Introduction to Statistical Computing and Visualization Using R

This book provides a foundational guide to statistical computing and visualization using R programming with an emphasis on practical data analysis skills that are directly applicable to diverse fields like finance, defence, health, and education. It uniquely combines a thorough explanation of basic constructs with advanced topics such as data visualization, statistical modelling, and probability, making it accessible yet comprehensive for learners across disciplines. This approach allows readers not only to build essential R skills but also to apply them to real-world scenarios, equipping students and professionals from various disciplines with versatile analytical tools. It offers a comprehensive yet approachable introduction for students and scholars from various disciplines using R.

- Includes practical and interactive elements such as quizzes, coding exercises, and hands-on projects to provide an engaging and effective learning experience for readers
- Provides complete code solutions to every problem presented, including detailed answers to even the most complex questions
- Presents case studies that can help contextualize the concepts covered in the book by showing how they are used in specific industries, fields, or contexts
- Offers application-based practical data analysis with cases in various fields and sectors, such as finance, healthcare, and marketing
- Focuses on best practices and efficient coding techniques, improving productivity and maintainability of R code

Introduction to Statistical Computing and Visualization Using R

Megha Rathi

CRC Press is an imprint of the
Taylor & Francis Group, an informa business
A CHAPMAN & HALL BOOK

Designed cover image: ShutterStock

First edition published 2025
by CRC Press
2385 NW Executive Center Drive, Suite 320, Boca Raton FL 33431

and by CRC Press
4 Park Square, Milton Park, Abingdon, Oxon, OX14 4RN

CRC Press is an imprint of Taylor & Francis Group, LLC

ISBN: 978-1-032-77281-3 (hbk)
ISBN: 978-1-032-78818-0 (pbk)
ISBN: 978-1-003-48938-2 (ebk)

DOI: 10.1201/9781003489382

Typeset in Palatino
by KnowledgeWorks Global Ltd.

Contents

Preface

As the amount of data continues to grow exponentially, there is a need for new and improved tools and techniques to effectively analyse and make sense of the overwhelming volume of information. In the past, there were two extremes of data analysis: lightweight methods that relied on tools like Tableau, Excel, and SPSS, and heavy-duty, high-performance analysis built using programming languages such as C++. As personal computers became more powerful, a middle ground emerged that combined both interactivity and robustness. An individual's exploratory analysis can easily be transformed into a foundation for advanced business processes. Statistical modelling and analysis help businesses and organizations increase their operational efficiency. Analysis is an essential process that involves the extraction, transformation, and analysis of data to uncover valuable insights and make informed decisions.

By analysing data, one can identify patterns, trends, and opportunities that would otherwise go unnoticed. This is the domain area of R. R is commonly used in the domain of data analysis, statistics, and machine learning, encompassing tasks such as data manipulation, exploratory data analysis, statistical modelling, data visualization, and machine learning algorithms. It is utilized in various fields such as academic research, business analytics, and data science, where large and complex data sets need to be analysed and interpreted. Furthermore, R is also employed for creating statistical software and packages, thereby extending its domain.

R was initially developed by Ross Ihaka and Robert Gentleman at the University of Auckland, New Zealand, in the mid-1990s. It was designed to be a free, open-source programming language that could provide a flexible and powerful tool for data analysis and statistical computing. The name "R" was derived from the initial of its creators' first names. Initially, it was called "S", which was developed by John Chambers and his colleagues at Bell Laboratories in the 1970s. However, Ihaka and Gentleman developed R as an open-source alternative to S, and hence they chose the letter "R" as a tribute to its predecessor.

In recent years, R has gained significant popularity as it has moved beyond academia and found widespread use in various industries, including but not limited to banking, marketing, pharmaceuticals, bioinformatics, genomics, finance, and data science. R programming offers a specialized, user-friendly environment for data analysis, and its open-source nature and integration capabilities make it a compelling choice for programmers seeking to work in data analysis and statistical computing. Programmers who work in these domains find R to be a more natural fit and prefer it over general-purpose languages. R has a large and active community of users and developers who have created many libraries and packages. This provides programmers with a wide range of tools and resources for data analysis and statistical computing. R can be easily integrated with other tools and programming languages, such as Python, SQL, and Hadoop. This makes it an ideal language for data analysts and scientists who work in heterogeneous environments.

The primary goal of this book is to enhance users' knowledge of R programming fundamentals and strengthen their basic understanding of statistics and analytical skills. By doing so, readers will be better equipped to predict future outcomes. This book offers a range of practice exercises, such as multiple-choice, short-answer, and long-answer questions. Notably, solutions are provided for all question types, enabling readers to verify their responses and enhance their comprehension of the material presented. This book is presented in a clear and user-friendly format, featuring step-by-step instructions and explanations. Additionally, it strikes a suitable balance between theoretical concepts and practical applications. This book contains real-world examples that demonstrate the practical applications of R in problem-solving.

Author Biography

Megha Rathi is currently serving as an assistant professor (senior grade) in the Department of Computer Science at Jaypee Institute of Information Technology (JIIT), Noida, India. She holds a Ph.D. in computer science from Banasthali University and has over 15 years of experience in teaching and research. Throughout her career, Dr. Rathi has contributed significantly to various research areas, including sustainable computing, data mining, data science analytics, health science, and machine learning. She has worked on key projects, such as the Xform generator at the National Informatics Centre (NIC), Delhi, and has experience in software development through her role as a project associate at the Indian Institute of Technology (IIT), Delhi. She has been actively involved in organizing special sessions at international conferences and has delivered invited talks. Additionally, she has organized numerous FDPs and workshops also guided numerous B.Tech/M.Tech/Ph.D. students, playing a pivotal role in shaping their research and academic development.

Introduction

Welcome to the world of R programming! This book serves as a comprehensive guide to learning the fundamentals of R programming and statistical analysis. Whether you are a beginner or an experienced programmer, this book has something for you. The use of R programming has become increasingly popular in various industries, such as finance, marketing, healthcare, and scientific research. This is due to R's wide range of statistical and graphical capabilities that enable researchers and analysts to efficiently manage and analyse large data sets.

In this book, you will learn the basics of R programming, including data types, functions, control structures, data structures, functions, and many more. A step-by-step procedure to install and set up R on various operating systems, as well as information on the integrated development environment (IDE) used for writing R code, is provided in this book. You will also gain an understanding of statistical concepts like central tendency measures, evaluation metrics, etc. This book includes information on how to use R for string handling and pattern recognition. The ability to handle strings and recognize patterns is crucial in data analysis and computational fields, as it facilitates the extraction, manipulation, and processing of textual data. The ability to handle strings and recognize patterns is crucial in data analysis and computational fields, as it facilitates the extraction, manipulation, and processing of textual data. This book includes numerous real-world examples and practice exercises to help reinforce your learning. Whether you are looking to enhance your data analysis skills, improve your understanding of statistics, or simply learn a new programming language, this book will provide you with a strong foundation in R programming and its practical applications.

The book provides guidance on programming best practices, including code organization, debugging techniques, and version control. This book enriches readers with a solid foundation in R programming and statistical analysis, preparing them for further learning and exploration.

Examples of real-world applications of R are included in the book, demonstrating how the language can be used to address problems in diverse fields like finance, healthcare, and social sciences. Complete details are provided to readers on utilizing R's visualization libraries to generate informative and impactful visualizations. It encompasses data manipulation techniques such as subsetting, merging, reshaping, and summarizing data frames.

R programming language is a powerful and flexible tool for data analysis and statistical computing. R aims to make complex statistical analysis accessible to a wide range of users, from beginners to experts, by providing a comprehensive set of built-in functions and libraries.

R is also designed to be extensible, allowing users to create their own functions and libraries to address specific analytical needs. Additionally, R is an open-source language, which means that it is free to use and modify, and has a vibrant community of developers and users who contribute to its ongoing development and improvement. Overall, the objective of R is to enable users to analyse, visualize, and model data efficiently and effectively, in a way that is accessible to both experts and novices.

Upon completing this book, readers should possess a comprehensive understanding of utilizing R for data analysis and modelling, including proficiency in essential statistical concepts and techniques. The goal of this book is to empower readers with the expertise and aptitude required to proficiently employ R in data analysis and modelling.

1

Introduction to R

Learning Objectives:

After reading this chapter, students and researchers will be able to:

- Understand the characteristics and features of the R programming language.
- Use Integrated Development Environment for R programming
- Understand basic concepts of R programming
- Discuss some information on R commands
- Outline concepts: vector, matrices, array, data frame, factor
- Summarize all the concepts

1.1 Introduction

Programming language R specifically intends for statistical computing and graphical visualization of data. It was developed by Ross Ihaka, Robert Gentleman, and their team at AT&T Bell Laboratories (Ihaka & Gentleman, 1996). R is globally used by data analysts, statisticians, and machine learning developers for the purpose of predictive modelling, data analysis, and building data mining models and applications (Paradis, 2022).

R is freely available under the Gnu's Not Unix General Public License and implementation of R programming is done in C, Fortran, and R itself. R is provided with the basic command line interface. R studio is an open-access integrated development environment (IDE) for R programming (Gardener, 2012).

R has a wide range of packages and functions available for all data mining tasks and mathematical computation. R programming is basically a collection of various packages that hold different methods and functions (Braun & Murdoch, 2021). These packages aid in all data mining tasks from data collection to data loading, data cleaning to data integration and transformation,

DOI: 10.1201/9781003489382-1

classification of data-to-data clustering, association analysis, and data storage with visualization facilities (Xia et al., 2018). R provides easy-to-use graphical interface for data-oriented and statistical tasks. R and its packages apply an ample number of mathematical and statistical techniques, including supervised and unsupervised learning models for predictive modelling, time-series analysis, association analysis, classification, clustering, and others. Most of the methods supplied by R are written in R itself, which help users to decide the particular algorithm for implementation (Matloff, 2011). For rigorous computational tasks, other programming languages like C, C++, Fortran, Java, and Python are utilized to manipulate R-objects directly. Another main edge of R over other statistical programming language is data visualization; R graphics packages are able to generate good quality graphs with dynamic and interactive visualization techniques for analysis and generating quality results from the data (Kronthaler & Zöllner 2021).

Listed below are a few key features of R programming:

1. R is a statistical programming language which includes programming constructs like other traditional programming languages. R includes effective programming features: conditional statements, iterative statements, user-defined functions, and input-output facilities.
2. R is designed for effective data storage, data loading, and data manipulation (i.e., overall it provides potent data-handling facilities).
3. R can work as a calculator and has a wide range of methods for vectors, arrays, lists, matrices, and data frames.
4. R is an integrated development tool for data miners, data analysts, and statisticians.
5. R is integrated with data visualization tools for impressive graphics and is perfect for producing reports.

1.2 R and RStudio Installation

R is a powerful, open-source tool available for statistical analysis and data visualization. R has the advantage of user-friendly graphical interface along with highly active and friendly support online. Reports created by R are extraordinarily extensible and employed by various high-level publications. R is available for Linux, MacOS, and Windows. The software can be downloaded from the Comprehensive R Archive Network (CRAN).

RStudio is an IDE for R (Okoye & Hosseini, 2024). It provides an enriched interactive environment for computation and incorporates the features like a

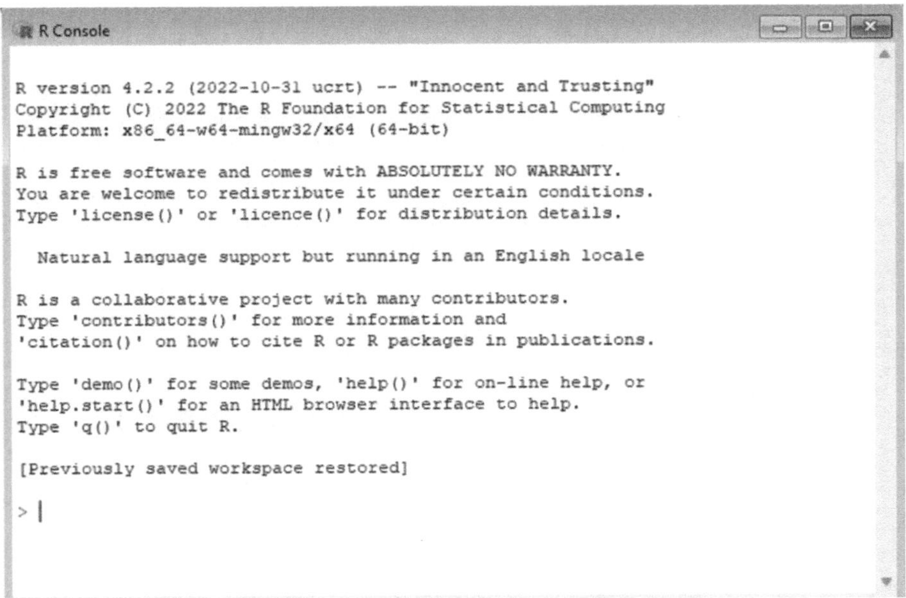

FIGURE 1.1
R GUI.

console, a window for writing and running scripts, a window for data visualizations, and R workspace.

Listed below are the series of steps to be carried to download R and RStudio in Windows (Gandrud, 2018):

1. R is easily downloadable from the source: http://cran.us.r-project.org/ (click on 'Download R for Windows' > 'base' > 'Download R 2.x.x for Windows')
2. Install R and set all settings equal to 'default' during installation.
3. Select R from the program list and start working on it. Figure 1.1 presents the snapshot of the R console. You can directly start writing R commands in the console.
4. Download RStudio from the source: http://rstudio.org/download/desktop and install it. Set all settings as 'default' during installation.
5. Open RStudio. Figure 1.2 presents the snapshot of RStudio. As seen, the window is divided into four parts. The upper-left window is for writing and running scripts, the lower-left window is for typing R commands, the upper-right window is to check all the connections and environment, and the lower-right window is for graphics. Also, you can check all the packages available, along with the data set description loaded for analysis (Verzani, 2011).

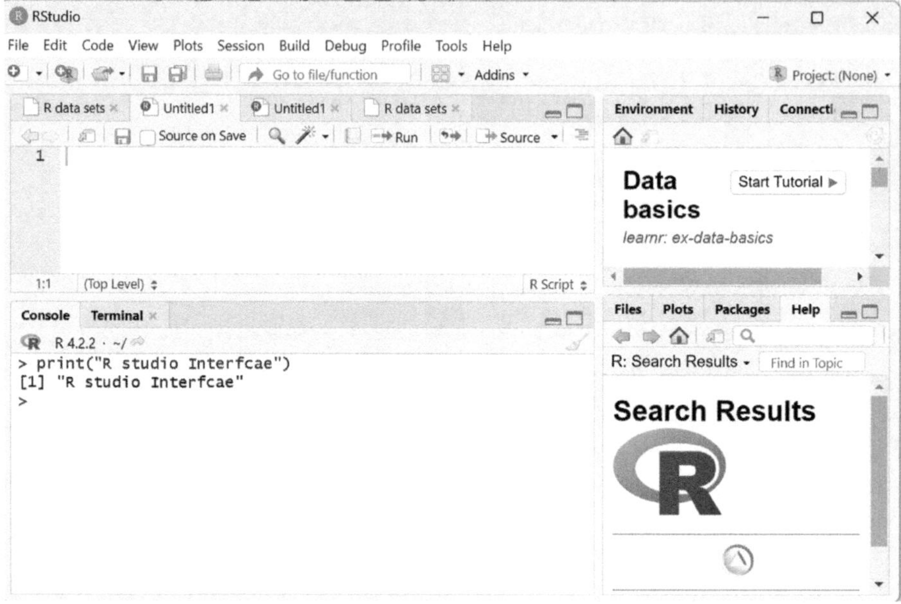

FIGURE 1.2
RStudio GUI.

6. R is extended by installing packages and packages are available for all statistical computation, data visualization, machine learning models, and data preprocessing.
7. Adding packages. A package may be installed by supplying its name to the function install.packages().

   For example, the 'Sandwich' package, which is used as a covariance matrix estimator can be downloaded and installed:

   R>install.packages("sandwich").

   The package functionality is available by:

   R>library("sandwich")

1.3 Why R?

1. **Open source and easily accessible**
   R is easily and freely accessible from CRAN and is licensed under GNU General Public License. R packages are easily obtainable under the same license and you can directly use them just after installation even for business and money-oriented applications (Gillespie & Lovelace, 2016).

2. **R is well-liked**
   According to the list published by Institute of Electrical and Electronics Engineers every year, R is a well-liked language for statisticians and data miners. The ranking given by IEEE was fifth in 2016, up from sixth in 2015. It is a tough job for a domain-precise language like R to be more accepted and favoured than a traditional programming language like C#. Trends for likeness are not only increasing for R but also for the domains like data analysis, data mining, machine learning, and data science, for which R is intended (Chambers, 2008).
3. **R is platform independent**
   R is uniformly designed to run on various distinct platforms such as Windows, Linux, and Mac. It is easy to port R workspaces from one platform to another. The ability to exchange data crossplatform is a predominant trait to have in today's computing era.
4. **Learning R will increase your chances of getting a job**
   Today, data scientists are being paid much higher salaries than traditional software developers. According to the survey conducted by O'Reilly Media in 2014, data scientists are paid a median of $98,000 worldwide. The figure is higher in the United States; around $144,000. Writing R scripts will not provide you a good opportunity; you have to be well versed in all advanced R concepts along with the knowledge of applying R methods and functions for data analysis and insights (Maronna et al., 2019).
5. **R applications to real-world problems**
   Adoption of R to solve real-world problems such as finance, marketing, supply chain, bioscience, and data storage and analysis is an indication of a programming language's potential (Kronthaler & Zöllner, 2021). Because of interactive user interface and simplicity, it has been adopted by companies globally for effective decision-making and data analysis. Table 1.1 lists a few ways industry stalwarts are using R and contributing to the R ecosystem.

TABLE 1.1

Industry-wise R Utilization

Company Name	Application
Google	Developed R mode guide for Google's R user circle
X	End-user experience monitoring
Ford	Social data analysis for effective decision-making for potential car designs
New York Times	Data analysis for journalism and for information graphics
Human Rights Data Analysis Group	War effect measurements
Microsoft	An extended and advanced Microsoft R server named Microsoft R Open is released

1.4 R Basic Programming Constructs

Command line is used to write basic commands in R. The user enters commands at the prompt (> by default) and each command is executed one at a time. R workspace is the current working environment of R and holds any user-defined data objects like vectors, array, matrices, data frames, lists (Das et al., 2022). R workspace is saved as an image, which is automatically loaded after R restarts again. Provided below is the basic information of R commands:

1. Ctrl+l: To clear the screen
2. q(): Quit command exits from the R console.
3. Source(): Source command executes from the current working directory.
4. Sink(): Sink command restore the output to given file or console.
5. Sink("record.lis"): stores the output to external file named record.lis.
6. R is case-sensitive language.
7. Either a semicolon or new line separates the R command.
8. # is used for comments in R.
9. Braces ({...}) integrate divergent elementary commands into a single composite expression.
10. For naming variables, all alphanumeric symbols are used in R as well as symbols '.' and '_' are utilized in naming conventions. Variable names always start either with the letter or '.', though variable name length is not restricted.
11. R provides unique prompt (by default, operator '+') for concurrent next expressions until the command completes at the end of a line. Approximately 4095 bytes can be entered on the command line console of R at a time.
12. R provides different escape sequences: \n, \r, \t, \v, \b, \f, and \a.
13. R is enriched with another feature: recall and re-execute previous commands. Vertical and horizontal arrow keys are utilized to achieve this functionality. Vertical arrow key scroll in both directions (i.e., forward and backward through R command history). Horizontal keys are then used to move the cursor within the command.

1.5 R Command Prompt

R is a command line–driven program. The user enters commands at the prompt (> by default) and each command is executed one at a time. R is an interpreted language that comes up with various built-in functions that

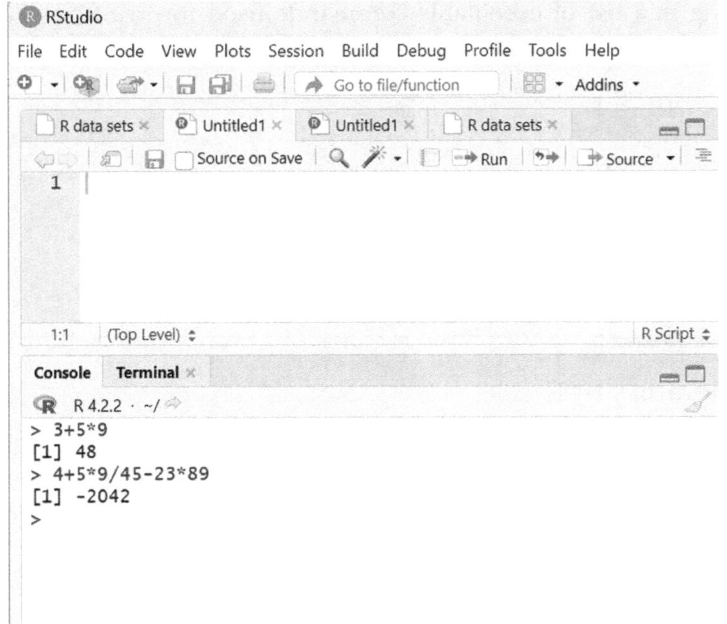

FIGURE 1.3
RStudio Command Prompt (Simple Calculator).

can be directly applied according to the need. R commands are entered in the console provided by R. R provides the ability to change the layout of the graphical user interface (GUI) according to user requirement by default commands written in red and results presented in blue. In contrast, in RStudio, all commands are written in blue and results are presented in black, and red is used to display error messages and warnings on the screen (Cirillo, 2016). R language can be directly used as a calculator. Simply type any arithmetic operation: 3+5*9 and press the Enter key; the computational result will be displayed on the next line. Figure 1.3 is a snapshot of the RStudio command line interface acting as a pocket calculator.

R can be directly used as a calculator. Just type 7+6 on the command line interface and the result will appear on next line after pressing the Enter key.

```
> 10+10
[1] 20
```

For the typed simple arithmetic command, the outcome is 20. Result [1] defines the 'initial and first requested data.' The calculation includes only a single element.

There are two methods to print data on the screen in R. First, to write a variable or object name directly; the second is to use the *print()* method.

Following is a list of executable commands used for printing data on the screen in R programming:

```
> data<-"INDIA"
> data
[1] "INDIA"
> print(data)
[1] "INDIA"
```

1.6 Operators in R

Operators play a significant role in mathematical, statistical, and logical computation. An operator notifies the compiler to perform a specific action for arithmetic manipulations. R language has abundant operators (Braun & Murdoch, 2021). Given below is the list of R operators:

- Arithmetic operators
- Relational operators
- Logical operators
- Assignment operators
- Miscellaneous operators

1.6.1 Arithmetic Operators

These operators are used to carry out mathematical operations such as addition and multiplication. Table 1.2 shows the arithmetic operators along with a description and example supported by R language.

1.6.2 Relational Operators

For comparison of data, relational operators are utilized. Table 1.3 provides a list of relational operators supported by R, along with a description and examples. An output of a relational operator is a Boolean value (i.e., True or False).

1.6.3 Logical Operators

Boolean operations such as AND and OR are carried out in R programming using logical operators. Table 1.4 presents a description and examples of logical operators supported by R. Operators AND (&) and OR (|) implement a bit-wise operation, generating an output of length greater than the operand. In contrast, logical operators AND (&&) and logical OR (||) implement only

TABLE 1.2

List of Arithmetic Operators with Examples

Operator	Description	Examples
+	Addition	> 1+1 [1] 2
–	Subtraction	> 20-10 [1] 10
*	Multiplication	> 10*10 [1] 100
/	Division	> 100/10 [1] 10
^	Exponent	> 2^4 [1] 16
%%	Modulus	> 19%%7 [1] 5
%/%	Integer Division	19%/%6 [1] 3

the first element of the operands, outputting this into a single length logical vector. By default, '0' is considered a Boolean value False and numbers other than zero are considered True.

1.6.4 Assignment Operators

For assigning values, assignment operators are used. Table 1.5 presents the description and examples of assignment operators supported by R language.

TABLE 1.3

List of Relational Operators with Examples

Operator	Description	Examples
<	Less than	> 4<6 [1] TRUE
>	Greater than	> 4>6 [1] FALSE
==	Equal to	> 4==6 [1] FALSE > 4==4 [1] TRUE
<=	Less than equal to	> 4<=5 [1] TRUE
> =	Greater than equal to	> 4>=4 [1] TRUE
!=	Not equal to	> 4!=9 [1] TRUE

TABLE 1.4

List of Logical Operators with Examples

<table>
<tr><th>Operator</th><th>Description</th><th>Examples</th></tr>
<tr><td>|</td><td>OR</td><td>> x<-2
> y<-3
> x|y
[1] TRUE</td></tr>
<tr><td>||</td><td>Logical OR</td><td>> x<-0
> y<-5
> x||y
[1] TRUE</td></tr>
<tr><td>&</td><td>AND</td><td>> x<-0
> y<-5
> x&y
[1] FALSE</td></tr>
<tr><td>&&</td><td>Logical AND</td><td>> x<-4
> y<-6
> x&&y
[1] TRUE</td></tr>
<tr><td>!</td><td>Logical NOT</td><td>> x<-4
> !x
[1] FALSE
> x<-0
> !x
[1] TRUE</td></tr>
</table>

TABLE 1.5

List of Assignment Operators with Examples

Operator	Description	Examples
<- or <<- or =	Left Assignment	> x<-5 > x [1] 5 > x=9 > x [1] 9 > x<<-5 > x [1] 5
-> or ->>	Right Assignment	> 10 -> x > x [1] 10 > 10->>x > x [1] 10

TABLE 1.6
List of Miscellaneous Operators with Examples

Operator	Description	Examples

:	Colon operator. Generates a sequence of numbers	`c<-1:10`
		`> c`
		`[1]  1  2  3  4`
		`5  6  7  8  9 10`

%in%	Used to check the existence of any element in the given set of data	`> a<-10`
		`> b<-20`
		`> c<-1:10`
		`> c`
		`[1]  1  2  3  4`
		`5  6  7  8  9 10`
		`> a%in%c`
		`[1] TRUE`
		`> b%in%c`
		`[1] FALSE`

%*%	Used for multiplying data with its transpose	`> a`
		`[1] 10`
		`> b`
		`[1] 20`
		`> c`
		`> a%*%b`
		`[,1]`
		`[1,]  200`

1.6.5 Miscellaneous Operators

For performing target specific computations, miscellaneous operators are required. They implement specific requirements, not just mathematical or logical arithmetic operations. Table 1.6 presents a list of miscellaneous operators along with their description and examples.

1.7 R Data Types

While developing any general or any specific application or just writing the code in any programming language, a placeholder is required to store data efficiently, and you will utilize this within the code. Most of the programming languages, such as C or C++, require data types to be preceded with placeholders to save data. Programmers are interested in saving data of different data types like integer, float, Boolean, character, etc. Depending upon the type of data operating system, this will allocate memory and determine what has to be saved in the area known as reserved memory. In this section, you will learn about all the data types available in R.

Provided below is the list of basic data types provided by R (Schmidt, 2011):

- Numeric
- Integer
- Complex
- Character
- Logical

1.7.1 Numeric Data Type

R provides a data type for decimal data known as numeric data. In R, data is directly assigned to variables without preceding it with the data type. In the assignment of decimal data to variable 'Y', make it numeric. R provides a built-in method class ( ) to check the class of data. We can check the data type of any data by supplying a variable name to function class ( ), as shown in the example given below:

```
> Y<-20.6
> Y
[1] 20.6
> class(Y)
[1] "numeric"
```

1.7.2 Integer Data Type

Another data type provided by R is 'integer.' Unlike other traditional programming languages, integer data in R takes the form '2L, 4L, or 0L.' R recognizes all the data with 'L' as an integer data type. In the example shown below, two variables hold data '4' and '4L' and, after using the function class ( ), it has been found that the class of 4L is an integer type and that 4 is a numeric type.

```
> x<-4
> y<-4L
> class(x)
[1] "numeric"
> class(y)
[1] "integer"
```

If you want to produce integer-type data of other data types, use method as.integer ( ). This method drives one data type value to an integer type like this:

```
> as.integer(5.67)
[1] 5
```

Method is.integer( ) is used to check whether data is an integer type or not.

```
> x<-56
> is.integer(x)
[1] FALSE
> y<-56L
> is.integer(y)
[1] TRUE
```

1.7.3 Complex Data Type

The pure imaginary value 'i' are used to define complex data types in R. Consider the following example of the complex data type in R:

```
> z<-4+7i
> z
[1] 4+7i
> class(z)
[1] "complex"
```

1.7.4 Character Data Type

String objects in R are represented by character data types. Single quotes (' ') and double quotes (" ") are used to define strings in R. Given below is an example that shows how to define strings:

```
> a<-"Hello World"
> b<-'Hi'
> class(a)
[1] "character"
> class(b)
[1] "character"
```

R provides the facility to convert any data object into a character by using the built-inn function as.character( ) and the code looks like:

```
> x<-34
> as.character(34)
[1] "34"
```

1.7.5 Logical Data Type

For comparative analysis of data, logical data type is used. An output is always represented as one of the two values (True or False) and the example code looks like:

```
> a<-67
> b<-89
> a> b
[1] FALSE
```

1.8 Data Structures in R

The data structure can be described as the determined form of structuring and saving the data. R programming aids five basic types of data structure, namely vector, matrix, list, data frame, and factor. In disparity to other programming languages like C and Java, the placeholders in R are not declared as a data type. The placeholders are assigned with R-objects and the data type of the R-object becomes the data type of the placeholder in R programming (Peng et al., 2020). Moreover, memory areas are reserved to hold variables that hold data. Data structures are the way to arrange and store data efficiently.

Unlike other general-purpose coding languages like C and Java, R variables are not declared by attaching a data type with the variable; rather, the variables are stored as objects in R and information from R-objects is further utilized to assign the data type of variable (Hillebrand & Nierhoff, 2015). This section will provide a brief overview of these data structures and the way to write these in R programming. Detailed descriptions of all these data structures are provided in Chapter 3. R programming provides the following basic data structures:

- Vector
- Matrix
- Array
- List
- Data frame
- Factor

1.8.1 Vector

A vector is the fundamental data structure provided by R and is actually comprised of data of the same type. Types of data stored within the vector can be logical, integer, double, character, complex, or raw. Vectors are described as arrangements of ordered collections of data elements. Function c( ) is used to create vectors in R. Since a vector contains a list of elements of the same data type, the c( ) method will try and constrain data elements to the same type if they are dissimilar. Following is a vector containing five data elements with values 2, 5, 6, 9, 0:

```
> a<-c(2,5,6,9,0)
> a
[1] 2 5 6 9 0
> typeof(a)
[1] "double"
> x <- c(5, 10.4, TRUE, "R Programming")
```

```
> x
[1] "5"              "10.4"              "TRUE"
[4] "R Programming"
> typeof(x)
[1] "character"
```

As you have seen in the above example, we have created two vectors named 'a' and 'x.' Vector 'a' consists of integer data only, so this vector is coerced to a higher type (i.e., double) and the second vector 'x' is coerced to character type because a character is the highest data type in that list.

1.8.2 Matrix

Another data structure used in R is 'matrix,' which is a two-dimensional data structure and can be created using the *matrix ()* function. You can provide a count of the total number of rows by 'nrow' and column by 'ncol,' which is passed as an argument to the *matrix()* function. Passing these arguments are optional as we can set the dimension of the matrix using length. This is the way of arranging data elements in a two-dimensional rectangular shape. A matrix contains data elements of the same type. A matrix differs from the vector in the sense that it contains an additional attribute 'dimension' to set the matrix *size.Attribute()* function, which is used to check the attributes of objects, and *dim()* function is used to check the matrix dimension. The syntax for matrix creation in R is:

matrix(data,nrow,ncol,byrow,dim)

Descriptions of the parameters passed inside the matrix method include the following:

- **data** is the data elements of a vector passed as input vectors.
- **nrow** is the count of the total number of rows.
- **ncol** is the count of the total number of columns.
- **byrow** is a logical value and it determines whether input data is filled row-wise or not. If TRUE, then the input vector elements are arranged by row.
- **dim** is the name assigned to the rows and columns.

In the given example, the matrix is created and elements are filled row-wise and in the subsequent example, the matrix is filled column-wise.

```
> # Elements are arranged sequentially by row.
> M<-matrix(c(10:21),nrow=4,byrow=TRUE)
> M
     [,1] [,2] [,3]
[1,]   10   11   12
[2,]   13   14   15
[3,]   16   17   18
[4,]   19   20   21
```

```
> # Elements are arranged sequentially by column
> M<-matrix(c(10:21),nrow=4,byrow=FALSE)
> M
     [,1] [,2] [,3]
[1,]   10   14   18
[2,]   11   15   19
[3,]   12   16   20
[4,]   13   17   21
```

1.8.3 Array

Arrays are the specific arrangement of data that can save data elements in more than two dimensions. An array is created using the *array* () function in R programming. Inside an *array ()* function, input data vectors are provided as the first parameter and dimension as the second parameter, which actually sets the dimension of an array. For instance, if you want to create an array with five columns and four rows, syntax is and the data is as follows:

```
> A<-array(4:30,dim=c(4,5,2))
> A
, , 1

     [,1] [,2] [,3] [,4] [,5]
[1,]    4    8   12   16   20
[2,]    5    9   13   17   21
[3,]    6   10   14   18   22
[4,]    7   11   15   19   23


, , 2

     [,1] [,2] [,3] [,4] [,5]
[1,]   24   28    5    9   13
[2,]   25   29    6   10   14
[3,]   26   30    7   11   15
[4,]   27    4    8   12   16
```

R vectors are all single or one-dimensional data objects; that is why while creating a vector, we provide only the length and no dimension as a parameter. The array exhibit property of a multidimensional data structure in contrast to vector and matrices. Matrices are also confined to two-dimensional structures. Matrices are a special type of two-dimensional array.

1.8.4 List

Lists are data structures provided in R programming to involve input elements of different data types. A list is dissimilar from a vector because a vector includes a similar type of data elements, but a list is comprised of the different types of data elements. The list () function is used to create a list in R. Lists are the unique type of object defined in R that hold dissimilar data elements such as numbers, strings, vectors, and other lists inside. A list can also hold a function or matrix as its data elements. The example shown below presents how to create a list in R, which is actually the combination of mixed data elements of strings, numbers, vectors, and logical values.

```
> list1 <- list("Hut", "Nest", c(2,12,111), FALSE, 81.99, 129.5)
> list1
[[1]]
[1] "Hut"

[[2]]
[1] "Nest"

[[3]]
[1]   2 12 111

[[4]]
[1] FALSE

[[5]]
[1] 81.99

[[6]]
[1] 129.5
```

In the result shown above, the subscript operator [[...]] is used for listing elements of the list. Data elements are sequenced as [[1]], [[2]], [[3]], and so on, where [1] denotes the first data element of the list. It is also possible to extract data elements from a specific position, as in the following example, where we extract the third member of the list with its second value. The third member of a list is a vector comprised of values 2, 12, and 111, and its second element is 12.

```
> list1[[3]][2]
[1] 12
```

1.8.5 Data Frame

A data frame is a table-like data structure provided by R that stores data in the form of a table or a two-dimensional arraylike structure in which each column contains attribute information, along with the type and each row contains a value for that attribute type. We can store divergent data in a data frame according to the need, such as numeric, factor, or character type. Data frames describe a column list where all columns specify the attribute information and are of the same length. Data frames are created in R using the function *data.frame()*.

Data frames are a powerful way of arranging data in tabular or spreadsheet form. It resembles a table, spreadsheet, or matrix. In most real-life applications, data needs to be stored in a tabular format, which is of great significance, as tabular data presents a clear view of the data. R provides a new concept to implement tables known as 'data frames.' Given below are some properties of the data frame:

- Column name should not be blank
- Distinctive row names
- Data types that can be stored in the data frame: numeric, character, or factor
- Size of each column is confined to the same length

Consider the instance of the given student table with the attributes Sid (student ID), Sname (name of the student), Age (age), and City (home city of the student).

Sid	**Sname**	**Age**	**City**
101	Amit	22	New Delhi
102	Neha	21	Hyderabad
103	Atharv	20	Noida
104	Misha	22	Mumbai

For creating the above tabular data storage structures in R, we create a data frame with four different columns, and for each column, we create a vector (i.e., four vectors are required of equal length for four different column names: Sid, Sname, Age, and City). The syntax to create the student table in R follows:

```
> stu.data<-
data.frame(sid=c(101,102,103,104),sname=c("Amit","Neha",
"Atharv","Misha"),age=c(22,21,20,22),city=c("NewDelhi",
"Hyderabad","Noida","Mumbai"))
```

The following result is generated after execution of the above code:

```
> stu.data
 sids   name age      city
1 101   Amit  22 New Delhi
2 102   Neha  21 Hyderabad
3 103 Atharv  20     Noida
4 104  Misha  22    Mumbai
```

1.8.6 Factor

Factors are R-objects reserved as integer vector values with the list of character values employed when the factor is presented. R uses the *factor()* method to create factors. A single parameter requires the list of vector values, which will be returned as vectors of factor values. For categorization of unique levels within the data, factors are used, which categorize and store the data at different levels. Both character and numeric data can be stored in factors. Factors are mostly implemented in the scenario when a column possesses a limited count of distinct values like gender categorization: 'Male' or 'Female.' An application includes data analysis for statistical or mathematical modelling (Giorgi et al., 2022). Factors are the structures that incorporate the finite number of distinct values; such data are often termed as *categorical data*. Shown below is the code to create a factor in R. In this example, initially a vector of categorical data is created that includes the categories of marital status type. Factor class can be checked using the function *is.factor()*. As seen, initially factors are not created, so the output is 'FALSE' and after creating the factor class it displays the output 'TRUE.'

```
data<-
c("Single", "Married", "Single", "Separated", "Single",
"Separated", "Married")
> data
[1] "Single" "Married" "Single" "Separated" "Single"
"Separated"
[7] "Married"
> print(is.factor(data))
[1] FALSE
> f_data<-factor(data)
> print(f_data)
[1] Single    Married   Single    Separated Single
Separated Married
Levels: Married Separated Single
> print(is.factor(f_data))
[1] TRUE
```

1.9 R-Objects

R actually acts by storing everything as an object. An object is an entity with independent existence. Placeholders are used to store a simple calculation, which is further stored as an object in R. R programming language is a kind of object-oriented programming language as it stores every calculation in the form of objects (Morandat et al., 2012). Workspaces are stored as R-objects in the database known as .Rdata in the current working directory of R. Lists, vectors, functions, and factors are all example of R-objects. Values can be assigned to R-objects using the symbol '<-' or '-> .' The simple command 5+35 is used to add two numbers and the result is stored in the variable 'sum.' In R-object, the assignment is required, and the variable sum actually holds the data object (Tippmann, 2015). We can assign an object, as shown in R:

```
> sum<-5+35
> sum
[1] 40
> 5+35-> sum
> sum
[1] 40
```

Listed below are some important points related to R-objects:

- R stores everything in the form of objects and provides a means for object assignment. Both left-side and right-side assignments are allowed in R.
- Objects can also be utilized in calculations. The following R command illustrates this:

```
> sum<-5+35
> sum
[1] 40
> solution<-(10.5+7)^2
> solution+sum
[1] 346.25
```

- R provides the facility to save the output of any calculation done with two or more objects. The following R command illustrates this:

```
> sol1<-solution/sum
> sol1
[1] 7.65625
```

- For retrieving the list of objects, the *ls()* function is used in R programming. Initially, when you start working on the R console, there are no stored objects, but after some time, when you write a few R commands, many objects are generated and stored. The following command illustrates how *ls()* functions.

```
> ls()
 [1] "a"        "A"          "data"      "emp.data" "f_data"
"list1"     "M"
[8]  "sol1"    "solution" "stu.data" "sum"       "x"
```

- For removing objects from memory, the *rm()* method is used in R programming. All objects are removed and you can also check it by the *ls()* function.

```
> rm(sol1)
> ls()
 [1] "a"          "A"          "data"     "emp.data"   "f_data"
"list1"     "M"
[8]  "solution" "stu.data" "sum"      "x"
```

The example shown above removes an object 'sol1' from memory. We can verify it by typing the *ls()* command and the output contains all the objects except 'sol1.' An error message will be displayed on the screen if you want to delete an object that does not exist.

1.10 Summary

- R is an open-source programming language specially designed for data analysis and data visualization.
- R is extended by incorporating several packages. R is a collection of so many packages that contain functionalities for statistical analysis. Packages are available for all data mining tasks from data loading to data preprocessing to data classification, data clustering, and graphical representation of data.
- Application areas include health care, finance, marketing, supply chain, social network analysis, and bioinformatics.
- R is an easy-to-operate interface and is a collection of add-on packages for statistical computation and analysis, which are two vital components of R programming.
- R packages are installed using the command install.packages() and includes various functions, examples, and help material. The *library()* method is used to load add-on packages.
- For R and RStudio installation for a distinct operating system, the official software downloading link is provided by CRAN (Comprehensive R Archive Network). CRAN provides approximately 8696 packages for different user needs (Jones et al., 2022).
- R language is case sensitive; for instance, 'M' and 'm' are different letters.
- R programming is a command line–driven programming language. R is an interpreted language and aids in statistical analysis, with the help of methods provided by R.

- R supports the following operators: arithmetic operators, relational operators, logical operators, assignment operators, and miscellaneous operators.
- Basic data types supported by R are integer, numeric, logical, character, and complex.
- In contrast to other traditional programming languages like C and Java, variable names in R are not declared by a preceding variable name with its corresponding data type like int a (example of variable declaration in C programming). Rather, variables are stored as objects, and metadata of R-objects are used to decide the data type of any variable or object.
- For efficient storage and retrieval of data, R supports data structures like vector, matrix, array, list, data frame, and factor.
- A vector is the rudimentary data storage structure in R that stores data elements of the same type. The *c()* function is used to create vectors in R.
- A matrix is a two-dimensional data structure and is created using the *matrix()* function. An array is also used to create a rectangular data representation, but the difference is that an array can be any dimension that is not restricted to a two-dimensional data structure. In R programming, the *array()* function is used to create arrays.
- The list is another data structure provided by R, which actually holds data of different types. The *list()* method is used to create lists in R.
- For storing data in the tabular or spreadsheet form, the data structure known as a 'data frame' is used in R. R creates a data frame using the method *data.frame()*.
- For categorization of data to different unique levels, factors are used and factors are created using the *factor()* function in R.
- R is an object-oriented programming language. R stores everything in the form of objects. Lists, vectors, functions, and variables are stored in the form of objects. The ls() method is used to retrieve the list of all objects stored in memory. For removing objects from memory, the *rm()* method is used.

EXERCISES

I. Multiple-Choice Questions

1. In 1991, R was created by Ross Ihaka and Robert Gentleman in the Department of Statistics at the University of _________.
   a. Auckland
   b. Oxford

c. California
d. Harvard

2. Find the wrong statement.
   a. R is a language for data analysis and visualization.
   b. One major constraint of the S language was that it was only available in a commercial package, S-PLUS.
   c. The initial version of S programming incorporates methods for statistical analysis.
   d. None of the above.
3. In which year was R version 1.0.0 released to the public?
   a. 2007
   b. 2010
   c. 2001
   d. 2000
4. In R programming, the following are types of basic data types, except which one?
   a. Character
   b. Logical
   c. Integer
   d. Data frame
5. We write the command x<-10 on the R console. What is the class of object 'x'?
   a. Double
   b. Numeric
   c. Integer
   d. Complex
6. What is the class of the following: x<-c(100, TRUE, A")?
   a. Integer
   b. Character
   c. Logical
   d. Raw
7. Which of the following is the default prompt for R command line interface?
   a. >
   b. <-
   c. =
   d. <

8. Find the incorrect statement.
   a. The grammar of the language determines whether an expression is complete or not.
   b. The operator used by R programming for assignment is <-.
   c. Comments are specified by the symbol ##.
   d. None of the above.
9. Out of the following functions, which is used to check whether an R-object is NULL or not?
   a. null()
   b. is.objnull()
   c. is.null()
   d. All of the above.
10. Which of the following codes is an example of explicit printing?
    a. x<-5
       x
    b. x<-5
       Print(x)
    c. x<-"auto printing"
       x
    d. None of the above.

II. Brief Questions

1. Describe R language. Also, provide five real-time applications of R programming language.
2. What are the major features of R programming?
3. What are the merits and demerits of R programming?
4. Why are packages used in R programming? How do you install the packages? Provide a list of a few packages used in R programming.
5. List all the operators used in R programming, along with an example.
6. How do you print data in R programming? Write a program to print the following list: ("cat", "tiger", "TRUE", c(12, 3, 4)).
7. Why is a matrix and an array used in R? Also cite the difference(s) between the two. Which is more advantageous for statistical modelling?

8. Apart from R, which technology is used for data analysis? Find the merit and demerit. Also, Compare and contrast both technologies.
9. What is coercion? In which data structure it is used? Write the function provided in R for coercion.
10. Explain the data frame. Write down the advantages and disadvantages of a data frame. Create a data frame of employees with attributes (emp_id, emp_name, salary, address, num_of_dependents).
11. Why are factors used in R programming? Describe using a suitable real-life example.
12. Describe a list with a suitable example.
13. Why is the operator '%*%' used in R? Explain with a suitable example.
14. What are R-objects? How are they created? Write a method to list all the stored objects. Also, define how objects are assigned in R.
15. Write the command to find out the class of any object in R.

III. Coding Exercise

1. Write a program to define two vectors: A (20, 45, 50) and B (100, 34, 56). What is the output of vector C, defined as A*B?
2. Write a program to generate an array of odd and even numbers.
3. Create a 4×3 matrix and fill the data column-wise.
4. Create a vector of length 10. Identify the position that contains even data.
5. Consider the two vectors: X=(10,20,30) and Y=(1,2,3,4,5,6). Find the value of X+Y.
6. Create a data frame named 'Customer' that stores information like Customer_id, Csutomer_name, Age, Address, and Income of five customers.
7. Execute the given command, [matrix(10:10,nrow=2,byrow=TRUE)] & [matrix(25:15,nrow=2,byrow=TRUE)], and answer the following questions:
    a. What is the total count of data elements generated by this command?
    b. What is the total number of columns?
    c. Is there any error in the above command? Or is there any warning produced while executing this command? If yes, explain. Also, alter the code to remove a warning.

8. Create the vector A={"aa", "bb", "cc", "dd", "ee"}. Write the R command to display output as {bb, cc, dd} from the given list of vectors.

References

Braun, W. J., & Murdoch, D. J. (2021). A first course in statistical programming with R. Cambridge University Press.

Chambers, J. (2008). Programming with R: The Basics. In Software for Data Analysis: Programming with R (pp. 37–78). Springer.

Cirillo, A. (2016). Rstudio for R statistical computing cookbook. Packt Publishing Ltd.

Das, B. K., Jha, D. N., Sahu, S. K., Yadav, A. K., Raman, R. K., & Kartikeyan, M. (2022). Introduction to R Software. In Concept Building in Fisheries Data Analysis (pp. 209–233). Singapore: Springer Nature Singapore.

Gandrud, C. (2018). Reproducible research with R and R studio. Chapman and Hall/CRC.

Gardener, M. (2012). Beginning R: The statistical programming language. John Wiley & Sons.

Gillespie, C., & Lovelace, R. (2016). Efficient R programming: A practical guide to smarter programming. O'Reilly Media, Inc.

Giorgi, F. M., Ceraolo, C., & Mercatelli, D. (2022). The R language: An engine for bioinformatics and data science. Life, 12(5), 648.

Hillebrand, J., & Nierhoff, M. H. (2015). Mastering RStudio–develop, communicate, and collaborate with R. Packt Publishing Ltd.

Ihaka, R., & Gentleman, R. (1996). R: A language for data analysis and graphics. Journal of computational and graphical statistics, 5(3), 299–314.

Jones, E., Harden, S., & Crawley, M. J. (2022). The R book. John Wiley & Sons.

Kronthaler, F., & Zöllner, S. (2021). Data analysis with RStudio. Springer.

Maronna, R. A., Martin, R. D., Yohai, V. J., & Salibián-Barrera, M. (2019). Robust statistics: Theory and methods (with R). John Wiley & Sons.

Matloff, N. (2011). The art of R programming: A tour of statistical software design. No Starch Press.

Morandat, F., Hill, B., Osvald, L., & Vitek, J. (2012). Evaluating the design of the R language: Objects and functions for data analysis. In ECOOP 2012–Object-Oriented Programming: 26th European Conference, Beijing, China, June 11–16, 2012. Proceedings 26 (pp. 104–131). Springer Berlin Heidelberg.

Okoye, K., & Hosseini, S. (2024). Introduction to R Programming and RStudio Integrated Development Environment (IDE). In R Programming: Statistical Data Analysis in Research (pp. 3–24). Singapore: Springer Nature Singapore.

Paradis, E. (2022). R Programming and Development from Basics to Advanced Topics. ISEM.

Peng, R. D., Kross, S., & Anderson, B. (2020). Mastering Software Development in R. Victoria, Canada: Leanpub. https://bookdown.org/rdpeng/RProgDA.

Schmidt, G. (2011). Relational mathematics (Vol. 132). Cambridge University Press.

Tippmann, S. (2015). Programming tools: Adventures with R. Nature, 517(7532), 109–110.

Verzani, J. (2011). Getting started with RStudio. O'Reilly Media, Inc.

Xia, Y., Sun, J., & Chen, D. G. (2018). Introduction to R, RStudio and ggplot2. In Statistical Analysis of Microbiome Data with R. ICSA Book Series in Statistics (pp. 77–127). Singapore: Springer. https://doi.org/10.1007/978-981-13-1534-3_4.

2

R Programming Concepts

Learning Objectives:

After studying this chapter, students and researchers will be able to:

- Understand the basic programming constructs of R
- Make decisions using R
- Use conditional statements and switch-case statements in R programming
- Use iterative statements in R programming
- Use built-in functions and user-defined functions in R
- Use R scripts
- Summarize all the concepts

2.1 The R Environment

Program constructs are the essential component of all programming languages. These constructs are utilized to control the order of execution (i.e., deciding which expressions are to be executed and which are not). Basically, three main basic constructs are provided by coding languages: sequence, selection, and iteration. R language also provides facility to concentrate on the sequence of execution of code. R provides basic constructs for looping, selection, and conditional statements, and also has the concept of creating user-defined functions, which makes coding much easier (Crawley, 2012). Functions, decision-making expressions, and iterative statements are the vital component of any coding technique.

R has another add-on concept known as 'scripts.' Script is a file that contains a set of R commands that can be saved in memory for further use. Scripts play a vital role in R programming because once written, scripts

DOI: 10.1201/9781003489382-2

can be used anywhere (Zuur et al., 2009). R is an interactive language for data analysis and all R commands and statements can be implemented as scripts. Scripts are useful when we are working on large projects with different modules and each module is repeatedly implemented in different scenarios. Instead of repeatedly writing the code, we can save it in the form of a script and run the script whenever required. Scripts are saved in R with an '.r' extension.

R programming provides various built-in functions (Alvaro, 2016). These functions can be directly implemented by passing values inside the function. For example, R provides built-in functions for all mathematical calculations such as finding square root, maximum value, mean value, absolute value, and round value (Pace, 2012). User-defined functions are required to minimize the line of code and make programming better and easier. The interpreter of R supervises the control of user-defined functions and once the function executes completely, it returns the control back to the interpreter.

R programming supports basic conditional statements such as *if, if.. else, nested if..else,* and switch-case statements for implementing decision-making (Albright et al., 2011). R makes use of iterative statements to iterate specific groups of statements. Loops supported by R include for, while, and repeat. This chapter discuss all the basic programming concepts of R in detail.

2.2 Decision-Making in R Programming

Like other general-purpose coding techniques, R language can also be used in a scenario where we want to implement conditions based on a specific requirement. For decision-making, *if, if..else,* and *switch-case* statements are used by R programming (Hafner, 2019). Conditional statements are required to verify a condition by executing a task in coding. Conditional or decision-making statements accomplish this by checking one or more conditions. Operators such as relational/logical can be amalgamated to check conditions in conditional statements. Conditional programming constructs are basic properties that work by executing different computations, depending upon whether a program-defined condition evaluates to True or False and is always attained by changing the execution flow based on a condition.

2.2.1 Conditional Statement *(if)*

The first conditional statement provided by R is the *if* statement. It executes by verifying one or more conditions and executes the set of statements accordingly depending upon whether the condition is evaluated as True or False. The syntax of an *if* statement is as follows:

```
if (test_condition)
{
        (Statements to be executed)
}
```

If test_condition evaluates to TRUE, all sets of expressions or statements are executed; otherwise, nothing gets executed.

if Statement Example in R

The given example provides an *if* conditional statement, which verifies whether a given number is odd or not:

```
> x<-5
> if(x%%2==1)
+ print("Odd Number")
[1] "Odd Number"
```

2.2.2 Conditional Statement *(if..else)*

Another conditional statement used in R programming is the *if..else* statement. It is actually an extension of the *if* statement, which checks the TRUE condition and does nothing if the condition is evaluated as FALSE (Alvaro, 2016). In the *if..else* statement, if the test_condition evaluates to FALSE, a set of statements is executed for the else part. The syntax of writing an *if..else* statement in R is as follows:

```
if (test_condition)
        {
                (Statements to be executed)
        } else
        {
                (Statements to be executed)
        }
```

If test_condition evaluates to TRUE, all sets of expressions or statements are executed; otherwise, all statements for the else part get executed by the program.

if..else Statement Example in R

```
> x<-246
> if(x%%2==0)
+ {print("Even Number")}else
+ {print("Odd Number")}
[1] "Even Number"
```

2.2.3 Conditional Statement *(Nested if..else)*

Nested if..else expressions are used in the scenario when two or more conditions need to be verified before making any decision. The user first checks the initial *if..else* test conditions and the further conditions. It is utilized to execute a block of expressions for more than two alternatives. The syntax of a *nested if..else* statement is as follows:

```
if (test_condition1)
        {
                (Statements to be executed)
        } else if(test_condition2)
        {
                (Statements to be executed)
        } else if(test_condition3)
        {
                (Statements to be executed)
        } else if(test_condition4)
        {
                (Statements to be executed)
        } else
        {
                (Statements to be executed)
        }
```

Out of the given set of statements, only one statement will be executed as determined by the test conditions.

Nested if..else Statement Example in R

```
> a<-10
> x<-100
> y<-20
> z<-1000
> if((x>y)&(x>z))
+ {print("x is greater")}else if(y>z)
+ {print("y is greater")}else
+ {print("z is greater")}
[1] "z is greater"
```

NOTE: The *readline()* function in R is used to enter a run-time one-line string in the R command line interface (Davies, 2016). The function *as.integer()* is used to typecast character data types to integers. Given below is an example showing the run-time input from the user.

readline() Example in R

```
> n<-readline("Enter any number:")
Enter any number:34
```

```
> if(n>0)
+ {print("Positive Data")}else
+ {print("Negative Data")}
Error in n%%2 : non-numeric argument to binary operator
```

As shown in the example, without using the function *as.integer()*, an error message will display on the screen because it takes the string data as its default data type, so a conversion from string to numeric type is required before doing any manipulation.

```
> n<-readline("Enter any number:")
Enter any number:34
> n<-as.integer(n)
> if(n>0)
+ {print("Positive Data")}else
+ {print("Negative Data")}
[1] "Positive Data"
```

2.2.4 Conditional Statement *(Switch Case)*

Switch case is another conditional programming construct utilized for decision-making. Switch-case expressions allow a variable condition to be tested against the possible case values. These conditional statements are generally implemented for multiple selections of conditional expressions. The syntax of a switch-case structure in R is as follows:

switch(test_condition,case1,case2,... ...caseN)

Switch-Case Rules:

- If the input expression is not in a string format, then it is coerced as an integer data type.
- An infinite number of cases can be embedded inside a single switch statement.
- If a range of integer data is between 1 and nargs()-1 (i.e., the maximum count of arguments), then the corresponding statement of the case test condition is executed and an output is returned.
- If output of an expression is String then, output is matched to data elements name.
- The default argument is not provided.
- In case it is not matched to any of the defined cases, then a NULL value is returned.

***Switch-Case* Example in R**

```
> Data<-switch(1,"A","B","C","D")
> Data
[1] "A"
> Data<-switch(10,"A","B","C","D")
```

```
> Data
NULL
```

Another Example

```
> Data<-readline("Enter any Numeric Data:")
Enter any Numeric Data:8
> Data<-as.integer(Data)
> S=switch (Data,"Jan","Feb","Mar","Apr","May","Jun","Jul","Aug",
"Sep","Oct","Nov","Dec")
> S
[1] "Aug"
```

2.3 Iterative Statements in R Programming

In statistical modelling, a situation may arise where coders want to execute a set of expressions together repeatedly. In such situations, iterative statements are required to execute a set of statements sequentially. Iterative statements minimize the lines of code. A conceptually iterative statement is a means to sequentially repeat expressions under a test condition. It automates some segments of code that require repetitive execution. R has three basic type of iterative programming constructs: *for, while,* and *repeat* loop (Paradis, 2022).

2.3.1 Iterative Statement *(for loop)*

A *for loop* is quite crucial in a situation when we wish to iterate over a list of data or a range of numbers. The *for loop* can be used to repeat a set of statements over a list, data frame, vector, matrix, or any other object (Trejo, 2017). The syntax of a *for loop* in R is as follows:

```
for (num in vector sequence)
{
        (Statements to be executed)
}
```

where 'num' is the iterative variable and the vector sequence is the vector limit range, the controlling condition (i.e., num) iterates during the loop up to this range.

***for loop* Example in R**

```
> for(val in 1:10)
+ {if(val%%2==0)print(paste(val,"Even Number"))}
[1] "2 Even Number"
[1] "4 Even Number"
[1] "6 Even Number"
[1] "8 Even Number"
[1] "10 Even Number"
```

***for loop* List Example in R**

```
> product<-list(itemtype=c("TV","Laptop","Washing Machine","Smart
Phones"),Price=c(30000,50000,25000,80000))
> for(val in product)
+ {print(val)}
[1] "TV"                "Laptop" "Washing Machine""Smart Phones"
[1] 30000 50000 25000 80000
```

NOTE: The *paste()* function is used to merge vectors after changing the data type to string. The *paste()* function combines vectors with a space separator. The syntax for the *paste()* function in R is:

paste(data...,sep="",collapse=NULL)

where data is the list of R objects to merge together:

sep is the symbol required to segregate two strings and,
collapse is an optional argument required to segregate the output.

***paste()* Example in R**

```
> paste('X',1:10,sep='')
[1] "X1"  "X2"  "X3"  "X4"  "X5"  "X6"  "X7"  "X8"  "X9"  "X10"
> paste(c('X','Y'),2:4,sep='_',collapse=' and ')
[1] "X_2 and Y_3 and X_4"
```

2.3.2 Iterative Statement *(while loop)*

The *while loop* repeatedly executes the given set of expressions until the specified test_condition is met. The important point about the *while* statement is that execution of the loop is not mandatory. It works by first checking the test_condition. If the condition evaluates to TRUE, then it starts the execution; otherwise, the loop body is skipped (Trejo, 2017). Here is the syntax of a *while loop* in R programming:

```
while (test_condition)
{
        (Statements to be executed)
}
```

***while loop* Example in R**

```
> num<-0
> while(num<10){
+ print(num)
+ num=num+1
+ }
[1] 0
[1] 1
```

```
[1] 2
[1] 3
[1] 4
[1] 5
[1] 6
[1] 7
[1] 8
[1] 9
```

2.3.3 Iterative Statement *(repeat loop)*

The *repeat loop* iterates the set of expressions again and again until the test criteria is met. It works the same as that of the *while loop*, except that no test condition is provided in the *repeat* loop. The *repeat loop* uses a *break* statement to exit the loop. If a *break* statement is not used, then epochs will take place an infinite number of times (Trejo, 2017). The syntax of a *repeat loop* in R is as follows:

```
repeat{
        (Statements to be executed)
                If(test_condition){
                break
                }
        }
```

***repeat loop* Example in R**

```
> num<-16
> repeat{
+ print(num)
+ num=num+2
+ if(num==30){break}
+ }
[1] 16
[1] 18
[1] 20
[1] 22
[1] 24
[1] 26
[1] 28
```

2.3.4 Loop Control Expressions

To alter the normal execution of iterative statements loop control structures are required. In R language, *break* and *next* control expressions are utilized to change the normal execution of program (Jones et al., 2022). Conceptually, control statements are implemented to reduce the size of memory required and to save the execution time. While altering the sequence of execution, all objects created within normal sequences are demolished. R supports the following two types of loop control expressions: *break* and *next.*

break Statement

The *break* control statement is used to exit from the normal execution of loop and transfer control to the expression immediately after the loop. It has the following two utilities in R programming:

- The loop is terminated after encountering a *break* statement and control is transferred to execute the next immediate expression after the loop.
- It is utilized to finish the case in the switch conditional statement.

The syntax of *break* in R is as follows:

```
if (test_condition)
{
        break
}
```

***break* Example in R**

```
> for(i in num){
+ if(i==3){print(paste("Taking exit from the loop where
i=",i))
+ break
+ }
+ print(paste("Values are:",i))
+ }
[1] "Values are: 1"
[1] "Values are: 2"
[1] "Taking exit from the loop where i= 3"
```

In the above example, the *break* statement is placed inside a *for loop* to check whether the value of i = 3 is true or not. If the test condition, which is i = 3, is FALSE, then the *break* statement is not executed and the result will be 1 and 2. Once the test condition evaluates to TRUE, the *break* statement is executed and control is transferred from the loop to the print statement, which is the next immediate statement after loop.

next Statement

The *next* control statement is used to skip the current iteration of the loop without exiting from the loop. On finding the next statement within the loop, control transfers to the next iteration and results in skipping the current iteration (Ren, 2016). Apart from the loop, the *next* control statement can also be used with conditional statements. The syntax is as follows:

```
if (test_condition)
{
        next
}
```

next Example in R

```
> x<-1:10
+ if(i==5){
+ next
+ }
+ print(i)
+ }
[1] 1
[1] 2
[1] 3
[1] 4
[1] 6
[1] 7
[1] 8
[1] 9
[1] 10
```

In the above example, the *next* control statement is placed inside a loop statement. If the test condition, which is i = 5, is TRUE, then execution of the current iteration will stop, as reflected in the output.

2.3.5 Vectorization

Vectorization is an alternative to iterative constructs in R programming. Vectorization is the method to change repeated statements on simple numeric data, known as scalars, to a single operation applied on vectors (Sudhamathy & Venkateswaran, 2019). A vector is the elementary arrangement of data that consists of similar kinds of data. Merging two vectors of equal length will result in a matrix. Therefore, a matrix is actually an amalgamation of horizontal and vertical vectors. Therefore, we can apply repeated operations on vectors with loops. Most of the iterative programming constructs can be made implicit by utilizing the concept of vectorization. Vectorization means several operations must occur and execute in parallel. The example shown below illustrates the concept of vectorization:

```
> # Addition of two vectors using for loop
> z<-numeric(length(x))
> for(i in 1:5)
+ {
+ z[i]<-x[i]+y[i]
+ }
> z
[1]  7  9 11 13 15

> # Addition of two Vectors after applying Vectorization
> x<-1:5
> y<-6:10
> z<-x+y
> z
[1]  7  9 11 13 15
```

In the above, R code is written to add two vectors. In the first code, a *for loop* is executed and in the second R, a vectorization operation applies for the addition of two vectors.

Another example of vectorization is in comparing logical data, as shown:

```
> x<-1:5
> x>=2
[1] FALSE  TRUE  TRUE  TRUE  TRUE
> x==8
[1] FALSE FALSE FALSE FALSE FALSE
```

Matrix operations are also vectorized. In this manner, R programming provides the easiest way to perform all matrix operations directly without employing iterative constructs, unlike other traditional programming languages. The code shown below reflects an example of how *matrix* operations can be vectorized:

```
> M1<-matrix(1:9,3,3)
> M1
     [,1] [,2] [,3]
[1,]    1    4    7
[2,]    2    5    8
[3,]    3    6    9

> M2<-matrix(rep(9:1),3,3)
> M2
     [,1] [,2] [,3]
[1,]    9    6    3
[2,]    8    5    2
[3,]    7    4    1

> # Multiplication of two matrices
> M1*M2
     [,1] [,2] [,3]
[1,]    9   24   21
[2,]   16   25   16
[3,]   21   24    9

> # True Matrix Multiplication
> M1%%M2
     [,1] [,2] [,3]
[1,]    1    4    1
[2,]    2    0    0
[3,]    3    2    0
```

2.4 R Scripts

A script is a text file used to store R commands for accomplishing a task. Scripts hold all the commands that are typed in the command line interface of R. Scripts are crucial when a program contain thousands of commands that are repeatedly required for data analysis (Pace, 2012). Data analysts can save all these commands in one place, known as scripts. Scripts can be stored in a current working directory of R and can be run at any time. Scripts are stored with the '.r' extension in R. While writing down the code in the interface of R is quite easy and simple, writing the code again and again is a tedious job. Also, statistical modelling requires complex commands and a group of R statements for further analysis. Writing difficult commands and procedures wastes time and is subject to typing errors. Repeating a set of coding expressions demands writing a complete set of code again. R programming provides the facility to overcome all these issues and scripts are the solution.

2.4.1 Creating Scripts in RStudio and R

RStudio provides a convenient way to create scripts (Ahmad et al., 2023). Simply click on the File icon and then select the New File option, which appears in the upper-left portion of RStudio. After clicking on New File, a new empty script is generated. A screenshot of creating new scripts in RStudio is presented in Figure 2.1.

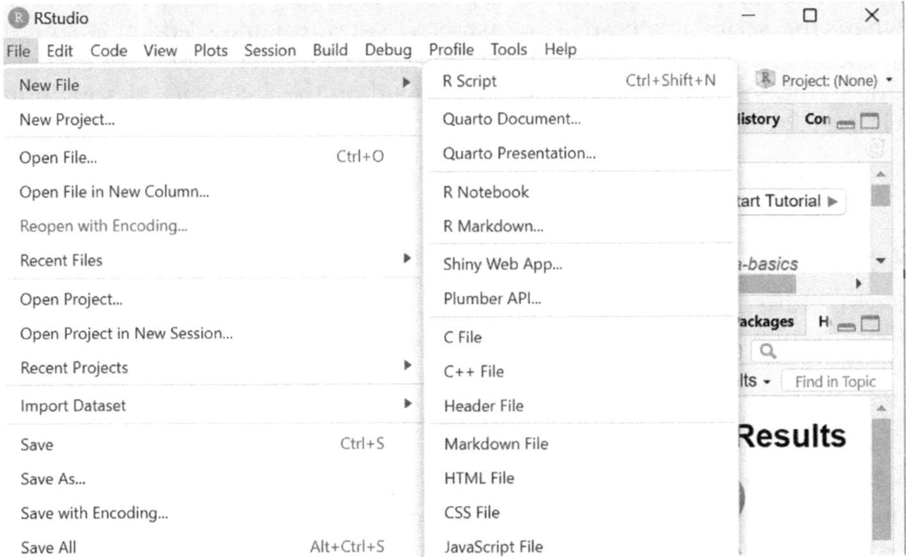

FIGURE 2.1
RStudio interface to create scripts.

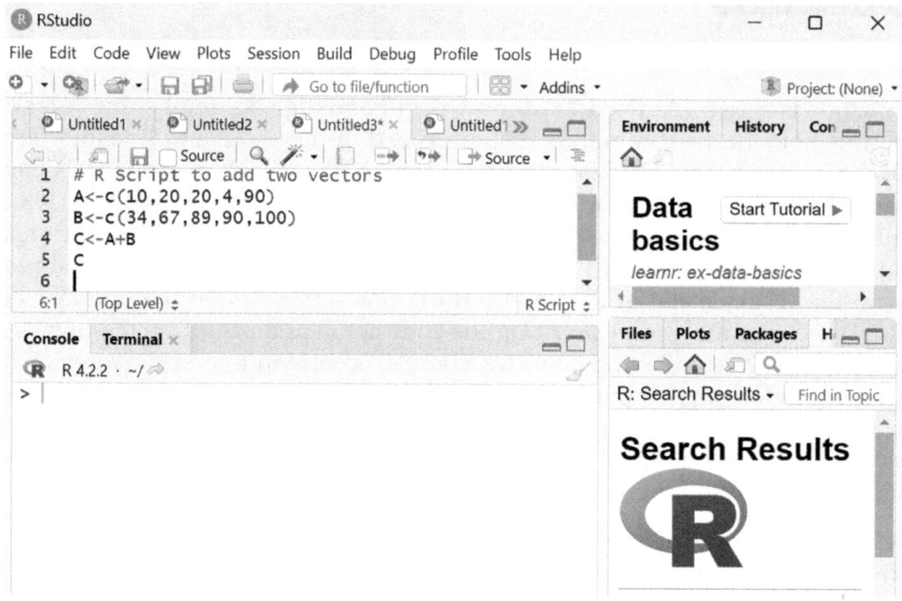

FIGURE 2.2
R Interface to create and open scripts.

For creating scripts in R, write the list of R commands in any editor, such as Notepad or text editor. For executing the script, the user can use the *source()* command to provide a complete path of the current workspace where the script is stored. For instance, a script named 'add.r' is stored at the following location: "C:/Users/megha.rathi/Documents". To execute this script, the following command is typed on the R console, as shown in Figure 2.2:

```
> source("C:\\Users\\megha.rathi\\Documents\\add.r")
```

2.4.2 Saving and Running Scripts in RStudio

R scripts are stored inside the current working directory of R. To know about the current working directory, the command is: *getwd()*.

```
> getwd()
[1] "C:/Users/megha.rathi/Documents"
```

If you want to change the current workspace of R command then *setwd()* method is used which helps in setting the new path.

After getting information about the current workspace, save the file and then run it whenever it is required by the end user.

After New File is open in the Editor panel, you can type R commands and the RStudio screen will look like Figure 2.3.

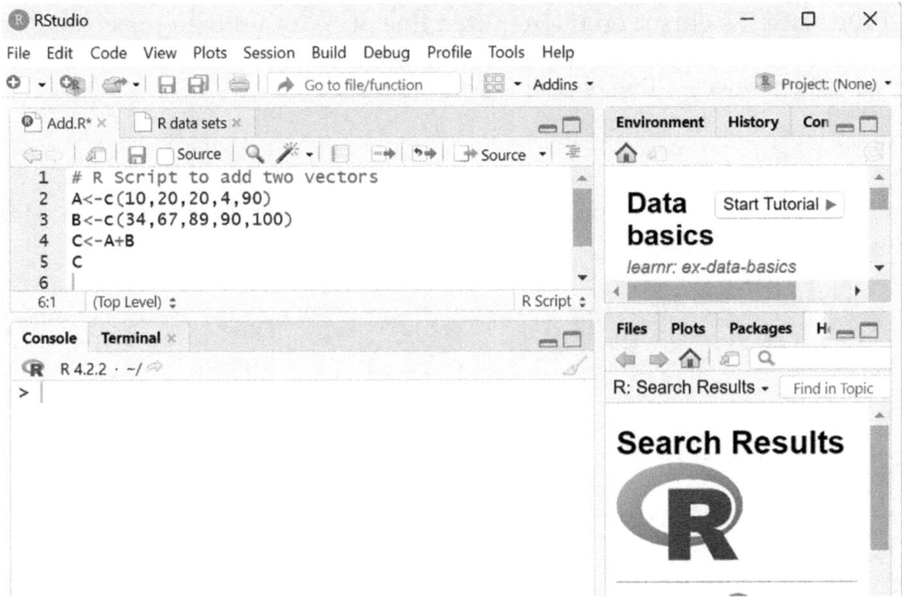

FIGURE 2.3
RStudio interface with editor.

Save your scripts by pressing the Save icon, as shown in Figure 2.4.

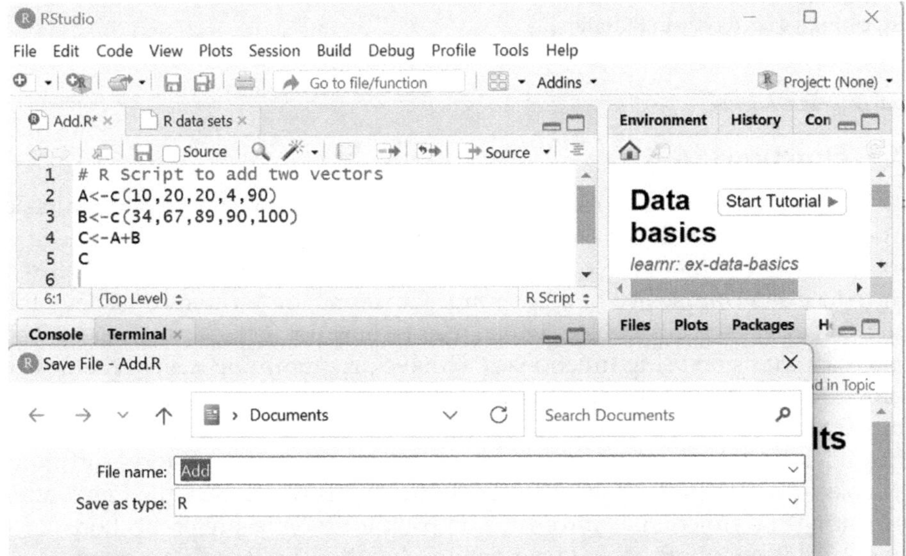

FIGURE 2.4
RStudio interface to save script (add.R).

Execute your code saved inside your scripts. The Run icon is provided in the RStudio interface. The Run icon can execute either the entire code of your script or the selected lines of code as per your requirement (Saiz & Rosteck, 2020). Place the cursor on the required line of code and then press the Run icon to execute the selected line of code. Figure 2.5 presents the screenshot of executing a selected line of script code in RStudio.

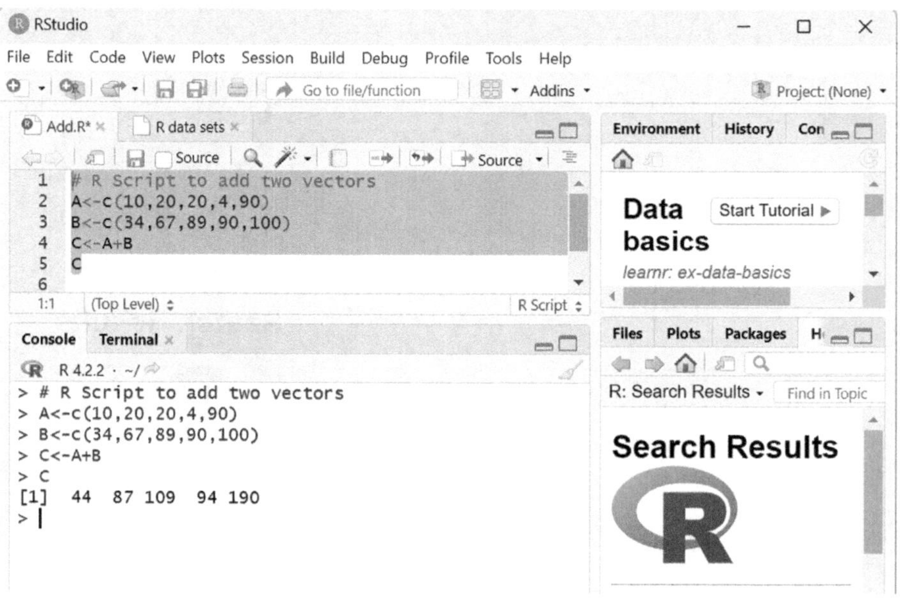

FIGURE 2.5
Executing selected code in RStudio.

2.5 Functions in R

A function is defined as group of statements to accomplish a predefined task. The main functionality of writing functions in any programming language is to avoid writing the same code again and again. Functions are also required to reduce the complexity of a program. Functions are written to achieve the solution of a specified problem which may or may not include parameters. The main emphasis on using functions is to make programming easier and to save time. Functions are basically of two types: built-in functions and user-defined functions (Lawrence & Verzani, 2018). R programming supports both types of functions. As R is an object-oriented programming language, it stores functions as objects. R interpreter passes control to functions whenever called and, after executing the function, control returns back to the interpreter. Sections 2.5.1 and 2.5.2 discuss built-in and user defined functions in R programming.

2.5.1 Built-in Functions in R

R language is comprised of several pre-defined built-in functions. A function is a set of working instructions to execute a task. R is a kind of coding language where every working instruction typed on the R console is a function. For example, the *c()* function is used to create a vector, which is a built-in function, and also to print data on the console, the *print()* function is used. To carry out a single task in R, built-in functions are utilized. These functions are implemented directly by passing suitable parameters. R programming language provides abundant sets of built-in functions for accomplishing the task in all domains from numbers, to strings, to data pre-processing, to data mining, to data visualization. Table 2.1 provides a list of some built-in functions used in R for different purposes (Jones et al., 2022).

TABLE 2.1

List of General Built-in Functions Along with Their Description

Function	Description
help()	Opens help page
append()	Appends data elements to a vector
identical()	Checks the equality of two objects
length()	Counts vector elements
range()	Minimum and maximum
rep(x,n)	n times data x is repeated
rev()	Reverses the order
sort()	Sorts data
tolower(),toupper()	Converts string to lower/upper case letters
unique()	Removes redundant data from the data set
getwd()	Returns current working directory of R
setwd()	Sets new working directory of R
seq(1,10,0.4)	Generates a sequence (1 -> 10, spaced by 0.4)
sequence()	Generates a vector of sequence
paste(data,sep="")	Concatenates string after using a separator symbol to separate two strings
ls()	Retrieves the list of all objects stored
rm()	Deletes object from memory
order(x)	Lists sorted element numbers of x

Figures 2.6 and 2.7 present an overview of how to use functions described in Table 2.1 in R programming.

```
> help()
> V<-c(10,20,30,40)
> append(V,50)
[1] 10 20 30 40 50
> V1=c(10,20,30,40,50)
> identical(V,V1)
[1] FALSE
> V
[1] 10 20 30 40
> length(V)
[1] 4
> range(V)
[1] 10 40
> rep(10,2)
[1] 10 10
> rev(V1)
[1] 50 40 30 20 10
> sort(V1)
[1] 10 20 30 40 50
> Str="hello"
> toupper(Str)
[1] "HELLO"
> tolower(Str)
[1] "hello"
> V3=c(10,10,34,45,78,90,90)
> V3
[1] 10 10 34 45 78 90 90
> unique(V3)
[1] 10 34 45 78 90
```

FIGURE 2.6
Execution of built-in function in R.

```
getwd()
[1] "C:/Users/megha.rathi"
> setwd("C:/Users/megha.rathi/Documents")
> seq(2,20,02)
[1]  2  4  6  8 10 12 14 16 18 20

> sequence(V)
 [1]  1  2  3  4  5  6  7  8  9 10  1  2  3
[14]  4  5  6  7  8  9 10 11 12 13 14 15 16
[27] 17 18 19 20  1  2  3  4  5  6  7  8  9
[40] 10 11 12 13 14 15 16 17 18 19 20 21 22
[53] 23 24 25 26 27 28 29 30  1  2  3  4  5
[66]  6  7  8  9 10 11 12 13 14 15 16 17 18
[79] 19 20 21 22 23 24 25 26 27 28 29 30 31
[92] 32 33 34 35 36 37 38 39 40
> paste(1:10,sep="")
[1] "1"  "2"  "3"  "4"  "5"  "6"  "7"  "8"
[9] "9"  "10"
> ls()
[1] "Str" "V"   "V1"  "V3"
> rm(V1)
> ls()
[1] "Str" "V"   "V3"
> order(V3)
[1] 1 2 3 4 5 6 7
>
```

FIGURE 2.7
Execution of built-in function in R.

Table 2.2 describes the list of mathematical functions used in R.

TABLE 2.2
List of Mathematical Functions Along with Their Description

Function	Description
sqrt()	Finds out the square root of a given number
sum()	Addition of numbers
log(x), log10()	Log(x) returns natural logarithm value and log(10) returns common logarithm
exp()	Finds out the exponential value
cos(),sin(),tan()	Hyperbolic functions
union(), intersect(), setdiff(), setequal()	Set operations
eigen()	Computes Eigen value and Eigen vector
ceiling()	Returns maximum value after rounding the numerical data
floor()	Returns minimum value after rounding off the numerical data

Figure 2.8 presents an overview of how to use mathematical functions in R programming.

```
> sqrt(16)
[1] 4
> sum(109090,6778)
[1] 115868
> log(10)
[1] 2.302585
> log10(10)
[1] 1
> exp(3)
[1] 20.08554
> cos(90)
[1] -0.4480736
> sin(90)
[1] 0.8939967
> tan(90)
[1] -1.9952
> v<-c(1,2,3,4,5)
> v1<-c(3,4,5,6,7,8,9)
> union(v,v1)
[1] 1 2 3 4 5 6 7 8 9
> intersect(v,v1)
[1] 3 4 5
> setdiff(v,v1)
[1] 1 2
> setequal(v,v1)
[1] FALSE
> ceiling(34.78)
[1] 35
> floor(34.55)
[1] 34
```

FIGURE 2.8
Execution of mathematical function in R.

Table 2.3 describes the list of statistical functions used in R.

TABLE 2.3

List of Statistical Functions Along with Their Description

Function	Description
lm()	Creates a linear model
glm()	For generalized linear model
cor.test()	Performs correlation test
cumsum(); cumprod(); cummin(); cummax()	Cumulative functions for vector
mad()	Finds out median absolute deviation
mean(x), weighted.mean(x), median(x), min(x), max(x), quantile(x),rnorm(), runif()	Produces random elements with Gaussian/ uniform distribution
sd()	Standard deviation
sample()	Random sample of data
summary(x)	Returns a summary of x: mean, min, max.
qqplot()	Draws quantile-quantile graph of data
var()	Finds out variance
ecdf()	Empirical cumulative distribution function
t.test()	Student's t-test

Table 2.4 describes the list of graphical functions used in R.

TABLE 2.4

List of Graphical Functions Along with Their Description

Function	Description
plot()	Draws a plot
curve(equation,add=T)	Plots an equation on curve
points(x,y)	Appends an additional set of points to an existing plot
arrows()	Plots arrows on a graph
abline()	Appends a straight line to an existing plot
lines()	Integrates points within the line
hist()	Draws a histogram
matplot()	Draws a column of matrices
pdf()	Graphs to a PDF file
png()	Graphs to a png file
jpeg	Graphs to a jpeg file
image()	Plots an image

2.5.2 User-Defined Functions in R

User-defined functions are a group of instructions organized together to accomplish the solution of a predefined problem. R provides the facility to create user-defined functions as per the user need. User-defined functions are also stored as objects. R interpreter transfers control to the calling function to complete its execution. Functions are created in R programming using the keyword *'function.'* The syntax of creating a function in R is as follows:

```
name_of_function<-function(arg1,arg2,.......)
{
        Function body
        return statements
}
```

Different components are provided in the syntax of functions and are defined as (Zamora Saiz et al., 2020):

- *Name_of_Function:* This is the actual name of the function and is saved in the R workspace in the form of an object with its name.
- *Arguments:* These are like variables. When a function is called in a real environment, values corresponding to each argument are passed. Parameters are optional. Also, an argument can take default values.
- *Function Body:* This is actually included in the group of statements to accomplish a predefined action. It describes the functionality of a function.
- *Return Value:* A return value is optional and is the last instruction in the function body. It specify the value returned from the function

A user can write user-defined functions in R. They are specific to the end-user requirement and, once coded, can be used several times and implemented like built-in functions in a program. Below is an example of how a function is created and used:

User-Defined Function Example in R

```
> # Creating First Function in R with no argument and no
return value
> func1<-function()
+ {
+ x<-10
+ y<-10
+ z<-x*y
+ print("Multiplication of two numbers:")
+ print(z)
+ }
> # Call Function
> func1()
[1] "Multiplication of two numbers:"
[1] 100
```

In the next example, a user-defined function is created with arguments but no return value:

```
> # Creating Function in R with arguments but no return value
> func2<-function(x,y)
+ {
+ z<-x*y
+ print("Multiplication of two numbers:")
+ print(z)
+ }
> # Function Calling
> func2(10,50)
[1] "Multiplication of two numbers:"
[1] 500
```

In another example, a user-defined function is created with arguments and return value:

```
> # Creating User Defined Function with argument and return
value
> func3<-function(x,y)
+ {
+ z<-x*y
+ print("Multiplication of two numbers:")
+ return(z)
+ }
> # Function Calling
> func3(34,45)
[1] "Multiplication of two numbers:"
[1] 1530
```

2.6 Programming Example: Conditional Statements, Loops, and Functions

In this section, we provide some useful programs to gain understanding of all the concepts studied in this chapter.

Program 1: R program to check whether a given number is Armstrong or not

Any number is an Armstrong number if the sum of the cube of its digits equals the number itself. For instance, 370 is an Armstrong number since 370 = 3*3*3 + 7*7*7 + 0*0*0.

Program 1 is implemented using the *if..else* conditional statement and *while loop*. As shown in the output 370 is identified as an Armstrong number. Figure 2.9 presents the algorithmic detail of the program. It illustrates all the steps used in the computation to find out an Armstrong number.

```
Console   Terminal ×
~/
> # R Program 1 (Check Armstrong Number)
> n<-as.integer(readline(prompt = "Enter the Number:"))
Enter the Number:370
> sum<-0
> temp<-n
> while(temp>0)
+ {
+ d<-temp%%10
+ sum<-sum+(d^3)
+ temp<-floor(temp/10)
+ }
> # Program Output
> if(n==sum){
+ print(paste(n,"Given number is Armstrong Number"))
+ }else{
+ print(paste(n,"Given number is not an Armstrong Number")
)}
[1] "370 Given number is Armstrong Number"
```

FIGURE 2.9
R program to check Armstrong number.

Program 2: R program to find out the sum of natural numbers

The sum of natural numbers is given by the formula: n*(n+1)/2. For example, if the value of n = 4 then sum = 4*(4+1)/2=10. Program 2 is implemented again using the *if..else* statement and *while loop*. Figure 2.10 presents the output along with the logic to find out the sum of natural numbers.

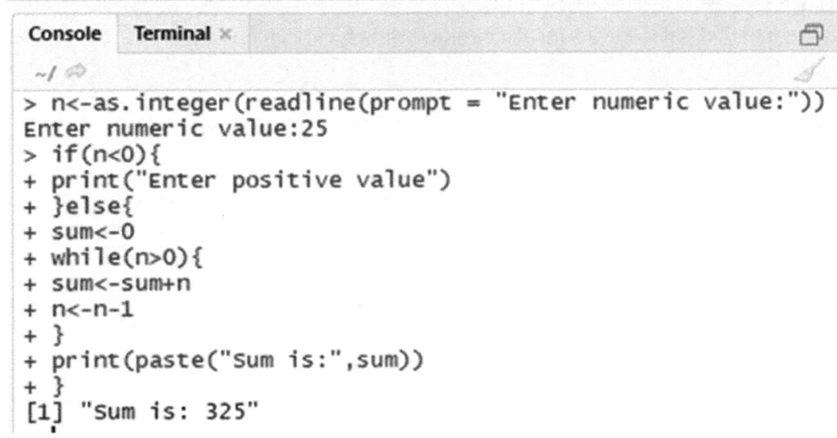

FIGURE 2.10
R program to find the sum of natural numbers.

Program 3: R program to find out the factorial of a number using recursion

Incorporating the concept of recursion makes the program easy, reduces the line of code, and is a clean way of coding. But sometimes it is difficult to write the logic for recursive programming because finding out the solution in a recursive manner is quite difficult. Recursive functions are memory intensive, since they call the function in an iterative manner.

The factorial of any number is given by the equation n*(n-1). In the program written below, we create a recursive function 'recursive_fact(x)' which will return the value until x equates to '0' When the value of x is zero, 1 is returned as the factorial of zero. Figure 2.11 presents the recursive factorial program.

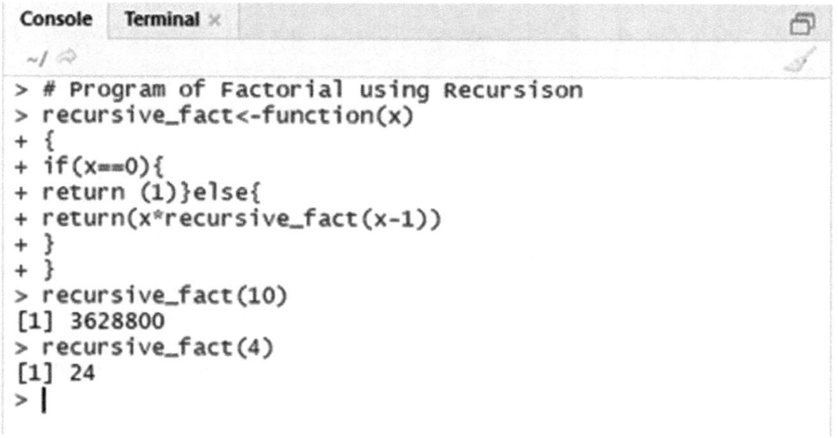

FIGURE 2.11
R program to find out the factorial of a number using recursion.

Program 4: R program to generate a Fibonacci sequence using recursion

The Fibonacci series is computed by summing up the last two elements. The initial two values are: 0 and 1. All other series if generated by adding the last data values. The equation to find out the Fibonacci series of 'n' terms is given by the formula: (n-1) + (n-2). For example, if we want to generate the Fibonacci sequence of 10 numbers, then the sequence is:

0,1,1,2,3,5,8,13,21,34.

Figure 2.12 presents the complete program along with the output to generate the Fibonacci sequence in a recursive manner.

```
># Function to generate Fibonacci series up to n terms
> fibonacci <- function(n) {
+     if (n <= 1) {
+         return(n)
+     } else {
+         return(fibonacci(n - 1) + fibonacci(n - 2))
+     }}
> # Function to check if the input is a positive integer
> is_positive_integer <- function(x) {
+     return(!is.na(as.numeric(x)) && as.numeric(x) > 0 && as.numer
ic(x) %% 1 == 0)
+ }
> # Function to print the full Fibonacci series up to n terms
> fibonacci_series <- function(n) {
+     cat("Fibonacci series up to", n, "terms:\n")
+     for (i| in 0:(n - 1)) {
+         cat(fibonacci(i), " ")}
+     cat("\n") # For a clean line break after the series}
> # Main program
> n <- readline(prompt = "Enter a positive integer: ")
Enter a positive integer: # Check if the input is valid
> if (is_positive_integer(n)) {
+     n <- as.numeric(n)
+     # Call the function to display the Fibonacci series
+     fibonacci_series(n)
+ } else {
+     cat("Please enter a valid positive integer.\n")}
Please enter a valid positive integer
> fibonacci_series(10)
Fibonacci series up to 10 terms:
0  1  1  2  3  5  8  13  21  34
```

FIGURE 2.12
R program to generate Fibonacci sequence using recursion.

Program 5: R program to generate a table of any number

'n' is the data input by the user. Then the iterative expression is used to generate a table of any number 'n.' The loop initializes from 1 and works till the value becomes 10; for generating a table command, this is as follows: print(paste(n,'X',i,':=',n*i)). Figure 2.13 presents the program.

```
Console  Terminal
R 4.2.2 · ~/
> # R program to generate table of any number
> n<-as.integer(readline(prompt = "Enter Data:"))
Enter Data:23
> for(i in 1:10)
+ {
+ print(paste(n,'x',i,'=',n*i))
+ }
[1] "23 x 1 = 23"
[1] "23 x 2 = 46"
[1] "23 x 3 = 69"
[1] "23 x 4 = 92"
[1] "23 x 5 = 115"
[1] "23 x 6 = 138"
[1] "23 x 7 = 161"
[1] "23 x 8 = 184"
[1] "23 x 9 = 207"
[1] "23 x 10 = 230"
```

FIGURE 2.13
R program to generate table of any number.

Program 6: R program to generate random data from standard deviations (SDs)

The R language provides functions to produce random data from the SD like uniform distribution, normal distribution, and binomial distribution. For checking the complete list of all SDs: type the command '? Distribution' on the R console. The details will be displayed on the lower-right portion of the RStudio interface, as shown in Figure 2.14.

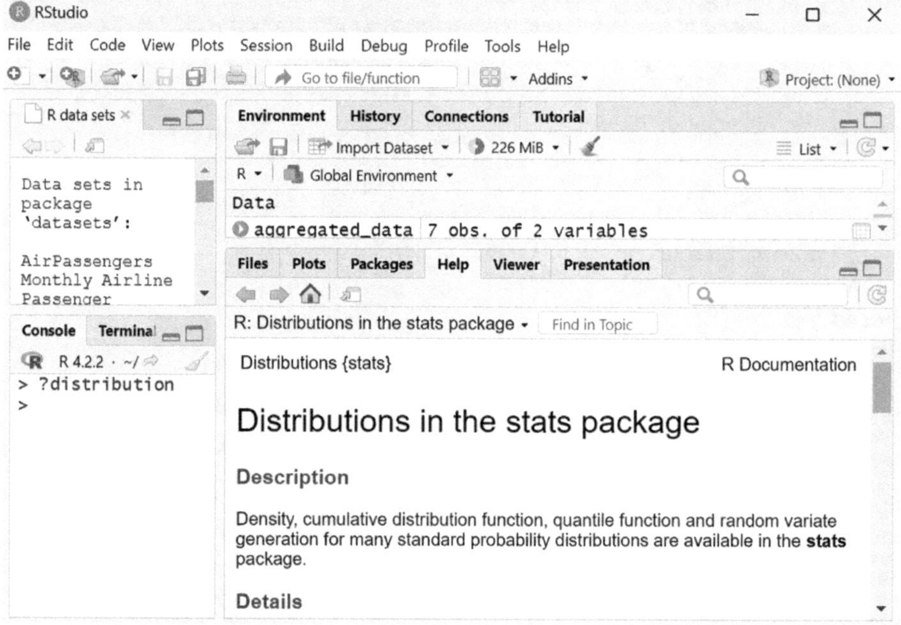

FIGURE 2.14
R command to generate the list of all distributions.

In R programming, all functions for random data start with 'r.' Consider the example for generating a random number from a uniform distribution and the available function is: *'runif()'* and for generating normal distribution function it is: *'rnorm().'*

In Figure 2.15, we have presented the output of random number generation for uniform and binary distribution. Binary distribution works by searching the occurrence of an even success. For an instance, tossing a coin has two possible outcomes: head or tail. R language provides the function *dbinom (data, size, probability)* and others to generate random binomial distribution. In the example, we have used the functions *runif()* and *dbinom()* for random number generation.

```
Console   Terminal ×
R 4.2.2 · ~/
> # Random number generation for Uniform Distribution
> runif(1)
[1] 0.1029247
> runif(2)
[1] 0.8998250 0.2460877
> runif(5,min=2,max=10)
[1] 2.336476 4.623366 9.636029 9.116315 7.542427
> #Random number generation for Binary Distribution
> # Function named dbinom() is used to generate probabilistic density
distribution
> data<-seq(50,100,by=2)
> dd<-dbinom(data,100,0.5)
> dd
 [1] 7.958924e-02 7.352701e-02 5.795840e-02 3.895256e-02
 [5] 2.229227e-02 1.084387e-02 4.472880e-03 1.559739e-03
 [9] 4.581053e-04 1.128170e-04 2.317069e-05 3.943369e-06
[13] 5.518672e-07 6.293223e-08 5.783981e-09 4.228163e-10
[17] 2.419003e-11 1.061697e-12 3.485735e-14 8.286361e-16
[21] 1.365543e-17 1.467975e-19 9.403635e-22 3.093301e-24
[25] 3.904861e-27 7.888609e-31
```

FIGURE 2.15
R program to generate uniform and binomial distribution.

Program 7: R menu–driven program for a simple calculator that can add data elements, find out the difference, and multiply and divide numbers using functions (Figure 2.16)

```
RStudio
File Edit Code View Plots Session Build Debug Profile Tools Help
Go to file/function   Addins
Console   Terminal ×
~/
> # R Program for Simple Calculator using Function
> sum<-function(a,b)
+ {
+ return(a+b)
+ }
> diff<-function(a,b)
+ {
+ return(a-b)
+ }
> mul<-function(a,b)
+ {
+ return(a*b)
+ }
> div<-function(a,b)
+ {
+ return(a/b)
+ }
> # Take Input from User
> print("SELECT ANY OPERATION:")
[1] "SELECT ANY OPERATION:"
> print("1. ADD")
[1] "1. ADD"
> print("2. DIFF")
[1] "2. DIFF"
> print("3. MULTIPLY")
[1] "3. MULTIPLY"
> print("4. DIVIDE")
[1] "4. DIVIDE"
> option=as.integer(readline(prompt = "Select any option [1/2/3/4]"))
Select any option [1/2/3/4]4
> data1<-as.integer(readline(prompt = "Enter first data"))
Enter first data23
> data2<-as.integer(readline(prompt = "Enter second data: "))
Enter second data: 45
> operator<-switch(option,"+","-","*","/")
> output <- switch(option, add(data1, data2), diff(data1, data2), mul(data1, data2), div(data1, data2))
> option
[1] 4
> operator
[1] "/"
> output
[1] 0.5111111
> |
```

FIGURE 2.16
R menu–driven program for a simple calculator.

2.7 Summary

- R language provides all the constructs for decision-making like other general-purpose programming languages. It supports *if*, *if..else*, and *nested if* conditional statements.
- The *if* conditional statement starts the execution if any of the supplied conditions become true. Conditions are also provided in the *if..else* statement; the only difference is that it works for false conditions as well.
- *Nested if..else* conditional statements are useful in a scenario when the user has a huge count of conditions that need to be validated before making any decision.
- R programming language supports the loops *for*, *while*, and *repeat*. These are three different styles of loops supported by R.
- The *for* loop constantly iterates either a single expression or a group of expressions and terminates when the criteria is met. The *for* loop is utilized when we want to repeatedly execute over a list of data.
- The *while loop* first checks the test condition and after verifying the condition, it starts the execution and constantly iterates groups of statements until the criteria is met.
- The last type of loop in R is *repeat* loop. It works similarly to the *while loop*; the only difference is that it has no test condition. The *repeat* loop repeatedly executes groups of expressions and finally turns into an infinite loop. So the loop terminating statement *'break'* is required to exit from the loop.
- To change the normal sequence of loop execution, we utilize loop control statements. R language supports two loop control structures: *break* and *next*. Loop control statements are executed and implemented to reduce the overall complexity as it exits early; thus, the result is reducing total memory and time required.
- *break* is used within the loop to alter the normal sequence of loop and programming control is transferred to the immediate next statement after loop. *break* is also used in a switch statement for ending the case.
- *next* is used to skip the current iteration of a loop without exiting from the loop. Once the *next* statement is encountered in the loop, programming control transfers to the succeeding iteration as a consequence, and the current iteration is skipped. Another usage of *next* is with *if..else* conditional statements.
- Vectorization is another very important programming concept in R programming. It is used to replace an iterative construct in R. For example, it provides the easiest way to add two matrices or vectors;

otherwise, a loop is required to perform both operations in other coding languages. With the help of vectorization, we can apply repeated operations on vectors; thus, replacing loops.

- A script in R is a text file used to store sets of statements or R commands together. An extension is used to save the script in R: '.r' For writing new script, you can use any editor like Notepad or text editor. The command used to run script in R is *'source.'*
- Scripts are saved in the current workspace of R. For extracting the path of the current workspace, the command is *'getwd().'* If you want to set a new path of the R workspace, the command is *'setwd().'*
- A function is described as a portion of code that is implemented to achieve a specific task. A function is a piece of code that executes on numerous inputs and generates output according to a specific function operation.
- Both built-in and user defined functions are supported by R programming. R has an abundant list of built-in functions for data analysis. It has functions available for statistics, string handling, data preprocessing, and data visualization.
- A user-defined function is created to meet the end-user requirement. They are created for a targeted specific need and a 'function' is used to create functions in R.

EXERCISES

I. Multiple-Choice Questions

1. Find out the total random number generated by the command: `d<-runif(4,5,55)`
   a. 5
   b. 4
   c. 9
   d. 55
2. Which of the following expressions is used to control the loop explicitly?
   a. exit
   b. for
   c. goto
   d. break

3. What will be the output of the following code?

```
> n<-c("x","y","z")
> for(i in 1:5)
+ {
+ print(n[i])
+ }
```

a.
```
[1] "x"
[1] "y"
[1] "z"
[1] "x"
[1] "y"
```
b.
```
[1] NA
[1] NA
[1] NA
[1] NA
[1] NA
```
c.
```
[1] "x"
[1] "y"
[1] "z"
[1] NA
[1] NA
```
d. None of the above

4. What will be the output of the following code?

```
> n<-5
> switch (1,2+2,mean(1:10),rnorm(5))
```

a. 5
b. 4
c. 7
d. 9

5. Which of the following codes can be used to avoid numeric problems such as taking the logarithm of a negative number?

a. if (any(x < 0)) y <- log(1+x) else y <- log(x)
b. if (any(x >= 0)) y <- log(1+x) else y <- log(x)
c. if (any(x <= 0)) y <- log(1+x) else y <- log(x)
d. None of the above

6. What will be the output of the following code?

```
> fun<-function(n=1)
+ {
+ var<-"Welcome"
+ for(i in seq_len(n))
+ {
```

```
+ cat(var)
+ }
+ c<-nchar(var)*n
+ c
+ }
> fun()
```

a. Welcome[1] 7
b. Welcome[1] 8
c. Welcome[1] 9
d. None of the Above

7. Which of the following arguments is used to print NULL?
   a. arg(paste)
   b. print(paste(arg))
   c. args(paste)
   d. All of the above
8. Which of the following is used to take runtime input from the console?
   a. read()
   b. paste()
   c. readline()
   d. cat()
9. Which of the following is not a built-in function in R?
   a. apply()
   b. mapply()
   c. sapply()
   d. All of the Above
10. What will be the output of the following code?

```
> n<-list(x=1:10,y=rnorm(20))
> lapply(n,mean)
```

a.
```
$x
[1] 5.0
$y
[1] -0.09012609
```

b.
```
$x
[1] 5.5
$y
[1] -0.08000000
```

c.
```
$x
[1] 5.5
$y
[1] -0.09012609
```

d.
```
$x
[1] 5.0
$y
[1] -0.010000
```

11. What will be the output of the following code?

```
> n<-list(x=1:10,y=rnorm(20))
> sapply(n,mean)
```

a.
```
         x           y
5.50000000 -0.09012609
```

b.
```
         x           y
5.00000000 -0.089677777
```

c.
```
         x           y
6.50000000 -0.09012609
```

d.
```
         x           y
7.50000000 -0.09012609
```

12. Which of the following functions is used to apply a function over the vector subset?
    a. sapply()
    b. tapply()
    c. mapply()
    d. vapply()
13. Which of the following functions is the same as that of sapply?
    a. apply()
    b. mapply()
    c. lapply()
    d. All of the above
14. What is the number of arguments supplied to the lapply function in R?
    a. 2
    b. 1
    c. 3
    d. 4

15. Which of the following code creates n samples of size 'size' with probability prob from the binomial?
    a. z <- rinom(n,size,prob)
    b. z <- rbinom(n,size,prob)
    c. z <- binom(n,size,prob)
    d. All of the above

II. Brief Questions

1. Describe functions. How many types of functions are supported by R programming? Explain with an example.
2. Describe the commonly used general built-in functions in R programming.
3. Describe the user-defined functions with examples.
4. Explain the *break* and *next* statements. Also, write down the differences between them.
5. Explain all the conditional statements used in R programming.
6. Describe the different types of loops used in R. Also, write down the difference in all the loops. Which one is the best and why?
7. Describe the benefits of scripts. Also, write command to run scripts in R programming.
8. Explain why *switch-case* statements are avoided most often. Also, describe switch case and in which scenario it is utilized in programming.

III. Coding Exercise

1. Write the R command to generate a sequence of odd and even numbers from 1 to 30.
2. Consider the given data: x(3,67,89,90,23,10,56,4,7,100,9). Write the R command to find the maximum and minimum value. Also, print the index (i.e., print the location of the maximum and minimum).
3. Write the R function to implement the concept of bubble sort.

4. Consider the given Zeller's formula:

   F = ([2.6m-0.2])+k+y+[y/4]+[c/4]-2c) mod 7, where [x] denotes the integer part of x (e.g., [9.5]=9).

   Zeller's congruence is used to return the day of the week, F, such that k = day, y= year, c= initial two digits of year, and m= month (where January is month 11 of the preceding year, February is month 12 of the preceding year, March is month 1).

   For example, Date: 2/2/1983 has m=12, k=2, c=19, and y=83.

5. Write a program to implement a quick sort algorithm. Also, test the result for a given vector: n (1,6,8,9,10,2,3,-9,56).
6. Create a function to accept a Customer data frame (CName, Gender, Job_profile, Cid) and print the first and second customer, as well as the Names and Cid of all the customers.
7. Create a function to print a number raised to another with one argument a default argument. Also, write down the function to print the class of input data.
8. Create the given series: 'D1' 'D2' 'D3' 'D4' 'D5.' Also, write R commands for the following: a) to display data elements 1 to 4, b) to display data elements other than 3 and 4.
9. Write a function f(x) such that x is the set of vectors: x=(x1,x2,x3, x4….,xn). Function f(x) returns the list of moving averages:

   x1+x2+x3/3, x2+x3+x4/3,……,xn-2+xn-1+xn/3. Test this function for the given data: f(c(1:3,4:1).

10. Write an R function that returns the real roots of the quadratic ax2 + bx + c. The function should take a, b, and c as arguments and return the appropriate messages if the values entered do not specify a quadratic or if there are no real roots. Use the function to determine the roots of $5x^2 - x - 8$.

References

Ahmad, W. M. A. W., Ghazali, F. M. M., & Yaqoob, M. A. (2023). Basic statistical analysis using RStudio Software. Penerbit USM.

Albright, S. C., Winston, W. L., Zappe, C. J., & Broadie, M. N. (2011). Data analysis and decision making (Vol. 577). South-Western/Cengage Learning.

Alvaro, F. (2016). R: Easy R programming for beginners, your step-by-step guide to learning R programming. CreateSpace Independent Publishing Platform.

Crawley, M. J. (2012). The R book. John Wiley & Sons.

Davies, T. M. (2016). The book of R: A first course in programming and statistics. No Starch Press.
Hafner, S. (2019). An introduction to R for beginners. Research gate.
Jones, E., Harden, S., & Crawley, M. J. (2022). The R book. John Wiley & Sons.
Lawrence, M., & Verzani, J. (2018). Programming graphical user interfaces in R. Chapman and Hall/CRC.
Pace, L. (2012). Beginning R: An introduction to statistical programming. Apress.
Paradis, E. (2022). R programming and development from basics to advanced topics. ISEM.
Ren, K. (2016). Learning R programming. Packt Publishing Ltd.
Saiz, Z., & Rosteck (2020). An introduction to data analysis in R. Springer International Publishing.
Sudhamathy, G., & Venkateswaran, C. J. (2019). R programming: An approach to data analytics. MJP Publisher.
Trejo, O. (2017). R programming by example: Practical, hands-on projects to help you get started with R. Packt Publishing Ltd.
Zamora Saiz, A., Quesada González, C., Hurtado Gil, L., Mondéjar Ruiz, D., Zamora Saiz, A., Quesada González, C., … & Mondéjar Ruiz, D. (2020). Introduction to R. In An Introduction to Data Analysis in R: Hands-On Coding, Data Mining, Visualization and Statistics from Scratch (pp. 9–67). Springer.
Zuur, A. F., Ieno, E. N., & Meesters, E. H. (2009). A beginner's guide to R (p. 150). New York: Springer.

3

Data Structures in R

Learning Objectives:

After studying this chapter, students and researchers will be able to:

- Understand the concept of data structures and utility of data structures in R programming
- Have an insight on various data structures used in R programming
- Discuss vectors along with all the operations
- Explain the concepts of matrix and array in R
- Discuss list and recursive list
- Explain the concept of data frames
- Explain the concept of factors
- Explain and illustrate all data structures along with different operations
- Briefly summarize all the concepts

3.1 Introduction

To build the pre-eminent model from R programming, users will have a great understanding of data structures and how to perform various operations on data structures (Mehta & Sahni, 2004). R programming also needs to utilize different variables to store data efficiently. Variables take some piece of memory to save data; when you initialize any variable, you book some memory location for that variable. Additionally, data structures are the way to organize data so that one can utilize it for further processing on data. R programming is an object-oriented coding technique; thus, all data structures are objects in R (Tanner-Smith et al., 2016). You cannot directly access

DOI: 10.1201/9781003489382-3

TABLE 3.1
List of Data Structures Used in R

Dimension	Homogeneous	Heterogeneous
One-dimensional	Vector	List
Two-dimensional	Matrix	Data Frame
N-dimensional	Array	

the memory locations of data objects; instead, all data objects are accessed via a variable that holds that object. It is crucial to gain better understanding of all data structures because these are required to perform day-to-day operations in R.

Data structures in R can be structured according to their dimensionality (i.e., one-dimensional, two-dimensional, or N-dimensional) and it also depends on whether they are homogeneous or heterogeneous. In R programming, for data analysis, five types of data structures (Rao et al., 2019) can be used, which are categorized as homogeneous (data stored is of the same type) and heterogeneous (different types of data are organized) data structures, are provided in Table 3.1.

R programming provides five basic types of data structures: (1) vector, (2) matrix, (3) array, (4) list, and (5) data frame for storing and executing different operations on data. Vector, matrix, and array are all examples of homogeneous data storage structures and they can store data of the same type, while list and data frame are examples of heterogeneous data storage structures that can store integer, real, Boolean, and string in one place (Prakash & Rao, 2016). For statistical modelling we are interested in a data structure which holds different types of data in one place, so in the context of data analysis and statistical modelling, list and data frame are of great use (Braun & Murdoch, 2021). Operations like addition, deletion, multiplication, splitting, and merging can be executed directly on all the above-mentioned data structures. This chapter discusses all these data structures in detail and the way to write and execute them in R programming.

3.1.1 Vectors

A vector is the most basic type of data structure used in R programming (Matloff, 2011). A vector is an example of a homogeneous data storage structure; that is, it holds data of same type. It can store data types such as integer, real, character, Boolean, complex, or raw. Data elements of vectors are often termed components.

c() is used to create vectors in R. Function length() is used to find out the total count of data elements stored within a vector. The typeof () function is used to find out the type of vector.

3.1.2 Vector Creation

Vectors are created in R using function *c()*. In the example shown below, a vector of a real data type is created. To check the type of vector, the typeof() function is used. If the vector is comprised of a mix of data, then the typeof() method will return the highest data type amongst all. The length() function is used to return the total count of data elements stored within a single vector (Cotton, 2013).

Vector Creation Example in R

```
> V<-c(12.56,78.9,45.7,88.9,90.0)
> V
[1] 12.56 78.90 45.70 88.90 90.00
> typeof(V)
[1] "double"
> length(V)
[1] 5
> V1<-c(10,89.9,TRUE,"R Programming")
> V1
[1] "10"              "89.9"              "TRUE"
[4] "R Programming"
> typeof(V1)
[1] "character"
```

Vector Creation Example Using ':' Operator and 'seq()' Function in R

```
> # Vector creation using ":"
> V<-1:10
> V
[1]  1  2  3  4  5  6  7  8  9 10
> typeof(V)
[1] "integer"
> length(V)
[1] 10
> #Vector creation using "seq()"
> V1<-seq(1,5,by=0.4)
> V1
[1] 1.0 1.4 1.8 2.2 2.6 3.0 3.4 3.8 4.2 4.6 5.0
> length(V1)
[1] 11
> typeof(V1)
[1] "double"
```

3.1.3 Accessing Vector Components

To access the components of a vector, indexing is used in R programming. Indexing is denoted as: [ ] brackets. It always starts with a value or position 1.

If you want to drop any component from the vector, use negative index. Other ways to represent indexing are TRUE, FALSE or 0 and 1.

***Accessing Vector Components* Example in R**

```
> # Indexing used to access Vector Elements
> Mon<-
c("jan","feb","mar","apr","may","jun","jul","aug","sep","oct",
"nov","dec")
> Mon[c(8,9,10)]
[1] "aug" "sep" "oct"
> # Accessing vector element using negative indexing
> Mon[c(-2,-5)]
  [1] "jan" "mar" "apr" "jun" "jul" "aug" "sep" "oct"
  [9] "nov" "dec"
> # Accessing element using 0/1
> Mon[c(0,0,0,1,0,0,1,1,1,1,0,1)]
[1] "jan" "jan" "jan" "jan" "jan" "jan"
> Mon[c(2,3,1,0,0,9)]
[1] "feb" "mar" "jan" "sep"
> Mon[c(0,-7)]
  [1] "jan" "feb" "mar" "apr" "may" "jun" "aug" "sep"
  [9] "oct" "nov" "dec"
> # Accessing data elements using Logical Index
> v<-c('a','b','c','d','e')
> b<-c(FALSE,TRUE,TRUE,FALSE,TRUE)
> v[b]
[1] "b" "c" "e"
```

3.1.4 Vector Operations

We can add, subtract, multiply, or divide two vectors producing the output as a vector. Other operations that we can apply on vectors are sort the vector, recycle the vector components, and merge two vectors (Wickham, 2019). Given below is the example showing the outcome of the vector after application of various different vector operations.

***Operation 1: Vector Arithmetic* Example in R**

```
> v1<-c(10,23,45,67,89,2)
> v2<-c(34,45,44,4,5,6)
> v1+v2
[1] 44 68 89 71 94  8
> v1-v2
[1] -24 -22   1  63  84  -4
> v1*v2
[1]  340 1035 1980  268  445   12
> v1/v2
[1]  0.2941176  0.5111111  1.0227273 16.7500000
[5] 17.8000000  0.3333333
```

Operation 2: Sort Vector Example in R

```
> v1<-c(10,23,45,67,89,2)
> v2<-c(34,45,44,4,5,6)
> sort(v1)
[1]  2 10 23 45 67 89
> sort(v2)
[1]  4  5  6 34 44 45
> r<-sort(v1,decreasing=TRUE)
> r
[1] 89 67 45 23 10  2
```

Operation 3: Recycle Vector Components Example in R

The application of an arithmetic operation to two unequal length vectors can lead to recycling the data elements of a shorter vector in order to complete the specified arithmetic operation. The example shown below recycles the data elements of vector 3 (v3) with length =2. v3=(10,11) and after applying the addition/subtraction operation, v3 will become: (10,11,10,11,10,11).

```
> v1
[1] 10 23 45 67 89  2
> v3<-c(10,11)
> v1+v3
[1] 20 34 55 78 99 13
> v1-v3
[1]  0 12 35 56 79 -9
```

Operation 4: Assign a Name to Vector Components Example in R

We can assign names to vector data elements, as shown below:

```
> v<-c("john","smith")
> v
[1] "john"  "smith"
> #Assign names
> names(v)<-c("First","Last")
> v
  First     Last
  "john" "smith"
> # Extracting the first member by name
> v["First"]
  First
  "john"
> # Reverse the order while displaying on screen
> v[c("Last","First")]
   Last   First
"smith"  "john"
```

Operation 5: Vector Modification Example in R

You can add, delete, or update the data elements in a vector after the creation of the vector in R:

```
> # Vector Modification & Deletion
> # Modify a Vector
> v<-c(-30,-20,-10,10,20,30)
> #Modify third element
> v[3]<-30
> v
[1] -30 -20  30  10  20  30
> #Modify all negative numbers
> v[v<0]<-100
> v
[1] 100 100  30  10  20  30
> v<-v[1:3]
> v
[1] 100 100  30
> # Vector Deletion
> v
[1] 100 100  30
> v<-NULL
> v
NULL
> v[1:3]
NULL
> # Adding Data elements to existing vector
> v
[1]  5  4  5  6  7  8 10
> v[8:9]<-c(67,78)
> v
[1]  5  4  5  6  7  8 10 67 78
```

3.2 Matrices

In R programming, a matrix is stored as an object and is defined as the arrangement of data stored in a two-dimensional rectangular shape. A matrix is another example of a homogeneous data structure (i.e., it contains all data elements of the same type). Arithmetic operations can be applied directly in R on a few matrix elements or on the complete matrix. In R, a matrix is created using the function matrix() (Wickham, 2019). The only difference between vector() and matrix() is that the vector does not have a dimension parameter. We have an additional parameter dimension to set the dimension of a matrix. The attributes() method is used to check all the attributes of a data object. The basic syntax to create a matrix in R is as follows:

matrix(data, nrow, ncol, byrow, dimension)

where *'data'* is the data to be inserted in the matrix,

'nrow' specify the total number of rows,

'ncol' specify the total number of columns,

'byrow' specify that data elements are arranged by row. It is a logical value if set FALSE arranges data elements by column and,

'dimension' is used to set the dimension of the matrix.

3.2.1 Matrix Creation

A matrix is the arrangement of data elements arranged into a fixed number of rows and columns. By using matrix() function we can generate a rectangular representation of data in R (Searle & Khuri, 2017). In order to set the dimension of a matrix, arguments must be passed to parameters 'nrow' and ncol.' Setting both dimensions is not mandatory. You can pass the value for one dimension and the other value is inferred from the length of data. Given below is the example of a matrix creation in R:

Matrix Creation Example in R

```
> M<-matrix(10:18,nrow=3,ncol=3,byrow="TRUE")
> M
      [,1] [,2] [,3]
[1,]    10   11   12
[2,]    13   14   15
[3,]    16   17   18
> M<-matrix(10:18,nrow=3,ncol=3,byrow="FALSE")
> M
      [,1] [,2] [,3]
[1,]    10   13   16
[2,]    11   14   17
[3,]    12   15   18
> M<-matrix(10:18,nrow=3,byrow="FALSE")
> M
      [,1] [,2] [,3]
[1,]    10   13   16
[2,]    11   14   17
[3,]    12   15   18
```

A matrix can also be created using a vector by setting the dim() attribute, as shown in the given example:

```
> # Matrix creation from a vector by using dim()
> M<-c(1,2,3,4,5,6,7,8,9)
> class(M)
[1] "numeric"
> dim(M)<-c(3,3)
> M
     [,1] [,2] [,3]
[1,]    1    4    7
[2,]    2    5    8
[3,]    3    6    9
```

Another method of matrix creation in R is by using the functions cbind() and rbind(). Following is an example that illustrates the creation of a matrix using cbind() and rbind():

```
> #Matrix creation using cbind() and rbind() function
> cbind(c(1,2,3),c(4,5,6),c(7,8,9))
     [,1] [,2] [,3]
[1,]    1    4    7
[2,]    2    5    8
[3,]    3    6    9
> rbind(c(10,11,12),c(13,14,15))
     [,1] [,2] [,3]
[1,]   10   11   12
[2,]   13   14   15
```

3.2.2 Accessing the Elements of a Matrix

Data elements can be accessed by specifying the column and row index. Provided below is the R code to access matrix data elements. To access the element of the m row and n column, the command is: M[m,n]. For example, to access the data element of 2 row and 2 column, we must type: M[2,2].

Access Data Elements of Matrix **Example in R**

```
> M
     [,1] [,2] [,3]
[1,]   10   13   16
[2,]   11   14   17
[3,]   12   15   18
> #Accessing 2 column 2 row
> print(M[2,2])
[1] 14
> #Accessing data element of 1 column and 2 row
> print(M[1,2])
[1] 13
```

```
> #Acess only the 3 row
> print(M[3,])
[1] 12 15 18
> #Access only the first column
> print(M[,1])
[1] 10 11 12
> print(M[1:2])
[1] 10 11
> # Access 1 and 3 column
> M[,c(1,3)]
     [,1] [,2]
[1,]   10   16
[2,]   11   17
[3,]   12   18
```

3.2.3 Matrix Operations

In this section, we will discuss the various matrix operations. We can apply various arithmetic operations on a matrix in R (Fieller, 2018). It is a mandate to use the same dimensional matrices in order to perform various operations on two matrices. All the operations on a matrix are discussed below with an example.

Operation 1: Matrix Arithmetic Operation **Example in R**

```
> M1<-matrix(1:9,nrow=3,ncol=3,byrow="TRUE")
> M1
     [,1] [,2] [,3]
[1,]    1    2    3
[2,]    4    5    6
[3,]    7    8    9
> M
     [,1] [,2] [,3]
[1,]   10   13   16
[2,]   11   14   17
[3,]   12   15   18
> # Matrix Sum
> sum<-M+M1
> sum
     [,1] [,2] [,3]
[1,]   11   15   19
[2,]   15   19   23
[3,]   19   23   27
> #Matrix Diff
> diff<-M-M1
> diff
     [,1] [,2] [,3]
[1,]    9   11   13
[2,]    7    9   11
[3,]    5    7    9
```

```
> #Matrix Multiplication
> pdt<-M*M1
> pdt
     [,1] [,2] [,3]
[1,]   10   26   48
[2,]   44   70  102
[3,]   84  120  162
> #Matrix Division
> div<-M/M1
> div
          [,1]  [,2]     [,3]
[1,] 10.000000 6.500 5.333333
[2,]  2.750000 2.800 2.833333
[3,]  1.714286 1.875 2.000000
```

Operation 2: Matrix Modification Example in R

It is possible to update the data elements stored within the matrix. You can add, delete, or modify the content of a matrix. The example shown below adds, deletes, and updates the data elements of any given matrix.

```
> M
     [,1] [,2] [,3]
[1,]    1    4    7
[2,]    2    5    8
[3,]    3    6    9
> # Altering the 2 row 2 column element
> M[2,2]<-15
> M
     [,1] [,2] [,3]
[1,]    1    4    7
[2,]    2   15    8
[3,]    3    6    9
> # Change all the elements with value less than 6
> M[M<6]<-20
> M
     [,1] [,2] [,3]
[1,]   20   20    7
[2,]   20   15    8
[3,]   20    6    9
> # Delete 3 column
> M=M[,-3]
> M
     [,1] [,2]
[1,]   20   20
[2,]   20   15
[3,]   20    6
> # Delete 3 row
> M=M[-3,]
> M
```

```
     [,1] [,2]
[1,]   20   20
[2,]   20   15
> M<-matrix(1:9,ncol=3)
> M
     [,1] [,2] [,3]
[1,]    1    4    7
[2,]    2    5    8
[3,]    3    6    9
> # Add row to the end
> rbind(M,c(10,11,12))
     [,1] [,2] [,3]
[1,]    1    4    7
[2,]    2    5    8
[3,]    3    6    9
[4,]   10   11   12
> # Insert row before 3 row
> rbind(M[1:2,],c(13,14,15),M[3,])
     [,1] [,2] [,3]
[1,]    1    4    7
[2,]    2    5    8
[3,]   13   14   15
[4,]    3    6    9
> # Add column
> cbind(M,c(20,20,20))
     [,1] [,2] [,3] [,4]
[1,]    1    4    7   20
[2,]    2    5    8   20
[3,]    3    6    9   20
> M
     [,1] [,2]
[1,]    1    5
[2,]    2    6
[3,]    3    7
[4,]    4    8
> #Changing dimension of Matrix
> dim(M)<-c(2,4)
> M
     [,1] [,2] [,3] [,4]
[1,]    1    3    5    7
[2,]    2    4    6    8
```

Operation 3: Transpose a Matrix Example in R

The function t() is used to find out how to transpose a given matrix.

```
> M
     [,1] [,2] [,3]
[1,]    1    4    7
[2,]    2    5    8
[3,]    3    6    9
```

```
> t<-t(M)
> t
     [,1] [,2] [,3]
[1,]    1    2    3
[2,]    4    5    6
[3,]    7    8    9
```

Operation 4: Summary Functions on a Matrix Example in R

In order to gain more insight of data, we must apply statistical or mathematical functions on data elements of a matrix. Analyzing data involves finding out the max, mean, min, median, and standard deviation. Listed below is the set of commands which presents the result of summary functions applied on data elements of a matrix.

```
> M
     [,1] [,2] [,3]
[1,]    1    4    7
[2,]    2    5    8
[3,]    3    6    9
> # Summary Functions
> r1<-M[1,]
> max(r1)
[1] 7
> r1
[1] 1 4 7
> max(M[3,])
[1] 9
> min(M[,2])
[1] 4
> mean(M[,1])
[1] 2
> median(M[2,])
[1] 5
> sd(M[2,])
[1] 3
> M
     [,1] [,2] [,3]
[1,]    1    4    7
[2,]    2    5    8
[3,]    3    6    9
> avg<-apply(M, 2,mean)
> avg
[1] 2 5 8
> a<-apply(M,1,mean)
> a
[1] 4 5 6
```

Operation 5: Other Miscellaneous Matrix Operations Example in R

Other matrix operation include: (1) to check the type of matrix, (2) multiplying data elements of a matrix by a scalar, (3) finding out the diagonal elements of a matrix, (4) identity of a matrix, (5) determinant of a matrix, (6) rank of a matrix, (7) total count of rows and columns and to find out the dimension of a matrix, (8) compute the row sum and the column sum, and (9) compute the row and column means.

```
> # Other Miscellaneous Operations on Matrix
> # To check Matrix
> M<-matrix(1:9,nrow=3)
> M
     [,1] [,2] [,3]
[1,]    1    4    7
[2,]    2    5    8
[3,]    3    6    9
> is.matrix(M)
[1] TRUE
> #Multiplication by Scalar
> S<-3
> S*M
     [,1] [,2] [,3]
[1,]    3   12   21
[2,]    6   15   24
[3,]    9   18   27
> # Diagonal Matrix
> D<-diag(M)
> D
[1] 1 5 9
> # Identity Matrix
> I<-diag(3)
> I
     [,1] [,2] [,3]
[1,]    1    0    0
[2,]    0    1    0
[3,]    0    0    1
> # Determinant of Matrix
> d<-det(M)
> d
[1] 0
> # Rank of Matrix
> rank<-qr(M)
> rank
$qr
           [,1]      [,2]          [,3]
[1,] -3.7416574 -8.552360 -1.336306e+01
[2,]  0.5345225  1.963961  3.927922e+00
[3,]  0.8017837  0.988693  1.776357e-15
```

```
$rank
[1] 2

$qraux
[1] 1.267261e+00 1.149954e+00 1.776357e-15

$pivot
[1] 1 2 3

attr(,"class")
[1] "qr"
> # Count of row and column
> nc<-ncol(M)
> nc
[1] 3
> nr<-nrow(M)
> nr
[1] 3
> # Matrix Dimension
> dim(M)
[1] 3 3
> # Column and row sum and means
> c<-colSums(M)
> c
[1]  6 15 24
> m<-colMeans(M)
> m
[1] 2 5 8
> r<-rowSums(M)
> r
[1] 12 15 18
> rowMeans(M)
[1] 4 5 6
```

3.3 Arrays

An array in R is the objects which save data in more than two dimensions. Data is actually stored in a rectangular layout like a matrix. To access data elements at the matrix level, row and column indices are utilized (Brass, 2018). The function array() is used to implement the concept of an array in R programming. An array can have more than two dimensions; actually, it is a vector with an additional attribute for specifying the dimension size. A matrix is an example of a two-dimensional array and a vector is an example of a one-dimensional array. The syntax to create an array in R is as follows:

array(data, dim=length(data), dimnames=NULL)

where *array()* is the method used to generate arrays in R, *'data'* is a vector or list of vectors provided to enter data elements of an array, *'dim'* attribute is used to set the size of an array, and *'dimnanes'* contains the named indices for all the dimensions. Its value can be set to NULL if we are not providing any name to the respective dimensions.

3.3.1 Array Creation

In R, array is created using the function array(). To create an array, vectors are fed as an input and dim() is used to set the dimension of an array. For example, if we want to create an array of two 3X3 matrices, then the command is: array(data,dim=c(3,3,2)) where data is the value of data elements to be inserted in the array, 3 is the total count of row, 3 is the total count of columns, and 2 is used to create two matrices (Chan & Chan, 2018).

Array Creation Example in R

```
> v1<-c(10,34,56,55,89,90)
> v2<-c(10,20,30)
> A<-array(c(v1,v2),dim=c(3,3,2))
> A
, , 1

     [,1] [,2] [,3]
[1,]   10   55   10
[2,]   34   89   20
[3,]   56   90   30

, , 2

     [,1] [,2] [,3]
[1,]   10   55   10
[2,]   34   89   20
[3,]   56   90   30

> a1<-array(1:30,dim=c(4,4,2))
> a1
, , 1
```

```
     [,1] [,2] [,3] [,4]
[1,]    1    5    9   13
[2,]    2    6   10   14
[3,]    3    7   11   15
[4,]    4    8   12   16

, , 2

     [,1] [,2] [,3] [,4]
[1,]   17   21   25   29
[2,]   18   22   26   30
[3,]   19   23   27    1
[4,]   20   24   28    2
```

3.3.2 Naming Columns, Rows, and Arrays in R

Providing names to each column and row is the easiest way to access the data elements of an array. In R, we can provide names to the columns and rows of an array by setting the parameters in a 'dimnames' attribute within the array() function. The code shown below illustrates how to name columns and rows.

Naming Array Row and Column **Example in R**

```
> v1
[1] 10 34 56 55 89 90
> v2
[1] 10 20 30
> cnames<-c("C1","C2","c3")
> rnames<-c("R1","R2","R3")
> anames<-c("A1","A2")
> A<-array(c(v1,v2),dim=c(3,3,2),dimnames =
list(rnames,cnames,anames))
> A
, , A1

   C1 C2 c3
R1 10 55 10
R2 34 89 20
R3 56 90 30

, , A2

   C1 C2 c3
R1 10 55 10
R2 34 89 20
R3 56 90 30
```

3.3.3 Accessing Array Elements

To access data elements, [] brackets are used. For example, if we wish to display the second row of the second matrix, the R command is: print(A[2,,2]).

***Access Data Elements of an Array* Example in R**

```
> A
, , A1

   C1 C2 c3
R1 10 55 10
R2 34 89 20
R3 56 90 30

, , A2

   C1 C2 c3
R1 10 55 10
R2 34 89 20
R3 56 90 30

> # Display second row of the second matrix
> print(A[2,,2])
C1 C2 c3
34 89 20
> # Display data elements of 1 row and 2 column of first
matrix
> print(A[1,2,1])
[1] 55
> # Display first matrix
> print(A[,,1])
   C1 C2 c3
R1 10 55 10
R2 34 89 20
R3 56 90 30
> # Display second matrix
> print(A[,,2])
   C1 C2 c3
R1 10 55 10
R2 34 89 20
R3 56 90 30
> # Access all the elements except 1 and 3 row in array 2
> print(A[-1,-3,2])
   C1 C2
R2 34 89
R3 56 90
```

3.3.4 Array Operations

In this section, we will discuss the various operations on an array (Beasley & Beasley, 2020). We can apply various arithmetic operations on an array in R. All the operations in an array are discussed below with an example.

Operation 1: Array Arithmetic Operation **Example in R**

```
> # Arithmetic Operations on an Array
>  A
, , A1

   C1 C2 c3
R1 10 55 10
R2 34 89 20
R3 56 90 30

, , A2

   C1 C2 c3
R1 10 55 10
R2 34 89 20
R3 56 90 30

> A2<-array(c(v3,v4),dim=c(3,3,2))
> A2
, , 1

     [,1] [,2] [,3]
[1,]   10   67   23
[2,]   30   78   55
[3,]   45   89   70

, , 2

     [,1] [,2] [,3]
[1,]   10   67   23
[2,]   30   78   55
[3,]   45   89   70

> A+A2
, , A1

    C1  C2  c3
R1  20 122  33
R2  64 167  75
R3 101 179 100
```

```
, , A2

    C1  C2  c3
R1  20 122  33
R2  64 167  75
R3 101 179 100

> A-A2
, , A1

   C1  C2  c3
R1  0 -12 -13
R2  4  11 -35
R3 11   1 -40

, , A2

   C1  C2  c3
R1  0 -12 -13
R2  4  11 -35
R3 11   1 -40

> A*A2
, , A1

     C1   C2   c3
R1  100 3685  230
R2 1020 6942 1100
R3 2520 8010 2100

, , A2

     C1   C2   c3
R1  100 3685  230
R2 1020 6942 1100
R3 2520 8010 2100

> A/A2
, , A1

         C1        C2        c3
R1 1.000000 0.8208955 0.4347826
R2 1.133333 1.1410256 0.3636364
R3 1.244444 1.0112360 0.4285714

, , A2

         C1        C2        c3
R1 1.000000 0.8208955 0.4347826
R2 1.133333 1.1410256 0.3636364
R3 1.244444 1.0112360 0.4285714
```

***Operation 2: Array Manipulation* Example in R**

R provides us the facility to alter the data elements of an array. Addition, deletion, and alteration are allowed in an existing array. Given below is the code to manipulate data elements stored in an array.

```
> A
, , A1

   C1 C2 c3
R1 10 55 10
R2 34 89 20
R3 56 90 30

, , A2

   C1 C2 c3
R1 10 55 10
R2 34 89 20
R3 56 90 30

> changing the value of 1 row and 3 column and set it to 100
> A[1,3,1]<-100
> A
, , A1

   C1 C2  c3
R1 10 55 100
R2 34 89  20
R3 56 90  30

, , A2

   C1 C2 c3
R1 10 55 10
R2 34 89 20
R3 56 90 30

> Deleting third row from an array
> A
, , 1

      [,1] [,2] [,3]
[1,]    10   67   23
[2,]    30   78   55
[3,]    45   89   70
```

```
, , 2

     [,1] [,2] [,3]
[1,]   10   67   23
[2,]   30   78   55
[3,]   45   89   70

> A=A[-3,,-3]
> A
, , 1

     [,1] [,2] [,3]
[1,]   10   67   23
[2,]   30   78   55

, , 2

     [,1] [,2] [,3]
[1,]   10   67   23
[2,]   30   78   55

> Deleting first column from an array
> A=A[,-1,]
> A
, , 1

     [,1] [,2]
[1,]   67   23
[2,]   78   55
[3,]   89   70

, , 2

     [,1] [,2]
[1,]   67   23
[2,]   78   55
[3,]   89   70
```

***Operation 3: Other Miscellaneous Operations on an Array* in R**

The following presents an example of how to sort the elements of an array.

```
> A
, , 1

     [,1] [,2] [,3]
[1,]   10   67   23
[2,]   30   78   55
[3,]   45   89   70
```

```
, , 2

     [,1] [,2] [,3]
[1,]   10   67   23
[2,]   30   78   55
[3,]   45   89   70

> apply(A,2,sort)
     [,1] [,2] [,3]
[1,]   10   67   23
[2,]   10   67   23
[3,]   30   78   55
[4,]   30   78   55
[5,]   45   89   70
[6,]   45   89   70
```

The example shown below presents another operation in arrays. The given example displays the sum of rows across all matrices:

```
> A
, , 1

     [,1] [,2] [,3]
[1,]   10   67   23
[2,]   30   78   55
[3,]   45   89   70

, , 2

     [,1] [,2] [,3]
[1,]   10   67   23
[2,]   30   78   55
[3,]   45   89   70

> r<-apply(A,c(2),sum)
> r
[1] 170 468 296
> r<-apply(A,c(1),sum)
> r
[1] 200 326 408
```

3.4 Lists

Another data structure used in R is list. Lists are the R objects which belong to the category of heterogeneous data storage structures (i.e., they store data of different types) (Wickham et al., 2023). A list contains data elements like numbers, characters, Boolean, vectors, or it can contain another list. A list can also store a matrix or any function as its member. Lists are implemented using the list() function in R programming. For data analysis, a data set comprises different data types; in order to load this data, we are interested in the storage structure that supports a heterogeneous data storage structure. A list is one such example which stores heterogeneous data in R for data analysis. Lists hold the ordered lists of different objects altogether. Ordered objects are known as elements or members of a list. Data elements of a list could be of any given type: character, numeric, vector, matrix, list, logical value, or it could be any function as well.

3.4.1 List Creation

List is a data storage structure that is comprised of a mixed type of data. The list() is used to create lists in R. The typeof() method is used to check the type of data structure used for data storage. To find out the count of data elements stored within the list function, length() is used. Consider an example where we created a list containing a character, vector, Boolean value, and numbers (Matloff, 2011).

***List Creation* Example in R**

```
> L<-list("Apple","Mango",c(20,30,40,50),FALSE,34.7,c("A","B"))
> L
[[1]]
[1] "Apple"

[[2]]
[1] "Mango"

[[3]]
[1] 20 30 40 50

[[4]]
[1] FALSE

[[5]]
[1] 34.7

[[6]]
[1] "A" "B"
```

Another example of creating a list is presented below:

```
> a=c(1,2,3)
> b=c("A","E","I","O","U")
> d=c(TRUE,FALSE)
> m<-matrix(10:13,2,2)
> x=list(a,b,d,m)
> x
```

```
[[1]]
[1] 1 2 3

[[2]]
[1] "A" "E" "I" "O" "U"

[[3]]
[1]  TRUE FALSE

[[4]]
     [,1] [,2]
[1,]   10   12
[2,]   11   13
```

3.4.2 Naming List Elements

A list is the composition of various different data types, so naming data elements of a list is essential. In this section, we will explain how to provide a name to the data elements of a list. In the previous example, we have created a list, and now we are going to assign unique names to data elements stored within the list. The *names()* function is used to assign unique labels to list data elements.

Naming List Elements Example in R

```
> names(l)=c("Numeric data","Vowels","Real Data","Matrix")
> l
$`Numeric data`
[1] 1 2 3

$Vowels
[1] "A" "E" "I" "O" "U"

$`Real Data`
[1]  TRUE FALSE


$Matrix
     [,1] [,2]
[1,]   10   12
[2,]   11   13
```

3.4.3 Accessing List Elements

The indexing technique organizes the data as ordered collection. Indexing is essential in a list as a list is the group of mixed data types. Like other data structures, [] brackets are used for indexing. The operator [[...]] is used to access list members in R. List members are sequenced as [[1]],….[[N]] where [[1]] represents the first member of the list and [[N]] represent the Nth member of the list. Extraction of data elements from a list is carried out with the help of the '$' operator or '[[.]]' subscript operator. The syntax for accessing the list members is defined as follows:

$$List\_name\$member\_name \ OR \ List\_name[[..]]$$

Accessing List Members Example in R

In the list shown below, we will access the first and third element and then display the content on the screen using the member name.

```
> l
$`Numeric data`
[1] 1 2 3

$Vowels
[1] "A" "E" "I" "O" "U"

$`Logical Data`
[1]  TRUE FALSE

$Matrix
     [,1] [,2]
[1,]   10   12
[2,]   11   13
> # Display first element of List
> print(l[1])
$`Numeric data`
[1] 1 2 3

> # Display third element of list
> print(l[3])
$`Logical Data`
[1]  TRUE FALSE

> # Display Data element by Names
> print(l$Matrix)
     [,1] [,2]
[1,]   10   12
[2,]   11   13
> print(l[c(1,2)])
$`Numeric data`
[1] 1 2 3
$Vowels
[1] "A" "E" "I" "O" "U"
```

In order to extract the list member directly, we can use [[]], as shown below:

```
> l[[2]][4]
[1] "O"
> l$`Numeric data`[3]
[1] 3
```

3.4.4 List Operations

A list is a heterogeneous data storage structure for holding the data of mixed data types and requires operations for data modelling and statistical modelling. Addition, deletion, extraction, and merging are some operations that can be applied to a list.

Operation 1: Adding and Deleting Elements of a List **Example in R**

To add a new member in the list, '$' or '[[]]' operators are used in R language. The syntax for adding a new data member is:

List$member_name<- Value OR *List[[index_value]]<-Value.*

For deletion of a data element, NULL needs to assigned to the data member that needs to be deleted from the list and the syntax is shown below:

List$member_name<-NULL OR *List[[index_value]]<-NULL*

```
> # Add New Data Member to the List
> l
$`Numeric data`
[1] 1 2 3

$Vowels
[1] "A" "E" "I" "O" "U"

$`Logical Data`
[1]  TRUE FALSE

$Matrix
     [,1] [,2]
[1,]   10   12
[2,]   11   13
> l$spec_char<-c("@","#","$","&")
> l[[6]]<-c("Mon","Tue","Wed")
> l
$`Numeric data`
[1] 1 2 3

$Vowels
[1] "A" "E" "I" "O" "U"

$`Logical Data`
[1]  TRUE FALSE
```

```
$Matrix
     [,1] [,2]
[1,]   10   12
[2,]   11   13

$spec_char
[1] "@" "#" "$" "&"

[[6]]
[1] "Mon" "Tue" "Wed"

> # Deleting element from the list
> l$`Logical Data`<-NULL
> l
$`Numeric data`
[1] 1 2 3

$Vowels
[1] "A" "E" "I" "O" "U"

$Matrix
     [,1] [,2]
[1,]   10   12
[2,]   11   13

$spec_char
[1] "@" "#" "$" "&"

[[5]]
[1] "Mon" "Tue" "Wed"
> l[[5]]<-NULL
> l
$`Numeric data`
[1] 1 2 3

$Vowels
[1] "A" "E" "I" "O" "U"

$Matrix
     [,1] [,2]
[1,]   10   12
[2,]   11   13

$spec_char
[1] "@" "#" "$" "&"
```

Operation 2: Merging of a List Example in R

R language provides us the facility to merge two lists. For merging two lists, R provides various built-in functions. *c()* and append() both can be used to merge two lists. Package 'Rcurl' (Lang, 2007) provides another built-in method, merge.list(), to merge two independent lists. The syntax to merge two lists is shown below using the functions *c()* and append():

C(list_1,list_2,list_3,.list_n) OR *append(list_1,list_2,list_3,...list_n)*

```
> L1<-list(a=c(10,20),b=c("A","B"))
> L2<-list(d=c("E","F"),e=c(TRUE,FALSE))
> c(L1,L2)
$a
[1] 10 20

$b
[1] "A" "B"

$d
[1] "E" "F"

$e
[1]  TRUE FALSE

> append(l,L1)
$`Numeric data`
[1] 1 2 3

$Vowels
[1] "A" "E" "I" "O" "U"

$Matrix
     [,1] [,2]
[1,]   10   12
[2,]   11   13

$spec_char
[1] "@" "#" "$" "&"

$a
[1] 10 20

$b
[1] "A" "B"
```

Operation 3: Change the Value of a List Element Example in R

Sometimes you need to alter the value stored for a list member. R provides a way to alter the data member value. As in the given example, we alter the special char value to '!.' The syntax to change the value is:

List$Member_name<-New_value OR *List[[index_value]]<-New_Value*

```
> l$spec_char<-"!"
> l
$`Numeric data`
[1] 1 2 3
```

```
$Vowels
[1] "A" "E" "I" "O" "U"

$Matrix
     [,1] [,2]
[1,]   10   12
[2,]   11   13

$spec_char
[1] "!"
> l[[2]][3]<-"M"
> l
$`Numeric data`
[1] 1 2 3

$Vowels
[1] "A" "E" "M" "O" "U"

$Matrix
     [,1] [,2]
[1,]   10   12
[2,]   11   13

$spec_char
[1] "!"
```

Operation 4: Converting a List to a Vector **Example in R**

A list can be changed to a vector for data analysis. We can apply all the vector operations on a list after converting the list to a vector. R language provides the method *unlist()* to convert a list to a vector. The following example presents how to convert a list to a vector:

```
> l
$`Numeric data`
[1] 1 2 3

$Vowels
[1] "A" "E" "M" "O" "U"

$Matrix
     [,1] [,2]
[1,]   10   12
[2,]   11   13

$spec_char
[1] "!"
```

```
> v<-unlist(l)
> v
Numeric data1 Numeric data2 Numeric data3
          "1"           "2"           "3"
      Vowels1       Vowels2       Vowels3
          "A"           "E"           "M"
      Vowels4       Vowels5       Matrix1
          "O"           "U"          "10"
      Matrix2       Matrix3       Matrix4
         "11"          "12"          "13"
    spec_char
          "!"
> # Another Example
> # Create two list and then convert them to vector
> L1<-list(10:20)
> L2<-list(50:60)
> L1
[[1]]
 [1] 10 11 12 13 14 15 16 17 18 19 20
> L2
[[1]]
 [1] 50 51 52 53 54 55 56 57 58 59 60
> v1<-unlist(L1)
> v2<-unlist(L2)
> v1
 [1] 10 11 12 13 14 15 16 17 18 19 20
> v2
 [1] 50 51 52 53 54 55 56 57 58 59 60
> # Addition of two vectors
> sum<-v1+v2
> sum
 [1] 60 62 64 66 68 70 72 74 76 78 80
```

3.4.5 R Predefined List

In R, a predefined list is available for displaying small and uppercase letters. Also, a list is available for displaying months. Lists are available for showing month in abbreviated form and also in non-abbreviated form (i.e., complete month names displayed on the screen).

Predefined List Example in R

```
> letters
 [1] "a" "b" "c" "d" "e" "f" "g" "h" "i" "j" "k" "l"
[13] "m" "n" "o" "p" "q" "r" "s" "t" "u" "v" "w" "x"
[25] "y" "z"
> LETTERS
 [1] "A" "B" "C" "D" "E" "F" "G" "H" "I" "J" "K" "L"
[13] "M" "N" "O" "P" "Q" "R" "S" "T" "U" "V" "W" "X"
[25] "Y" "Z"
```

```
> month.abb
 [1] "Jan" "Feb" "Mar" "Apr" "May" "Jun" "Jul" "Aug"
 [9] "Sep" "Oct" "Nov" "Dec"
> month.name
 [1] "January"   "February"  "March"     "April"
 [5] "May"       "June"      "July"      "August"
 [9] "September" "October"   "November"  "December"
```

3.4.6 Recursive List

R language has additional features of embedding a list within another list (i.e., a list can be considered recursive data storage objects). The data structures vector and array are example of atomic objects in R, while a list is recursive in nature (Walker & Morrisett, 2000). Data objects can be either atomic or recursive but never both. Two functions are available in R to check whether data objects are atomic or recursive in nature. The functions *is.atomic()* and *is.recursive()* are used for data object types.

Recursive List Example in R

```
> L1<-list(1:10)
> L1
[[1]]
 [1]  1  2  3  4  5  6  7  8  9 10
> is.atomic
function (x) .Primitive("is.atomic")
> is.atomic(L1)
[1] FALSE
> is.recursive(L1)
[1] TRUE
> # Create Recursive List
> L2<-list(20:30,L1)
> L2
[[1]]
 [1] 20 21 22 23 24 25 26 27 28 29 30
[[2]]
[[2]][[1]]
 [1]  1  2  3  4  5  6  7  8  9 10
> # Another example of Recursive List
> L<-c(list(A=1,B="A",C=list(D="M",E="TRUE")))
> L
$A
[1] 1

$B
[1] "A"
```

```
$C
$C$D
[1] "M"

$C$E
[1] "TRUE"
```

3.5 Data Frame

Another heterogeneous type of data storage structure in R is data frames. A data frame is used to store data in a tabular layout. It is a list of data elements, but the only difference is that it contains data elements of the same size. It is a two-dimensional arrangement of data objects and is considered a specialized version of a list (Petersohn et al., 2020). Each data element forms the attribute name (i.e., column and value of attribute forms the rows in a data frame).

A data frame is basically utilized to save data in order to outline the major components of a data set and store data in a rectangular shape like a table or spreadsheet. Each row contains an instance of data and each column contains a vector for each attribute type. While creating a data frame, the following points must be addressed:

1. Data elements could comprise different types of data (i.e., it can contain data of numeric, string, and logical types)
2. Each column comprises the same type of data
3. The column name should not be empty
4. Row names should be unique
5. Each column should contain same number of data elements

To create a data frame in R programming, the function data.frame() is used (Kabacoff, 2022). We can also verify that a given object is a data frame or not by using the function class(). Table 3.2 outlines the structure of a data frame.

TABLE 3.2

Student Data Frame

R_NO	Name	Age	City	CGPA
101	Atharv	20	Noida	9.5
102	Rishav	22	New Delhi	8.9
103	Ananya	21	Bangalore	7.8
104	Amit	23	Mumbai	6.7
105	Aman	22	Dehradun	7.1

The table shown below presents the data frame of 'Student' with attributes: R_NO, Name, City, and CGPA and each row correspond to the measurement or value for each attribute type.

3.5.1 Data Frame Creation

The function data.frame() is used to create data frames in R programming. It generates a tightly coupled set of data elements which possess the similar characteristics as that of matrices and lists. The syntax to create a data frame in R is given below (Wickham et al., 2023):

data.frame(data,row.names=NULL,check.rows=FALSE, chacek.names=TRUE,fix.empty.names=TRUE,stringAsFactors())

where *'data'* is the value against each column attribute and *'row.names'* is a string which specifies the names of each row. This argument can be set equal to NULL; *'check.rows'* if the value is set to TRUE, then it verifies the rows for uniformity of length and names; *'fix.empty.names'* takes a logical value as input and automatically generates names for unnamed arguments. *'stringAsFactors'* is used to convert strings to factors and takes a logical value as input.

Data Frame Creation **Example in R**

```
> # Data Frame Creation Named:"Student"
> Student<-
data.frame(R_N0=c(101,102,103,104,105),Name=c("Atharv",
"Rishav","Ananya","Amit","Aman"),Age=c(20,22,21,23,22),
City=c("Noida","New Delhi","Bangalore","Mumbai","Dehradun"),
CGPA=c(9.5,8.9,7.8,6.7,7.1))
> Student
  R_N0   Name Age      City CGPA
1  101 Atharv  20     Noida  9.5
2  102 Rishav  22 New Delhi  8.9
3  103 Ananya  21 Bangalore  7.8
4  104   Amit  23    Mumbai  6.7
5  105   Aman  22  Dehradun  7.1

> # Another Data Frame "Customer"
> cust<-
data.frame(cid=c(1:4),cname=c("Naina","Amit","Sneha","Surbhi"),
income=c(20000,45908,78900,78000),purchase_date=as.Date(c("2010-01-
01","2018-10-24","2009-06-07","1999-04-30")),stringsAsFactors =
FALSE)
> cust
  cid  cname income purchase_date
1   1  Naina  20000    2010-01-01
2   2   Amit  45908    2018-10-24
3   3  Sneha  78900    2009-06-07
4   4 Surbhi  78000    1999-04-30
```

3.5.2 Accessing Data from a Data Frame

To extract data from a data frame, indexing is used. [] Brackets are used to specify a particular row and column index and these indices are further utilized to extract specific data from a data frame. A data frame is a two-dimensional data structure. The left-hand subscript data is used to access a row and the right-hand subscript data is used to access a column. For example, D [3,4] specifies the third row and fourth column of data frame 'D.' Given below is the R code to access data from data frames:

***Access Data from a Data Frame* Example in R**

```
> Student
  R_NO   Name Age      City CGPA
1  101 Atharv  20     Noida  9.5
2  102 Rishav  22 New Delhi  8.9
3  103 Ananya  21 Bangalore  7.8
4  104   Amit  23    Mumbai  6.7
5  105   Aman  22  Dehradun  7.1

> # Accessing data from Data Frame
> # Extract specific column

> out<-data.frame(Student$R_NO,Student$Name)
> out
  Student.R_NO Student.Name
1          101       Atharv
2          102       Rishav
3          103       Ananya
4          104         Amit
5          105         Aman

> # Extract specific Rows
> r<-Student[1:3,]
> r
  R_NO   Name Age      City CGPA
1  101 Atharv  20     Noida  9.5
2  102 Rishav  22 New Delhi  8.9
3  103 Ananya  21 Bangalore  7.8
> c<-Student[,1:3]
> c
  R_NO   Name Age
1  101 Atharv  20
2  102 Rishav  22
3  103 Ananya  21
4  104   Amit  23
5  105   Aman  22
```

```
> # Access 2nd and 3rd row along with 1st and 2nd column
> r<-Student[c(2,3),c[1,2]]
> r
[1] New Delhi Bangalore
5 Levels: Bangalore Dehradun Mumbai ... Noida
```

3.5.3 Naming Data Frame Components

For assigning row and column names of data frame names () and rownames (), built-in functions are used in R language. The syntax to provide row and column names is:

names(Data_Frame)<-c("col1_name","col2_name",... ....,"coln_name")
rownames(Data_Frame)<-c("row1_name","row2_name",... ....,"rown_name")

***Naming Rows and Columns of a Data Frame* Example in R**

```
> a<-c(1,2,3)
> b<-c("A","B","C")
> d<-c("#","$","&")
> D<-data.frame(a,b,d)
> D
  a b d
1 1 A #
2 2 B $
3 3 C &
> # Assigning row and column names
> names(D)<-c("Numbers","Character","Special Character")
> rownames(D)<-c("Data 1:","Data 2:","Data 3:")
> D
        Numbers Character Special Character
Data 1:       1         A                 #
Data 2:       2         B                 $
Data 3:       3         C                 &
```

3.5.4 Data Frame Operations in R

A data frame is a table-like data structure where data is stored in the form of rows and columns. Its structure is similar to that of a matrix, table, or spreadsheet, where each row corresponds to one instance and each column represents a metric of one attribute. A data frame supports various operations such as addition, deletion, sorting, merging, and update for data analysis. To build data-oriented machine learning models, all such operations are required. In this section, we will discuss various data frame operations along with examples.

Operation 1: Adding Columns and Rows to a Data Frame

A data frame can be extended by adding additional rows and columns. To add data elements to a data frame you have to append rows and columns with values. For appending rows and columns of a data frame, R provides operators [$ and [[]], along with built-in functions rbind() and cbind(). cbind() and rbind() are used to attach rows and columns to a data frame. The syntax to append rows and columns in R is as follows:

Data_Frame<-cbind(Data_Frame,"column_value")
Data_Frame<-rbind(Data_Frame,"row_value")

Adding Rows and Columns to a Data Frame **Example in R**

```
> Student
  R_NO   Name Age      City CGPA
1  101 Atharv  20     Noida  9.5
2  102 Rishav  22 New Delhi  8.9
3  103 Ananya  21 Bangalore  7.8
4  104   Amit  23    Mumbai  6.7
5  105   Aman  22  Dehradun  7.1

> # Adding column and row
> Hobby<-c("Dancing","Basket Ball","Cricket","singing","Video
Games")
> Student<-cbind(Student,Hobby)
> Student
  R_NO   Name Age      City CGPA       Hobby
1  101 Atharv  20     Noida  9.5     Dancing
2  102 Rishav  22 New Delhi  8.9 Basket Ball
3  103 Ananya  21 Bangalore  7.8     Cricket
4  104   Amit  23    Mumbai  6.7     singing
5  105   Aman  22  Dehradun  7.1 Video Games

> rbind(Student,list(106,"neha",20,"Noida",7.8,"Dancing"))
   R_NO   Name Age      City CGPA       Hobby
1   101 Atharv  20     Noida  9.5     Dancing
2   102 Rishav  22 New Delhi  8.9 Basket Ball
3   103 Ananya  21 Bangalore  7.8     Cricket
4   104   Amit  23    Mumbai  6.7     singing
5   105   Aman  22  Dehradun  7.1 Video Games
6   106   neha  20     Noida  7.8     Dancing
```

Operation 2: Deleting Data Frame Elements

Data frame components can be deleted in a similar way as we delete the data elements of a list. For example, if we want to delete any column, then set the value of that column equal to NULL and rows are deleted through reassignments (i.e., we provide an index of a row within the bracket to delete it: [-2,]. This was to delete the second row from the data frame. The syntax to delete rows and columns is as follows:

Data_Frame$Col_name<-NULL
Data_Frame<-Data_Frame[-row_index]

Deleting Rows and Columns from a Data Frame Example in R

```
> Student
  R_NO   Name Age      City CGPA
1  101 Atharv  20     Noida  9.5
2  102 Rishav  22 New Delhi  8.9
3  103 Ananya  21 Bangalore  7.8
4  104   Amit  23    Mumbai  6.7
5  105   Aman  22  Dehradun  7.1

> # Delete column CGPA
> Student$CGPA<-NULL
> Student
  R_NO   Name Age      City
1  101 Atharv  20     Noida
2  102 Rishav  22 New Delhi
3  103 Ananya  21 Bangalore
4  104   Amit  23    Mumbai
5  105   Aman  22  Dehradun

> # Delete 4th Row
> Student<-Student[-4,]
> Student
  R_NO   Name Age      City
1  101 Atharv  20     Noida
2  102 Rishav  22 New Delhi
3  103 Ananya  21 Bangalore
5  105   Aman  22  Dehradun
```

Operation 3: Merging Data Frames

The merging operation is applied on a data frame to merge rows and columns of two data frames. This operation is very much required for data analysis, as we have to load and integrate data from two or more sources in order to prepare data for analysis. In R language, the function to merge two data frames is *merge()*. The syntax to merge two data frames in R is:

> *merge(data_frame_1,data_farme_2,by=intersect(names (data_frame_1),names(data_frame_2)),by.x=by,by.y=by,all=FALSE,all. x=all,all. y=all,….)*

where *'data_frame_1 & data_frame_2'* specifies the two data frames which need to be merged; *'by,by. x,by. y'* specifies the common value of rows and columns; *'all'* specifies the test condition and contains a Boolean value of either TRUE or FALSE; *'all. x'* specifies the test condition for the left outer join and contains a Boolean value of either TRUE or FALSE; *'all. y'* specifies the test condition for the right outer join and contains a Boolean value of either TRUE or FALSE; and *'….'* specifies other optional arguments.

Merging Two Data Frames Example in R

```
> DF1<-data.frame(Spec_char=c("#","&"),Real_data=c(78.9,89.7))
> DF2<-data.frame(Spec_char=c("@","%"),Real_data=c(88.9,69.5))
> MDF<-merge(DF1,DF2)
> MDF
[1] Spec_char Real_data
<0 rows> (or 0-length row.names)
> MDF<-merge(DF1,DF2,all.x = TRUE,all.y = TRUE)
> MDF
  Spec_char Real_data
1         #      78.9
2         &      89.7
3         %      69.5
4         @      88.9
```

Operation 4: Converting Objects into a Data Frame

The function *as.data.frame()* is used to change different objects to data frames. The syntax for converting an R-object to a data frame is:

DF_name<-as.data.frame()

Converting Objects to Data Frames Example in R

```
> M1<-matrix(1:4,2,2)
> M1
     [,1] [,2]
[1,]    1    3
[2,]    2    4
> M<-as.data.frame(M1)
> M
  V1 V2
1  1  3
2  2  4
```

Operation 5: Changing the Content of a Data Frame

To alter the data content of a data frame, specify the row and column that needs alteration and set the new updated value as shown in the example. In this example, we alter the Age attribute and set its value equal to 21 for the first row of a data frame. The syntax to alter the content of a data frame is as follows:

Data_Frame[row_index,"col_name"]<-New_value

Changing Content of Data Frames Example in R

```
> # Altering the content of Data Frame
Student[1,"Age"]<-21
> Student
```

```
  R_NO   Name Age      City
1  101 Atharv  21     Noida
2  102 Rishav  22 New Delhi
3  103 Ananya  21 Bangalore
5  105   Aman  22  Dehradun

> # Another Example
> Student[2:3,"Age"]<-20
> Student
  R_NO   Name Age      City
1  101 Atharv  21     Noida
2  102 Rishav  20 New Delhi
3  103 Ananya  20 Bangalore
5  105   Aman  22  Dehradun
```

3.5.5 Data Frame Useful Functions

Various built-in functions provided in R language can be applied directly on data frames. These built-in functions are utilized according to the problem and in the implementation of various real-life applications. Table 3.3 presents an overview of a few useful built-in functions for a data frame:

Data Frame Built-In Functions Example in R

```
> # Data Frame Student
> Student
 R_NO   Name Age      City
  101 Atharv  21     Noida
  102 Rishav  20 New Delhi
  103 Ananya  20 Bangalore
  105   Aman  22  Dehradun
```

TABLE 3.3

Data Frame Built-In Functions

Function_Name	Description
str()	Describes the structure of a data frame
summary()	Describes the statistical summary of a data frame
names()	Extracts column names
head()	Displays first six rows of a data frame
tail()	Displays last six rows of a data frame
subset()	Generates a subset of a data frame
length()	Finds out the length of a data frame
unique()	Displays unique data of a data frame
duplicated()	Displays the Boolean value if any repeated data is found in a data frame
tapply()	Applies a function to each cell
split()	Divides into subgroups

```
> # Getting Structure of Data Frame
> str(Student)
'data.frame':       4 obs. of  4 variables:
$ R_NO: num  101 102 103 105
$ Name: Factor w/ 5 levels "Aman","Amit",..: 4 5 3 1
$ Age : num  21 20 20 22
$ City: Factor w/ 5 levels "Bangalore","Dehradun",..: 5 4 1 2

> # Displaying Summary of Data Frame
> summary(Student)
     R_NO             Name          Age
Min.   :101.0   Aman  :1   Min.   :20.00
1st Qu.:101.8   Amit  :0   1st Qu.:20.00
Median :102.5   Ananya:1   Median :20.50
Mean   :102.8   Atharv:1   Mean   :20.75
3rd Qu.:103.5   Rishav:1   3rd Qu.:21.25
Max.   :105.0              Max.   :22.00
        City
 Bangalore:1
 Dehradun :1
 Mumbai   :0
 New Delhi:1
 Noida    :1

> # Displaying Column Names
> names(Student)
[1] "R_NO" "Name" "Age"  "City"

> # Displaying first six rows
> head(Student)
  R_NO   Name Age      City
1  101 Atharv  21     Noida
2  102 Rishav  20 New Delhi
3  103 Ananya  20 Bangalore
5  105   Aman  22  Dehradun

> # Displaying last six rows
> tail(Student)
  R_NO   Name Age      City
1  101 Atharv  21     Noida
2  102 Rishav  20 New Delhi
3  103 Ananya  20 Bangalore
5  105   Aman  22  Dehradun
```

Since the data frame 'Student' comprises five rows in total, it displays the same outcome after applying the *head()* and *tail()* operations. We use a built-in data set 'mtcars' provided in R to display the outcome of *head()* and *tail ()* operations. As you have seen, different rows are displayed: head() displays the first six rows while tail() shows the last six rows.

```
> head(mtcars)
                   mpg cyl disp  hp drat    wt  qsec
Mazda RX4         21.0   6  160 110 3.90 2.620 16.46
Mazda RX4 Wag     21.0   6  160 110 3.90 2.875 17.02
Datsun 710        22.8   4  108  93 3.85 2.320 18.61
Hornet 4 Drive    21.4   6  258 110 3.08 3.215 19.44
Hornet Sportabout 18.7   8  360 175 3.15 3.440 17.02
Valiant           18.1   6  225 105 2.76 3.460 20.22
                  vs am gear carb
Mazda RX4          0  1    4    4
Mazda RX4 Wag      0  1    4    4
Datsun 710         1  1    4    1
Hornet 4 Drive     1  0    3    1
Hornet Sportabout  0  0    3    2
Valiant            1  0    3    1
> tail(mtcars)
                mpg cyl  disp  hp drat    wt qsec vs
Porsche 914-2  26.0   4 120.3  91 4.43 2.140 16.7  0
Lotus Europa   30.4   4  95.1 113 3.77 1.513 16.9  1
Ford Pantera L 15.8   8 351.0 264 4.22 3.170 14.5  0
Ferrari Dino   19.7   6 145.0 175 3.62 2.770 15.5  0
Maserati Bora  15.0   8 301.0 335 3.54 3.570 14.6  0
Volvo 142E     21.4   4 121.0 109 4.11 2.780 18.6  1
               am gear carb
Porsche 914-2   1    5    2
Lotus Europa    1    5    2
Ford Pantera L  1    5    4
Ferrari Dino    1    5    6
Maserati Bora   1    5    8
Volvo 142E      1    4    2
> # Display the length of Data Frame
> length(mtcars)
[1] 11
> length(Student)
[1] 4
> unique(Student)
  R_NO   Name Age      City
1  101 Atharv  21     Noida
2  102 Rishav  20 New Delhi
3  103 Ananya  20 Bangalore
5  105   Aman  22  Dehradun
> duplicated(Student)
[1] FALSE FALSE FALSE FALSE
> subset(Student,,2)
    Name
1 Atharv
2 Rishav
3 Ananya
5   Aman
> order(subset(Student,,2))
[1] 4 3 1 2
```

```
> # tapply() on another in builtin data frame named "iris"
Levels: M T S
> Data<-head(iris)
> Data
  Sepal.Length Sepal.Width Petal.Length Petal.Width
1          5.1         3.5          1.4         0.2
2          4.9         3.0          1.4         0.2
3          4.7         3.2          1.3         0.2
4          4.6         3.1          1.5         0.2
5          5.0         3.6          1.4         0.2
6          5.4         3.9          1.7         0.4
  Species
1  setosa
2  setosa
3  setosa
4  setosa
5  setosa
6  setosa
> tapply(iris$Sepal.Length, iris$Species, mean)
    setosa versicolor  virginica
     5.006      5.936      6.588

> # split() example on IRIS data set
> R<-Data$Sepal.Length
> s<-split(Data,R)
> s
$`4.6`
  Sepal.Length Sepal.Width Petal.Length Petal.Width
4          4.6         3.1          1.5         0.2
  Species
4  setosa

$`4.7`
  Sepal.Length Sepal.Width Petal.Length Petal.Width
3          4.7         3.2          1.3         0.2
  Species
3  setosa

$`4.9`
  Sepal.Length Sepal.Width Petal.Length Petal.Width
2          4.9           3          1.4         0.2
  Species
2  setosa

$`5`
  Sepal.Length Sepal.Width Petal.Length Petal.Width
5            5         3.6          1.4         0.2
  Species
5  setosa
```

```
$`5.1`
  Sepal.Length Sepal.Width Petal.Length Petal.Width
1          5.1         3.5          1.4         0.2
  Species
1  setosa

$`5.4`
  Sepal.Length Sepal.Width Petal.Length Petal.Width
6          5.4         3.9          1.7         0.4
  Species
6  setosa
```

3.6 Factors in R

Factors in R programming are the data objects which can have limited counts of unique values. Unique values actually belong to categorical data (i.e., variable can belong to one of the given categories of data). Factors are best suited to create a statistical model as saving data as factors ensures the behaviour of modelling methods uses categorical data without any error. For building statistical models, unique sets of values are used to represent subgroupings of data like social class, cancer stage, and diagnosis of disease (Fabrigar & Wegener, 2012). This data is fed as an input using codes and such data is referred as factors in R programming.

3.6.1 Factor Creation

Factors are the objects which can store data as unique levels and categorize it into different unique categories (Toby, 2006). Factors are able to store both numeric and string data. They are applicable in a scenario where a data set consists of a limited number of unique levels. For example, the attribute gender can have one of the possible values of either 'Male,' or 'Female.' Another example of categorical data is 'Marital Status' and different levels are single, married, separated, divorced, or widowed. It is required to have beforehand knowledge of all the unique levels within the data set. The syntax to create a factor in R is as follows:

$$factor(x = character(), levels, labels = levels, exclude = NA,$$
$$ordered = is.ordered(x), nmax = NA)$$

where *'x'* is the data with unique levels,

'levels' specifies the unique set of values,

'labels' assigns names to a unique set of values,

'exclude' group of values to be excluded while creating unique levels,
'ordered' to check unique levels should be defined in the given order and,
'nmax' maximum limit of unique levels.

***Factor Creation* Example in R**

```
> # Factor Example
> Gender=c("Male","Male","Female","Male","Female")
> F<-factor(Gender)
> F
[1] Male     Female Male   Female
Levels: Female Male
>Day=c("Sunday","Monday","Wednesday","Sunday","Thursday","Monday",
"Saturday","Friday","Tuesday","Friday")
> D<-factor(Day)
> D
 [1] Sunday    Monday    Wednesday Sunday    Thursday
 [6] Monday    Saturday  Friday    Tuesday   Friday
7 Levels: Friday Monday Saturday Sunday ... Wednesday

> # To display data in Tabular layout
> table(D)
D
   Friday    Monday  Saturday    Sunday  Thursday
        2         2         1         2         1
  Tuesday Wednesday
        1         1
```

3.6.2 Access Factor Elements

Accessing factor data elements is similar to that of other data structures like vectors and data frames. [] Brackets are used to access components. An example shown below illustrates accessing factor data components.

***Accessing Factor Elements* Example in R**

```
> D
 [1] Sunday    Monday    Wednesday Sunday    Thursday
 [6] Monday    Saturday  Friday    Tuesday   Friday
7 Levels: Friday Monday Saturday Sunday ... Wednesday
>
> # Access Factor Components
> D[4]
[1] Sunday
7 Levels: Friday Monday Saturday Sunday ... Wednesday
> D[c(1,5)]
[1] Sunday   Thursday
7 Levels: Friday Monday Saturday Sunday ... Wednesday
```

```
> D[-4]
[1] Sunday    Monday    Wednesday Thursday  Monday
[6] Saturday  Friday    Tuesday   Friday
7 Levels: Friday Monday Saturday Sunday ... Wednesday
> D[1:3]
[1] Sunday    Monday    Wednesday
7 Levels: Friday Monday Saturday Sunday ... Wednesday
```

3.6.3 Factor Updation

To update the content of a data frame, specify the data element position and set it to the new value. The syntax to update the content of factors is as follows:

$$Factor[index\_value] <-New\_value$$

To alter the order of levels, first create the factor and then use the *levels()* argument to reorder the level of unique values within the data. The example shown below presents how to modify the order of level.

Factor Updation Example in R

```
> # To alter the content of second element of Factor
> D[2]<-"Monday"
> D
 [1] Sunday    Monday    Wednesday Sunday    Thursday
 [6] Monday    Saturday  Friday    Tuesday   Friday
7 Levels: Friday Monday Saturday Sunday ... Wednesday

> # To change the levels of factor
> data<-
c("mon","tue","mon","wed","thu","tue","fri","sat","sun","fri")
> f<-factor(data)
> f
  [1] mon tue mon wed thu tue fri sat sun fri
Levels: fri mon sat sun thu tue wed
> # Change the level order
> new<-factor(f,levels =
c("mon","tue","wed","thu","fri","sat","sun"))
> new
  [1] mon tue mon wed thu tue fri sat sun fri
Levels: mon tue wed thu fri sat sun
```

3.6.4 Factor Useful Functions

Various built-in functions are provided that we can directly apply to factors. A list of all such functions are described in Table 3.4.

TABLE 3.4
Factor Built-In Functions

Function_Name	Description
is.factor()	Checks the class of data object
gl()	Produces factor levels
as.factor()	Changes any attribute to a factor
as.ordered()	Displays sorted data
is.ordered()	Checks whether data is sorted or not

Factors Built-In Functions Example in R

```
> D
 [1] Sunday    Monday    Wednesday Sunday    Thursday
 [6] Monday    Saturday  Friday    Tuesday   Friday
7 Levels: Friday Monday Saturday Sunday ... Wednesday
> D1<-c(1,2,3,4,5)
> is.factor(D1)
[1] FALSE
> is.factor(D)
[1] TRUE
> as.factor(D1)
[1] 1 2 3 4 5
Levels: 1 2 3 4 5
> as.ordered(D)
 [1] Sunday    Monday    Wednesday Sunday    Thursday
 [6] Monday    Saturday  Friday    Tuesday   Friday
7 Levels: Friday < Monday < Saturday < ... < Wednesday
> as.ordered(D1)
[1] 1 2 3 4 5
Levels: 1 < 2 < 3 < 4 < 5
> is.ordered(D)
[1] FALSE
> is.ordered(D1)
[1] FALSE
> v<-gl(3,4,labels=c("M","T","S"))
> v
 [1] M M M M T T T T S S S S
```

3.7 Programming Example

In this section, we provide some useful programs to gain the understanding of all the concepts studied in this chapter. Figures 3.1–3.4 present the output of Programs 1–4.

Program 1: R code to generate the given matrix pattern using the outer function:

	[1]	[2]	[3]	[4]	[5]
[1]	0	9	8	7	6
[2]	1	0	9	8	7
[3]	2	1	0	9	8
[4]	3	2	1	0	9
[5]	4	3	2	1	0

SOLUTION:

```
Console   Terminal ×
R 4.2.2 · ~/
> outer(2:6,2:6,"-")%%10
     [,1] [,2] [,3] [,4] [,5]
[1,]    0    9    8    7    6
[2,]    1    0    9    8    7
[3,]    2    1    0    9    8
[4,]    3    2    1    0    9
[5,]    4    3    2    1    0
```

FIGURE 3.1
Matrix pattern generation using outer function.

Program 2: R program to find out the sum of a given mathematical expression:

$$2+\frac{4}{2}+\left(\frac{48}{24}\right)+\left(\frac{4812}{246}\right)+\cdots+\left(\frac{48\ldots40}{24\ldots20}\right)$$

SOLUTION:

```
> 2+sum(cumprod(seq(4,40,b=4)/seq(2,20,b=2)))
[1] 2048
```

Program 3: Create a 5 × 5 matrix of random integers chosen from 10, 11,: : :, 20 by executing the following two lines of code:

set.seed(50)
M <- matrix (sample(10:20, size=25, replace=T), nr=5)

a. Display the count in each row with a value greater than 15.
b. Display the total count of rows which contain exactly three occurrences of the number 17.

c. Display those pairs of columns whose total (over both columns) is greater than 35. The answer should be a matrix with two columns; so, for example, the row (2; 3) in the output matrix means that the sum of columns 2 and 3 in the original matrix is greater than 35. Repeating a column is permitted.

SOLUTION:

```
Console   Terminal ×
~/
> set.seed(50)
> M <- matrix( sample(10:20, size=25, replace=T), nr=5)
> M
     [,1] [,2] [,3] [,4] [,5]
[1,]   17   10   14   17   16
[2,]   14   17   12   19   12
[3,]   12   17   17   14   17
[4,]   18   10   10   10   19
[5,]   15   11   13   11   13
> apply(M, 1, function(f){sum(f>15)})
[1] 3 2 3 2 0
> which( apply(M,1,function(s){sum(s==17)==3}) )
[1] 3
> c<-colSums(M)
> which( outer(c,c,"+")>150, arr.ind=T )
     row col
[1,]   1   1
[2,]   5   1
[3,]   1   5
[4,]   5   5
> |
```

FIGURE 3.2
R Code of Program -3.

Program 4:

Write a function in R with one parameter: 'Input Matrix.' The function should generate the new matrix which is same as the original, except for every odd data element, whose value is doubled. Consider the given example, if the input matrix is:

```
     [1]  [2]  [3]
[1]    2    2    4
[2]    5    8    9
[3]    1    2    3
```

Output matrix:

```
     [1]  [2]  [3]
[1]    2    2    4
[2]   10    8   18
[3]    2    2    6
```

Program 5:

Create a data frame that stores some basic information of a PC, such as the configuration of a PC from at least five companies. Apply length(), str(), summary(), duplicated(), and unique() functions on this data frame.

SOLUTION:

```
Console  Terminal
R 4.2.2 · ~/
> M<-matrix(data=c(2,2,4,5,8,9,1,2,3),nrow=3,byrow = T)
> M
     [,1] [,2] [,3]
[1,]    2    2    4
[2,]    5    8    9
[3,]    1    2    3
> Fn<-function(M)
+ {
+ M[M%%2==1]<-2*M[M%%2==1]
+ M
+ }
> Fn(M)
     [,1] [,2] [,3]
[1,]    2    2    4
[2,]   10    8   18
[3,]    2    2    6
```

FIGURE 3.3
R Code of Program -4.

```
Console  Terminal
R 4.2.2 · ~/
> PC=data.frame(Sno=c(1,2,3,4,5),Company=c("Lenovo","HP","Dell","Apple","Asus"),Configuration=c("i7 14th generation","i5 13th gener
ation","i5 12th generation","i7 14th generation","i5 14th generation"))
> PC
  Sno Company      Configuration
1   1  Lenovo i7 14th generation
2   2      HP i5 13th generation
3   3    Dell i5 12th generation
4   4   Apple i7 14th generation
5   5    Asus i5 14th generation
> length(PC)
[1] 3
> summary(PC)
      Sno       Company          Configuration
 Min.   :1   Length:5           Length:5
 1st Qu.:2   Class :character   Class :character
 Median :3   Mode  :character   Mode  :character
 Mean   :3
 3rd Qu.:4
 Max.   :5
> duplicated(PC)
[1] FALSE FALSE FALSE FALSE FALSE
> unique(PC)
  Sno Company      Configuration
1   1  Lenovo i7 14th generation
2   2      HP i5 13th generation
3   3    Dell i5 12th generation
4   4   Apple i7 14th generation
5   5    Asus i5 14th generation
>
```

FIGURE 3.4
R Code of Program -5.

3.8 Summary

- A vector is a simple data structure which contains homogeneous data elements. Data types can be numeric, character, Boolean, and complex. To generate a sequence of a vector, the ':' operator is used.
- The c () function is used to create a vector in R. The typeof() method is used to check the vector type.
- For accessing data elements, indexing is used. Indexing in vectors starts from 1. Negative indexing outputs all the data elements except the specified one. Negative and positive indices cannot mix for accessing data elements.
- The following are the vector operations in R: addition, deletion, updation, sorting.
- A matrix is the arrangement in which we store data objects in a two-dimensional layout and is a kind of homogeneous data structure. The matrix() function is used to create a matrix in R.
- The following operations can be applied to a matrix: addition, subtraction, multiplication, division, transpose, updation, inverse, determinant, Eigen values, and inverse.
- One limitation with a matrix is that it cannot store data in two dimensions only. Arrays are used to store data with dimension greater than two. For example, A(3,3,3) creates three rectangular structures of size 3 × 3 (i.e., three rows and three columns). The array() is used to generate an array.
- To assign column and row names in matrices and arrays, the dimnames() parameter is used.
- The function apply() applies mathematical or any other predefined method over data elements of an array or matrix. It returns a list or a vector or an array after applying the predefined function.
- The lapply() function is applied for mathematical or any other operation over data objects and outputs the list of data elements of the same size.
- The sapply() function applies the predefined function or method over an array or matrix margin. The wrapper class of lapply() is sapply().
- A list is an example of a heterogeneous data storage structure as it holds objects of different types (i.e., a list can contain integers, doubles, characters, and Boolean). A list presents systematic and ordered arrangements of data objects. The list() method is used to create lists in R language.
- Addition, deletion, updation, extraction, and merging are some operations that can be applied to a list.

- R supports conversion of one data object structure to another structure according to the need of the user. For example, as.data.frame() converts a data object into a data frame type. Similarly, as.list() changes any structure to a list-type structure.
- Recursion is the process of calling data objects itself and data objects could be any type, such as list, function, or vector. The concept of recursion is implemented to reduce the line of code in programming. R programming provides the concept of recursion in a list. A recursive list is used to embed one list within another list.
- A data frame is the arrangement of data in tabular form. A data frame is also able to hold data of different types. It stores data in a structure such as a table, matrix, or spreadsheet, where each column represents attributes and each row represents one instance corresponding to that attribute.
- To create a data frame, the function data.frame() is used. One constraint of a data frame is that it is mandatory to set the same length for all columns.
- To specify a row name of a data frame, the rownames() built-in method is used and for adding and deleting columns and rows, the cbind() and rbind() methods are used. Addition, deletion, updation, extraction, and merging are some operations that can be applied to a data frame.
- Factors are used to provide categorization within the data and save unique categories as levels. Factors are able to hold both numeric and string-type data objects. The factor() is the keyword used to create a factor in R.
- The application of a factor includes analysis of data, building a statistical model, and many more.

EXERCISES

I. Multiple-Choice Questions

1. The following R code is given to create a vector 'V' of length 4. How would you access the third element?

   ```
   V<-c(30,40,67,78)
   ```

   a. V[2]

   b. V(2)

   c. V[3]

   d. V(3)

2. The following R code is given to create a matrix 'M' of size 3×3. How would you subset the second column?

```
M<-matrix(nrow=3,ncol=3,1:9)
```

   a. M[,2,]
   b. M[, , 2]
   c. M[2,2]
   d. M[,2]

3. What would be the outcome of the following command to add two vectors?

```
v1<-2:5
v2<-10:12
v1+v2
```

   a. 12 14 16 15
   b. 12 14 16 5
   c. 12 14 16 NA
   d. Error

4. Consider the given vector: `v<-c(2,3,4,6,7,8,19,68,90)`. Write the R code to update all the values greater or equal to 10, with a new value of 1000.
   a. v[v]>=10=1000
   b. v[v>=10<-1000]
   c. v[v>=10]<-1000
   d. All of the Above

5. Which function in R language iterates over a list and evaluates a function on each data object?
   a. apply
   b. sapply
   c. lapply
   d. tapply

6. Name the function which is similar to lapply.
   a. tapply
   b. sapply
   c. apply
   d. rapply

7. Which of the given functions returns a vector as an output?
   a. tapply
   b. lapply
   c. apply
   d. sapply

8. Which function is used to calculate statistical measures like mean, max, min, and median?
   a. lapply
   b. mapply
   c. sapply
   d. tapply
9. Which of the given functions need a margin as an input parameter?
   a. lapply
   b. apply
   c. sapply
   d. tapply
10. Identify the incorrect statement.
    a. Semantics of operator "$" is identical to operator "[[".
    b. Operator "$' in R is used to extract data objects of a data frame or a list.
    c. Operator "[" outputs objects of a class identical to the original.
    d. None of the above.
11. Which of the given functions is used to output the day of the week?
    a. dayweek
    b. date
    c. weekdays
    d. time
12. What is the output of the given code.

```
> m<-matrix(10:13,2,2)
> m1<-matrix(rep(20,30),2,2)
> m*m1
```

a.

```
     [,1] [,2]
[1,]  200  240
[2,]  220  260
```

b.

```
     [,1] [,2]
[1,]  220  200
[2,]  240  260
```

c.

```
     [,1] [,2]
[1,]  240  200
[2,]  260  220
```

d.

```
      [,1] [,2]
[1,]   200  240
[2,]   200  240
```

13. What would be the output of the following code?

```
L<-list(A=7:10,B=0.9,C="India")
a<-"A"
l[[a]]
```

a. 6,7,8,9

b. 7,8,9,10

c. 7.9,8.9,9.9,10.9

d. 7777

14. What would be the total count of rows generated by the given data frame?

```
D<-data.frame(a=rnorm(5),b=runif(10))
```

a. 15

b. 10

c. 5

d. 0

15. What would be the outcome of the given R command?

```
A<-10:15
A>11
```

a. 12 13 14 15

b. TRUE TRUE TRUE TRUE

c. FALSE FALSE

d. FALSE FALSE TRUE TRUE TRUE TRUE

II. Brief Questions

1. What are the advantages of using a vector and a matrix for statistical computation?
2. Explain the concept of an array. Also, write the command to create two arrays of size 4×4.
3. What is indexing? Also, state the purpose of using indexing in R programming.
4. Explain a recursive list with one real-time example.
5. Explain what a data frame is with its characteristics. Write the syntax to add columns and rows to an existing data frame.

6. Write the importance and difference amongst the following functions: apply(), lapply(), sapply(), and tapply().
7. What is the advantage, disadvantage (if any), and difference between a data frame and a list?
8. Write, with suitable example, the entire list of functions of a data frame.
9. Write the application of the following data structure in data analytics and statistical computation: vector, matrix, data frame, and list.
10. Explain in detail with a suitable example the difference between the lapply() and sapply() functions.
11. What is merging used for? List the name of the data structures in R on which we can apply the merging operation.
12. Define a factor with a suitable example. Also, write the application of a factor.
13. What operations can we perform on a list? Provide details with a suitable example.
14. Explain the difference between the cbind() and rbind() functions.

III. Coding Exercise

1. Create the given vectors:
   a. (10; 11; 12; : : : ; 29; 30)
   b. (30; 29; : : : ; 2; 1)
   c. (10; 11; 12; : : : ; 29; 30; 29; 28; : : : ; 2; 1)
   d. (14; 16; 10) and assign it to the name temp.
   e. (14; 16; 10; 14; 16; 10; : : : ; 14; 16; 10) where there are 10 occurrences of all data objects.
   f. (14; 16; 10;14; 16; 10; : : : ; 14; 16; 10) where there are 11 occurrences of 14, 10 occurrences of 16, and 11 occurrences of 3.
2. Generate a vector of the values of $e^x\cos(x)$ at x = 2,2.2,….4.
3. Calculate the following:

   a. $$\sum_{i=20}^{90}(i^4 + 3i^2)$$

   b. $$\sum_{i=10}^{20} \frac{3^i}{i} + \frac{4^i}{i^2}$$

4. Create two vectors, 'x' and 'y,' of random integers starting from 0,1....to 500. Integers are chosen with a replacement and have a length=100.
5. For the vectors 'x' and 'y' generated in question 4, answer the following:
   a. Find the values in both vectors which are greater than 400.
   b. What are the index positions in x and y of the values which are > 400?
   c. Find the values in 'y' which correspond to the values in 'x' which are > 400.
   d. Find the count of values in 'x' which are within 300 of the minimum value of the terms in 'x.'
   e. How many data elements in 'y' are divisible by 4?
   f. Arrange the data elements in the vector 'y' in the order of increasing values in vector 'x.'
6. Create the given matrix and answer the following questions:

```
       [1]   [2]   [3]
[1]      1    10    10
[2]     10    -5    10
[3]     -5     1     1
```

   a. Replace the first column of matrix 'M' by the sum of the first and third columns.
   b. Compute the value of M^TM, where T stands for transpose a matrix.
   c. Replace the value of matrix 'M' by 2, where the difference of columns and rows equals 1.
7. Create the following matrix pattern using the 'outer' function:

   a.

```
             [1]  [2]  [3]  [4]  [5]
[1]      0    -1    -2    -3    -4
[2]      1     0    -1    -2    -3
[3]      2     1     0    -1    -2
[4]      3     2     1     0    -1
[5]      4     3     2     1     0
```

   b.

```
             [1]  [2]  [3]  [4]  [5]
[1]      2     3     4     5     6
[2]      3     4     5     6     7
[3]      4     5     6     7     8
[4]      5     6     7     8     9
[5]      6     7     8     9    10
```

c.

```
                [1]  [2]  [3]  [4]  [5]
[1]      0     1     0     1     0
[2]      1     0     1     0     1
[3]      0     1     0     1     0
[4]      1     0     1     0     1
[5]      0     1     0     1     0
```

8. Create a 6 × 6 matrix of random integers chosen from 90, 91,: : :, 100. Also, write R code for the following:
   a. Display the count in each row with a value greater than 99.
   b. Display the total count of rows which contain exactly two occurrences of data equals to 95?
   c. Display those pairs of columns whose sum (over both columns) is greater than 500. The answer should be a matrix with two columns; so, for example, the row (2; 3) in the output matrix means that the sum of columns 2 and 3 in the original matrix is greater than 35. Repeating a column is not permitted.
9. Create the function Dayweek(day,mon,year), which outputs the day of the week using Zeller's congruence equation, which is given by:

$$\text{F} = \left([2.6M - 0.2] + N + Y + \left[\frac{Y}{4}\right] + \left[\frac{A}{4}\right] - 2A \right) mod7$$

   where [X] is the integer value of X (e.g., [9.5] = 9) and

   N = Day of month

   Y = Year

   A = First two digits of the year

   M = Month (where January represents 11 of the preceding year, February is month 12 of the preceding year, March is month 1, and so on.)

   E.g.: Date = 12/02/2010

   In this date, M = 12, N = 12, A = 20, and Y = 10.
10. Generate a sorted vector of alphabets which include 10 random letters of uppercase and 10 of lowercase, incorporating both into a single vector.
11. Create the following list: v<-list(a = 10:15, b = 20:26, c = 30:37).
    a. Write a program to display the length of all the list elements. Also, calculate the sum of each element.
    b. Add an element to the list called 'newcolumn' that is a vector of 10 numbers that range from 90 to 100.
12. Write an R program to check the distribution frequency of gender data which is the combination of two unique values: Male and Female. Also, check the cumulative frequency of this data.

13. Create the given data frame and answer the following questions:

Cid	Cname	Income	City
C01	Amit	56000	Delhi
C02	Neha	80000	Noida
C03	Aman	45000	Noida
C04	Sneha	56000	Hyderabad

   a. Display the details of customer name along with income level.
   b. Write the output after executing the following command:
   `df[c(1,2),c(3,4)]`
   c. Add the column Age (34,56,45,30) to the Customer data frame and display the details of customer with the maximum age.
   d. Get the details of all customers whose income is greater than 50,000 and age is greater than 30.

14. Create a list with following data items: X (a sequence of integers ranging from 10 to 20), Y(random sample of 6 uppercase letters), Z(3 × 3 matrix of sequence 1:9).
   a. Display all the values of data object 'X.'
   b. Display the value stored within the matrix at position [2, 2]. Also, output the value of the second column.
   c. Find the length of all the data stored within the list. Also, check whether all the data objects are of the same length or not.
   d. Add the new data object known as 'rnum' to the list which is a collection of five standard normal distribution elements.

References

Beasley, R. E., & Beasley, R. E. (2020). Array Operations. In Essential ASP. NET Web Forms Development: Full Stack Programming with C#, SQL, Ajax, and JavaScript (pp. 221–240). Springer.

Brass, P. (2008). Advanced data structures (Vol. 193). Cambridge: Cambridge University Press.

Braun, W. J., & Murdoch, D. J. (2021). A first course in statistical programming with R. Cambridge University Press.

Chan, B. K., & Chan, B. K. (2018). Data analysis using R programming. In Biostatistics for Human Genetic Epidemiology (pp. 47–122).

Cotton, R. (2013). Learning R: A step-by-step function guide to data analysis. O'Reilly Media, Inc.

Fabrigar, L. R., & Wegener, D. T. (2012). Exploratory factor analysis. Oxford University Press.

Fieller, N. (2018). Basics of matrix algebra for statistics with R. Chapman and Hall/CRC.

Kabacoff, R. (2022). R in action: Data analysis and graphics with R and Tidyverse. Simon and Schuster.

Lang, D. T. (2007). R as a Web Client–the RCurl package. Journal of statistical software, 10(2), 1–42.

Matloff, N. (2011). The art of R programming: A tour of statistical software design. No Starch Press.

Mehta, D. P., & Sahni, S. (2004). Handbook of data structures and applications. Chapman and Hall/CRC.

Petersohn, D., Macke, S., Xin, D., Ma, W., Lee, D., Mo, X., ... & Parameswaran, A. (2020). Towards scalable dataframe systems. arXiv preprint arXiv:2001.00888.

Prakash, P. K. S., & Rao, A. S. K. (2016). R data structures and algorithms. Packt Publishing Ltd.

Rao, T. R., Mitra, P., Bhatt, R., & Goswami, A. (2019). The big data system, components, tools, and technologies: A survey. Knowledge and information Systems, 60, 1165–1245.

Searle, S. R., & Khuri, A. I. (2017). Matrix algebra useful for statistics. John Wiley & Sons.

Tanner-Smith, E. E., Tipton, E., & Polanin, J. R. (2016). Handling complex meta-analytic data structures using robust variance estimates: A tutorial in R. Journal of developmental and life-course criminology, 2, 85–112.

Toby, B. H. (2006). R factors in Rietveld analysis: How good is good enough?. Powder diffraction, 21(1), 67–70.

Walker, D., & Morrisett, G. (2000, September). Alias types for recursive data structures. In International Workshop on Types in Compilation (pp. 177–206). Berlin, Heidelberg: Springer Berlin Heidelberg.

Wickham, H. (2019). Advanced R. Chapman and Hall/CRC.

Wickham, H., Çetinkaya-Rundel, M., & Grolemund, G. (2023). R for data science. O'Reilly Media, Inc.

4

String Handling in R

Learning Objectives:

After studying this chapter, students and researchers will be able to:

- Handle and process strings in R
- Use string manipulation in R
- Use packages and functions for string handling
- Use string manipulation in a 'stringr' package
- Use pattern matching
- Use regular expression in R
- Briefly summarize all string handling concepts

4.1 Introduction

R language provides us one important type of data to handle and process text strings, which is character. Character string objects are held by the 'character' class. Data written inside single or double quotes is treated as a string in R. Externally, users can use any quotes for strings, but internally R language employs double quotes to hold character data. This chapter helps researchers and students to get insights on the concepts of string handling. Applications include analysis with character type data. The concept of strings can be applied for statistical text analysis (Ionescu et al., 2016). R is not a kind of scripting programming language like Python, Perl, or Ruby. For data analysis, manipulation of a string is required for better analysis. R provides many string functions and operations for statistical text analysis (Welbers et al., 2017).

A data set is the composition of all data types like numeric, Boolean, and character. Any data set contains column and row headers which are character

DOI: 10.1201/9781003489382-4

strings. In order to deal with the following cases, string handling and processing is required:

- Replacing one variable name or value with other
- Letter conversion like lower- to uppercase or vice versa
- Changing the value of specified text within your data
- Handling XML and HTML data
- Text modification in Excel and CSV
- Searching for a specific pattern
- Regular expression creation

4.2 String Creation in R

A character is one of the important types of data in R language. Character strings are presented using either single or double quotes (Sanchez, 2013). For example, *'R Programming Language'* OR *"R Programming Language."* Apart from using single/double quotes for string creation, R provides the character() function to generate strings. The character() function creates an object of 'character' type. To create NULL strings, use single/double quotes (' ' or " ") without any text; for example, NULL String= " ".

The following points are required to create strings in R:

- Single or double quotes are mandated at the beginning or end of the string. They cannot be used inside one another.
- Single quotes can be embedded within another string, starting and ending with double quotes.
- Double quotes can be embedded within another string, starting and ending with single quotes.
- Single quotes cannot be embedded within another string, starting and ending with single quotes.
- Double quotes cannot be embedded within another string, starting and ending with double quotes.

String Creation Example in R

```
> # String Creation example in R
> s<-"String Creation example"
> r<-'R Programming'
> s
[1] "String Creation example"
> r
```

```
[1] "R Programming"
> ss<- "single ' quote in double"
> ss
[1] "single ' quote in double"
> sr<-'double "in single'
> sr
[1] "double \"in single"
```

4.2.1 NULL String

The NULL string is created by using single/double quotes consecutively. For example:

N< ""

***Null String Creation* Example in R**

```
> NULLstr<-""
> NULLstr
[1] ""
> class(NULLstr)
[1] "character"
```

4.2.2 String Creation Using character()

The function to create a character vector in R language is character(). The character() function will generate a sequence of characters after specifying the length of the vector. For generating a sequence of characters using the character() function, the first step is to create a NULL character object. Then, append new elements by supplying its index value. The following example illustrates how to create a character vector using the character() function.

***Character Vector Creation* Example in R**

```
> cvector<-character(10)
> cvector
[1] "" "" "" "" "" "" "" "" "" ""
> length(cvector)
[1] 10
> # Add First element
> cvector[1]="R"
> # Add second element
> cvector[2]="Programming"
> cvector
[1] "R"         "Programming"
[3] ""          ""
[5] ""          ""
[7] ""          ""
[9] ""          ""
> cvector[10]="Statistical"
```

```
> cvector
[1] "R"          "Programming"
[3] ""           ""
[5] ""           ""
[7] ""           ""
[9] ""           "Statistical"
```

4.2.3 NULL Character Vector

To generate a NULL character vector in R, the function character() is used. To generate a basic string vector, the character() function is utilized and the length of its argument is zero. One major difference in the NULL string and NULL character vector is that both have different lengths. As shown in the example, the length of the NULL character vector is '0' while that of the NULL string is '1'.

Null Character Vector Creation **Example in R**

```
> nullchr<-character(0)
> nullchr
character(0)
> class(nullchr)
[1] "character"
> length(nullchr)
[1] 0
> length(NULLstr)
[1] 1
```

4.2.4 is.character() and as.character() Functions

The is.character() function is a generic method to test the object data of a character type. The as.character() function is another generic method to create an object of a character type.

is.character() **and** ***as.character()*** **Example in R**

```
> str="R Programming"
> n=2+2
> is.character(str)
[1] TRUE
> is.character(n)
[1] FALSE
> class(str)
[1] "character"
> class(n)
[1] "numeric"
> # Converting Numeric to Character type
> s=as.character(n)
> s
[1] "4"
> class(s)
[1] "character"
```

4.3 String Manipulation in R

With the increase in the volume of text data, compatible and built-in generic methods are required to extract valuable information from text data. Earlier, when data size was smaller, we could use and extract relevant information using Control+F, Control+C, and Control+V, but these approaches are slow and erroneous. In analytics of character type data, these traditional approaches will not work. For text analytics, we require techniques which are faster, convenient, and more reliable for data extraction and manipulation. R provides a complete set of functions for string manipulation (Boehmke, 2016). Table 4.1 presents a brief summary of string functions for manipulation (Zhang et al., 2017).

TABLE 4.1

String Functions

Function_Name	Description	Syntax
paste()	Merges characters or strings	paste(..., sep = " ", collapse = NULL)
nchar()	Counts characters in a given string	nchar(str)
tolower()	Converts text to lowercase	tolower(str)
toupper()	Converts text to uppercase	toupper(str)
length()	Extracts the length of a string	length(str)
substring()	Extracts parts of a string	substring(str,start,stop)
format()	Formats strings and numeric data	format(str, digits, nsmall, scientific, width, justify = c("left", "right", "centre", "none"))
print()	Prints a string	print(str)
noquote()	Prints a string with no quotes	noquote(str)
cat()	Concatenates a string with another string	cat(..., file = "", sep = " ", fill = FALSE, labels = NULL, append = FALSE)
sprintf()	Prints formatted text data	sprintf(format_data, ...)
toString()	Converts data objects to a string	toString(x)
casefold()	Used for case folding	casefold(x, upper)
abbreviate()	Assigns an abbreviation to a given string	abbreviate(str, minlength = 4, dot = FALSE, strict = FALSE, method = c("left.keep", "both. sides"))
substr()	Extracts parts of a given string	substr(str, first, last)
chartr()	Translates a character	chartr(old_str, new, x)
strsplit()	Splits the elements of a string	strsplit(str,splitting_condition)

4.3.1 Elementary String Manipulation Functions

1. **paste() function:** This is one of the major generic functions which is widely used to generate strings. paste() takes one or more data objects as an input and changes an object to a character type and then appends it with another string. The syntax of the paste() function is as follows:

   paste(data,sep= " "collapse=NULL)

   where data is the input data objects, sep= " " is used to separate strings, and argument collapse is an optional argument which is used to specify whether or not we want all the data elements to be collapsed.

   Example

```
> # paste() example of data objects of different length
> paste("R",1:10,sep="-")
[1] "R-1"   "R-2"   "R-3"   "R-4"   "R-5"
[6] "R-6"   "R-7"   "R-8"   "R-9"   "R-10"
> #paste with collapse
> paste("R",1:5,sep="-",collapse = "")
[1] "R-1R-2R-3R-4R-5"
```

2. **cat() function:** Concatenation of strings in R is implemented using the cat() function. The syntax is provided below:

   cat(data, file = "", sep = " ", fill = FALSE, labels = NULL, append = FALSE)

   If we input data objects of numeric or complex types, it automatically converts it into a character type. Strings are merged by default with space as a separator, and the user can alter it with a 'sep' argument.

   Example

```
> str="Statistical Modeling"
> cat(str)
Statistical Modeling
> cat(str,str)
Statistical Modeling Statistical Modeling
> cat
> cat(str,str,sep="::-->")
Statistical Modeling::-->Statistical Modeling
> # Dispaly last four months using cat()
> cat(month.name[9:12],sep="-")
September-October-November-December
```

3. **format() function:** This function is used to improve the look and feel aspect of printed data. The syntax of format() is as follows:

   format(str, digits, nsmall, scientific, width, justify = c("left", "right", "centre", "none"))

The arguments of the format() functions are:

- width: specify the width of string
- trim: TRUE value is set to remove the space
- justify: can take three possible values: 'left,' 'right,' 'centre,' and 'none'
- digits: specify the total digits
- scientific: is used for scientific notation. FALSE values are set for standard notation and TRUE values are set for scientific notation.
- big.mark: For printing large data in a sequenced manner, we use big.mark or big.interval. big.mark presents output in a comma separated way.

Example

```
> format(56.78765544,digits = 2)
[1] "57"
> format(56.78765544,digits = 4)
[1] "56.79"
> format(c("Data","Mining","Book"),width=6,justify ="centre")
[1] " Data " "Mining" " Book "
> format(c("Data","Mining","Book"),width=6,justify = "left")
[1] "Data  " "Mining" "Book  "
> format(c("Data","Mining","Book"),width=6,justify = "right")
[1] "  Data" "Mining" "  Book"
> format(c("Data","Mining","Book"),width=6,justify = "none")
> format(8768767897,big.mark = ",")
[1] "8,768,767,897"
```

4. **nchar() and length() functions: The** nchar() function is used to count the number of characters in a given string. length() is used to retrieve the total number of strings in the input data.

Example

```
> nchar("R Programming Book")
[1] 18
> nchar(c("R","Programming","Book"))
[1]  1 11  4
> length(c("R","Programming","Book"))
[1] 3
> length("R Programming Book")
[1] 1
```

5. **tolower() and toupper() functions:** Both are used for case conversion. tolower() converts data from upper- to lowercase and toupper() performs vice versa of tolower() (i.e., it converts data from lowercase to uppercase).

Example

```
> str
[1] "Statistical Modeling"
> toupper(str)
[1] "STATISTICAL MODELING"
> tolower(str)
[1] "statistical modeling"
```

6. **casefold() and chartr() functions:** The wrapper method of tolower() and toupper() is casefold(). By default, the casefold() method converts entire strings to lowercase. But if the user wants data in uppercase, then the TRUE value will be set for the upper argument. The syntax is:

   casefold(str,upper=FALSE)

   For character translation, the chartr() method is used. The syntax is as follows:

   chartr(str,old_value,new_value)

   The chartr() function will take three parameters. First: Input String, Second: Old_value, and Third: New_value. Let's understand the concept of the chartr() method by this example. Consider the string: "america".

   Suppose we want to replace 'a' with 'A' in the given string, then the output is:

   "AmericA"

   Example

```
> #casefold() example
> casefold("R Programming Workshop")
[1] "r programming workshop"
> casefold("R Programming Workshop",upper = TRUE)
[1] "R PROGRAMMING WORKSHOP"
> chartr("r","R","r programming workshop")
[1] "R pRogRamming woRkshop"
> # Multiple Translation Example
> chartr("rso","$%&","r programming workshop")
[1] "$ p$&g$amming w&$k%h&p"
```

7. **abbreviate():** Another principal string manipulation function is abbreviate(). The abbreviate() method is used to assign abbreviations to input strings. Its structure has the following parameters:

   abbreviate(str, minlength = 4, dot = FALSE, strict = FALSE, method = c("left.keep", "both.sides"))

   'str' contains the string object that we wish to abbreviate.

Example

```
> mon=month.name[4:8]
> mon
[1] "April"  "May"   "June"   "July"
[5] "August"
> # Abbreviation by default
> a1=abbreviate(mon)
> a1
April   May   June    July    August
"Aprl" "May" "June"  "July"  "Agst"
> #Abbreviation with minimum length
> a2=abbreviate(mon,minlength = 2)
> a2
April   May   June   July    August
"Ap"   "My"  "Jn"   "Jl"     "Ag"
> #with method parameter
> a3=abbreviate(mon,minlength = 3,method="both.sides")
> a3
April   May   June   July    August
"Apr"  "May" "Jun"  "Jly"    "Ags"
```

8. **substring() and substr() functions:** Both methods are used to extract and replace subparts of the input string. The syntax is provided below:

   substring(str, start, stop)
   substr(str, start, stop)

Example

```
> substring("xyzdfghhj",2,5)
[1] "yzdf"
> substr("DHJJKLLLL",2,5)
[1] "HJJK"
> # Replace 4th letter with $
> str=c("India","Australia","Canada","Mexico")
> substr(str,4,4)<-"$"
> str
[1] "Ind$a"    "Aus$ralia" "Can$da"
[4] "Mex$co"
> #Replace 3 and 4 letter with happy face
> substr(str, 3, 5) <- ":)"
> str
[1] "In:)a"    "Au:)ralia" "Ca:)da"
[4] "Me:)co"
```

4.3.2 Elementary Set Functions for String Manipulation

R language has exclusive functions for executing set operations on string vectors. We can apply all set functions like set difference, set union, set equality, and set intersection (Boehmke, 2016) on text data. Table 4.2 provides the list of set functions that we can directly apply on text data.

TABLE 4.2

Set Functions

Function_Name	Description
setdiff()	Finds out the difference in two string vectors
union()	Merges two string vectors
setequal()	Tests the equality of two sets
intersect()	Finds common elements of two sets
identical()	Tests whether two sets are exactly equal or not
%in%()	Checks whether the set contains a specific data object
sort()	Sorts data
is.element()	Tests elements within the set

Set Functions Example for String Manipulation

```
> s1=c("R","Programming","statistical","modelling")
> s2=c("R","workshop","statistical","computation")
> setdiff(s1,s2)
[1] "Programming" "modelling"
> union(s1,s2)
[1] "R"            "Programming"  "statistical"
[4] "modelling"    "workshop"     "computation"
> s3=c("R"," statistical "," Programming ","modelling")
> setequal(s1,s2)
[1] FALSE
> setequal(s1,s3)
[1] TRUE
> identical(s1,s2)
[1] FALSE
> identical(s1,s3)
[1] FALSE
> intersect(s1,s2)
[1] "R"            "statistical"
> e1="R"
> e2="Programming"
> is.element(e1,s3)
[1] TRUE
> e3="Maths"
> is.element(e3,s3)
[1] FALSE
> e1%%s3
> e1 %in% s3
[1] TRUE
> e3 %in% s2
[1] FALSE
> sort(s3)
[1] " Programming " " statistical "
[3] "modelling"     "R"
```

```
> sort(s3,decreasing = TRUE)
[1] "R"                "modelling"
[3] " statistical " " Programming "
```

4.4 String Handling Packages in R

String manipulation is a problem where the user is supposed to process the given text data and then further use that processed data; for example, searching for any pattern in the given data, shifting each character of a string by 'N' positions, reversing a string, reversing a string by a character, and substring sorting. Two important packages are available in R for string processing: 1) 'stringi' package and 2) 'stringr' package (Wickham & Bryan, 2023). Both packages are used for string manipulation and text processing.

4.4.1 'stringi' package

This package is used to provide text data processing facilities. Characteristics include methods that are very fast, consistent, easy to use, correct, and portable. The 'stringi' package incorporates a library named 'International Components for Unicode (ICU)' which makes this package platform independent, like Java and Python (Gagolewski, 2022). Applications include case mapping, string concatenation, string metaphrasing, pattern generation and searching, random generation of string, parsing, normalization of unicode, date and time formatting, and many more. Following commands are used to download 'stringi' package for string processing:

```
> install.packages("stringi")
> library(stringi)
```

Table 4.3 provides a list of a few available functions in the 'stringi' library.

TABLE 4.3

Useful Functions of 'stringi' Package

Function_Name	Description	Syntax
stri_enc_get	Extracts or sets default character encoding in 'stringi'	*stri_enc_set(enc)* where 'enc' is the encoding name
stri_duplicated	Determines repeated data element	*stri_duplicated(str, fromLast = FALSE, ..., opts_collator = NULL)* where 'str' is string vector, 'fromLast' is a logical value which indicates whether the search performs from last or not, and 'opts_collator' is named the ICU collator list

(Continued)

TABLE 4.3 ***(Continued)***

Useful Functions of 'stringi' Package

Function_Name	Description	Syntax
stri_length	Determines the string length	*stri_length(str)* where 'str' is the string vector
stri_count	Counts the pattern occurrence in a given string	*stri_count(str, ..., regex, fixed, coll, charclass)* where 'str' is the string vector, '....' specifies any additional arguments, and 'regex, fixed, coll, charclass' specify the search pattern
stri_enc_detect	Determines character data, character set, or encoding	*stri_enc_detect(str, filter_angle_brackets = FALSE)* where 'str' is the string vector and 'filter_angle_brackets' is a logical value indicating whether or not filtering is enabled
stri_enc_isutf16be	Checks whether a given input stream is valid UTF(16LE/16BE/32LE/32BE)	*stri_enc_isutf16be(str)* where 'str' is the string vector
stri_opts_fixed	Creates a list with a predefined pattern	*stri_opts_fixed(case_insensitive = FALSE, overlap = FALSE, ...)* where 'case_insensitive' is used to enable case matching, 'overlap' is used to enable determination of overlapping matches, and '...' specifies additional arguments if required
stri_datetime_add	Updates date and time by adjoining a fixed time unit	*stri_datetime_add(time, value = 1L, units = "seconds", tz = NULL, locale = NULL* where 'time' specifies the time object, 'value' is the fixed time unit to be added, 'units' is single string comprised of 'day,' 'weeks,' 'months,' 'years,' 'minutes,' 'seconds,' 'milliseconds,' or 'hours,' 'tz' specifies the default time zone, and 'locale' is used to specify the default locale
stri_timezone_info	Provides information regarding the time zone	*stri_timezone_info(tz = NULL, locale = NULL, display_type = "long")* where 'tz' specifies the default time zone, 'locale' is used to specify the default locale, and 'display_type' is used to present the output in any of the given forms: 'long,' 'generic_long,' and 'short.'

TABLE 4.3 ***(Continued)***

Useful Functions of 'stringi' Package

Function_Name	Description	Syntax
stri_list2matrix	Converts a list to a character matrix	*stri_list2matrix(x, byrow = FALSE, fill = NA_character_, n_min = 0)* where 'x' is alist of vectors, 'byrow' is used for setting the transpose of a matrix, 'fill' is a single string, and 'n_min' specifies the minimum count of columns and rows in a matrix.
stri_opts_regex	Adjusts the behaviour of a regular expression matcher	*stri_opts_regex(case_insensitive, comments, dotall, literal, multiline,unix_lines, uword, error_on_unknown_escapes, ...)* where 'case_insensitive' is used to enable case matching, 'comments' includes comments, 'dotall' logical value indicating line terminator, 'literal' is for considering literal as a complete pattern, 'multiline' administers the behaviour of '$' & '^', 'unix-line' line ending in unix, 'unword' is word boundaries (Unicode), and 'error' generates an error message.
stri_wrap	Wraps text for paragraph formatting	*stri_wrap(str, width = floor(0.9 * getOption("width")),cost_exponent = 2, simplify = TRUE, normalize = TRUE, indent = 0,exdent = 0, prefix = "", initial = prefix,whitespace_only = FALSE, use_length = FALSE, locale = NULL)* where 'width' specifies the number of words per line, 'cost_exponent' will select an algorithm for text wrapping, 'indent' provides indentation of the initial line, and 'exdent' provides indentation to the consecutive lines.
stri_subset	Extracts subdata that matches a particular pattern	*stri_subset(str,regex)* where 'str' is the string vector and 'regex' represents the search pattern.
stri_width	Determines the width of text data	*stri_width(str)*
stri_split	Splits text data based on the pattern	*stri_split(str, ..., regex, fixed, coll, charclass)* where 'str' is the string vector and 'regex,fixed,coll,charclass' represents the search pattern.
stri_match_all	Extracts data based on complete pattern matching	*stri_match_all(str, ..., regex)* where 'str' is the string vector and 'regex' represents the search pattern.
stri_unique	Extracts unique text data	*stri_unique(str, ..., opts_collator = NULL)* where 'opts_collator' is a named ICU collator list.

(Continued)

TABLE 4.3 *(Continued)*

Useful Functions of 'stringi' Package

Function_Name	Description	Syntax
stri_read_lines	Reads text data from a file	*stri_read_lines(fname, encoding = "auto", locale = NA, fallback_encoding = stri_enc_get())* where 'fname' is the input file, 'encoding' is used for encoding, 'locale' is used to specify locale value to be used, and 'fallback_encoding' is used if the encoding technique fails.
stri_join_list	Merges a string vector within a list	*stri_join_list(x, sep = "", collapse = NULL)* where 'x' is the string vector, 'sep' is the separating symbol, and 'collapse' is an optional output separator
stri_rand_strings	Used for random text generation	*stri_rand_strings(n, length, pattern = "[A-Za-z0-9]")* where 'n' is the number of random strings to generate, 'length' is the length of each generated string, and 'pattern' is a regular expression defining the set of characters to use in the random strings.

Example

 # Determination of Duplicated Elements

```
stri_duplicated(c("india","america","canada","NA","canada","mexico",
"india"))
[1] FALSE FALSE FALSE FALSE TRUE FALSE TRUE
stri_duplicated(c("india","america","canada","NA","canada","mexico",
"india"),fromLast = TRUE)
[1] TRUE FALSE TRUE FALSE FALSE FALSE FALSE
stri_duplicated_any(c("india","america","canada","NA","canada",
"mexico","india"))
[1] 3
```

 # Count the Length of Data Elements

```
>stri_length(c("india","america","canada","NA","canada","mexico",
"india"))
[1] 5 7 6 2 6 6 5
```

 # Count Pattern in the Given String

```
> str="R Programming Workshop"
> stri_count(str,fixed = "Programming")
[1] 1
> stri_count(str,regex="R+")
[1] 1
> stri_count(str,regex="r+")
[1] 3
```

Character Set and Language Determination

```
> data<-c("raw","10000","#$tt")
> stri_enc_detect(data)
[[1]]
        Encoding Language Confidence
1     UTF-8                  0.15
[[2]]
        Encoding Language Confidence
1     UTF-8                  0.15
2 UTF-16BE                   0.10
3 UTF-16LE                   0.10
[[3]]
        Encoding Language Confidence
1 ISO-8859-1        no        0.75
2 ISO-8859-2        hu        0.75
3       UTF-8                 0.15
4   UTF-16BE                  0.10
5   UTF-16LE                  0.10
```

Pattern Matching

```
> stri_detect_fixed("may","MAY")
[1] FALSE
> stri_detect_fixed("may","MAY",opts_fixed =
stri_opts_fixed(case_insensitive         = TRUE))
[1] TRUE
> #OR
> stri_detect_fixed("may","MAY",case_insensitive = TRUE)
[1] TRUE
```

Date and Time Functions

```
> dt<-stri_datetime_now()
> dt
[1] "2019-04-01 17:01:16 IST"
> stri_datetime_add(dt,units = "months")<-2
> dt
[1] "2019-06-01 17:01:16 IST"
> stri_datetime_add(dt,-2,units = "months")
[1] "2019-04-01 17:01:16 IST"
>  stri_datetime_add(stri_datetime_create(1983,  8,  25),  1,
units="years")
[1] "1984-08-25 12:00:00 IST"
>  stri_datetime_add(stri_datetime_create(1983,  8,  25),  1,
units="years",   locale="@calendar=hebrew")
[1] "1984-09-13 12:00:00 IST"
> stri_timezone_info()
$ID
[1] "Asia/Calcutta"

$Name
[1] "India Standard Time"

$Name.Daylight
[1] NA
```

```
$Name.Windows
[1] "India Standard Time"

$RawOffset
[1] 5.5

$UsesDaylightTime
[1] FALSE
```

 # Conversion of List to Matrix

```
> stri_list2matrix(list(c("May","June"),c("Mon","Tue")))
       [,1]   [,2]
[1,] "May"  "Mon"
[2,] "June" "Tue"
> stri_list2matrix(list(c("May","June"),c("Mon","Tue")),byrow = TRUE)
       [,1]  [,2]
[1,] "May" "June"
2,] "Mon" "Tue"
> stri_list2matrix(list(c("May","June"),c("Mon","Tue")),
fill = "",n_min=5)
       [,1]   [,2]
[1,] "May"  "Mon"
[2,] "June" "Tue"
[3,] ""     ""
[4,] ""     ""
[5,] ""     ""
```

 # Text Wrapping

```
> cat(stri_wrap(str,10,0.0),sep="\n")
R
Programming
Workshop
> cat(stri_wrap(str,10,0.0),sep="-")
R-Programming-Workshop
> cat(stri_pad(stri_wrap(str), side='both'), sep="\n")
       R Programming Workshop
```

 # Substring Extraction from the Given String

```
> stri_subset_regex(str,"R+",negate = TRUE)
character(0)
> stri_subset_regex(str,"R+")
[1] "R Programming Workshop"
> stri_subset_regex(str,"^[0-9]+$")
character(0)
> stri_subset_regex(str,"^[0-9]+$",negate = TRUE)
[1] "R Programming Workshop"
```

 # Splitting Text Data

```
> str
[1] "R Programming Workshop"
> stri_split_fixed(str," ")
[[1]]
[1] "R"             "Programming" "Workshop"
```

```
> stri_split_fixed("X*Y*Z","*")
[[1]]
[1] "X" "Y" "Z"

> stri_split_fixed("X*Y*Z","*",omit_empty = TRUE)
[[1]]
[1] "X" "Y" "Z"

> stri_split_fixed("X*Y*Z","*",n=2)
[[1]]
[1] "X" "Y*Z"

> stri_split_fixed("X*Y*Z","*",n=1)
[[1]]
[1] "X*Y*Z"

> stri_split_fixed("X*Y*Z","*",n=3)
[[1]]
[1] "X" "Y" "Z"

> stri_split_fixed("X*Y*Z","*",n=3,tokens_only = TRUE)
[[1]]
[1] "X" "Y" "Z"

> stri_split_charclass("R Programming Workshop","\\p{WHITE_SPACE}")
[[1]]
[1] "R"             "Programming" "Workshop"
```

 # Extract Data Based on Match Cases

```
> stri_match_first_regex(c("xyz", ":xyz", ":xyz:"),
"^(:)?([^:]*)(:)?$")
[,1]     [,2] [,3]  [,4]
[1,] "xyz"    NA    "xyz" NA
[2,] ":xyz" ":" "xyz" NA
[3,] ":xyz:" ":""xyz" ":"
> stri_match_all_regex(c("xyz", ":xyz", ":xyz:"), "^(:)?([^:]*)(:)?$")
[[1]]
[,1]   [,2] [,3]   [,4]
[1,] "xyz" NA    "xyz" NA

[[2]]
[,1]    [,2] [,3]   [,4]
[1,] ":xyz" ":"  "xyz" NA

[[3]]
[,1]     [,2] [,3]   [,4]
[1,] ":xyz:" ":"  "xyz" ":"
```

 # Sort Character Vector

```
> str=c("hello","bad","good","lion","zebra","apple")
> stri_sort(str)
[1] "apple" "bad" "good" "hello" "lion"
[6] "zebra"
> stri_sort(str,locale="sk_SK")
[1] "apple" "bad" "good" "hello" "lion"
[6] "zebra"
```

```
> stri_sort(str,locale="pl_PL")
[1] "apple" "bad" "good" "hello" "lion"
[6] "zebra"
```

#Merging String in a List

```
> l=list(c("HI","Hello"),c("Car","Bike"))
> l
[[1]]
[1] "HI"  "Hello"

[[2]]
[1] "Car" "Bike"
> stri_join_list(l,sep="",collapse=NULL)
[1] "HIHello"  "CarBike"
> stri_join_list(stri_extract_all_regex(c("back front bacon",
"789 111", "spam 789 sausage"), "\\p{L}+"),sep=",")
[1] "back,front,bacon" NA
[3] "spam,sausage"
```

 # Random String Generation

```
> A=1
> stri_rand_strings(A,10,"[0-9]")
[1] "3880082163"
> stri_rand_strings(A,5,"[0-9]")
[1] "32799"
> stri_rand_strings(A,5,"[aA-zZ]")
[1] "vnKIz"
> A=3
> stri_rand_strings(A, sample((A/2):(A+1), (A/2), replace=TRUE),
'[a-zA-Z0-9]')
[1] "0z" "Qw" "ip"
```

 # To Extract a Unique and Duplicated Element

```
> d<-tail(iris)
> d
Sepal.Length Sepal.Width Petal.Length Petal.Width  Species
145             6.7          3.3            5.7          2.5 virginica
146             6.7          3.0            5.2          2.3 virginica
147             6.3          2.5            5.0          1.9 virginica
148             6.5          3.0            5.2          2.0 virginica
149             6.2          3.4            5.4          2.3 virginica
150             5.9          3.0            5.1          1.8 virginica
> stri_unique(d$Sepal.Length)
[1] "6.7" "6.3" "6.5" "6.2" "5.9"
> stri_unique(c("\u0202", stri_trans_nfkd("\u0202")))
[1] "Â"
> stri_duplicated(d)
[1] FALSE FALSE FALSE FALSE FALSE
> stri_duplicated(d$Species)
[1] FALSE  TRUE  TRUE  TRUE  TRUE  TRUE
> stri_duplicated(d$Sepal.Length)
[1] FALSE  TRUE FALSE FALSE FALSE FALSE
```

4.4.2 'stringr' Package

'stringr' is another package in R for string processing and manipulation (Wickham & Bryan, 2023). It's an uncomplicated, compatible package for string operations. All methods and parameters are compatible, and all available built-in functions are able to deal with 'NA' values and a NULL vector, and the response from one method can be easily fed as an input to another method. Character vector is the chief component in data analysis and modelling and play a vital role in data cleaning, integration, and processing tasks. This package provides a rich set of functions for working with character objects. 'stringr' is the extension of the 'stringi' package. The main emphasis of 'stringr' is to provide functions for string manipulation, while 'stringi' is a rich library which has functions for almost all string operations. Four different core functions provided in 'stringr' are as follows:

- String Manipulation: Functions available for text data manipulation
- Trimming: Functions to add and remove whitespaces within text
- Locale: Functions for locale operations
- Pattern Matching: Methods to generate and match given pattern

The following commands are used to download 'stringr' package for string related operations:

```
install.packages("stringr")
library(stringr)
```

Table 4.4 provides a list of a few available functions in the 'stringr' library.

TABLE 4.4

Useful Functions of the 'stringr' Package

Function_Name	Description	Syntax
str_c	Used for string concatenation	*str_c(..., sep = "", collapse = NULL)* where '...' is the character vector, 'sep' is used to separate two strings, and 'collapse' is used to merge input character objects to one string.
str_length	Returns the length of input character vector	*str_length(str)* where 'str' is the string vector.
str_dup	Duplicates and merges strings	*str_dup(str, times)* where 'str' is the string vector and times' specify the count of duplication required.
str_trim	Deletes spaces form a string	*str_trim(str, side = c("both", "left", "right"))* where 'str' is the string vector and 'side' specifies the side from which space needs to be removed.

(Continued)

TABLE 4.4 ***(Continued)***

Useful Functions of the 'stringr' Package

Function_Name	Description	Syntax
str_sub	Used to find and replace a substring	*str_sub(str, start = 1L, end = -1L)* where 'str' is the string vector, 'start' is the starting index, and "end" is the last index.
str_pad	Used for text data padding	*str_pad(str, width, side = c("left", "right", "both"), pad = " ")* where 'str' is the string vector and 'width' is the padding width.
str_count	To find the count of match in character data	*str_count(str, pattern = "")* where 'str' is the string vector and 'pattern' is the pattern to look for.
str_wrap	Wraps text and formats text	*str_wrap(str, width = 80, indent = 0, exdent = 0)* where 'str' is the string vector, 'width' specifies the target line width, 'indent' is the indentation of the initial line, and 'exdent' indents the following consecutive lines.
str_extract	Fetches the matching pattern from the given string	*str_extract(str, pattern)* where 'str' is the string vector and 'pattern' is the matching pattern to look for.
word()	Extracts a word from the given string	*word(str, start = 1L, end = start, sep = fixed(" "))* where 'str' is the string vector, 'start' is the index of the first word, 'end' is the index of the last word, and 'sep' is the separator between words.
str_to_upper() str_to_lower()	Converts the case of a given string	*str_to_lower(str, locale = "en") OR* *str_to_upper(string, locale = "en")* where 'str' is the string vector and 'locale' specifies the locale to be used for translations.
str_trunc()	Trims the given string	*str_trunc(str, width, side = c("right", "left", "center"),ellipsis = "...")* where 'str' is the string object, 'width' specifies the width of a given string, and 'side, ellipsis' specifies the position and content to be truncated.
str_split()	Splits the given string	*str_split(str, pattern, n = Inf, simplify = FALSE)* where 'str' is the string vector, 'pattern' represents the search pattern, 'n' represents the count of split text, and 'simplify' is used to present output as a character type of vector or matrix.
str_order() str_sort()	First function generates the ordering of data objects and the second function sorts the data content in ascending/ descending order	*str_order(data, decreasing = FALSE)* *str_sort(data, decreasing = FALSE)* where 'data' is the data object required to perform ordering/sorting and 'decreasing' is used to display the output in decreasing order if set equal to TRUE.

Example

 # String Concatenation

The function for a string concatenation in 'stringr' is str_c(). It acts like the paste() function. The main difference between the str_c() and paste() function is that str_c() automatically deletes NULL character objects. Another difference is that str_c uses ' ' as a default separator, and paste () uses white space.

```
> library(stringr)
> str
[1] "hello" "bad"   "good"   "lion"   "zebra"
[6] "apple"
> str_c(str,1:6,sep = "-")
[1] "hello-1"  "bad-2"  "good-3"  "lion-4"
[5] "zebra-5" "apple-6"
> str_c(str,1:6,collapse = "")
[1] "hello1bad2good3lion4zebra5apple6"
> str_c(str,1:6,sep="-",collapse = "")
[1] "hello-1bad-2good-3lion-4zebra-5apple-6"
> str_c(LETTERS,"is:","...")
[1] "Ais:..." "Bis:..." "Cis:..." "Dis:..."
[5] "Eis:..." "Fis:..." "Gis:..." "His:..."
[9] "Iis:..." "Jis:..." "Kis:..." "Lis:..."
[13] "Mis:..." "Nis:..." "Ois:..." "Pis:..."
[17] "Qis:..." "Ris:..." "Sis:..." "Tis:..."
[21] "Uis:..." "Vis:..." "Wis:..." "Xis:..."
[25] "Yis:..." "Zis:..."
> str_c("May","June",NULL,"JULY",character(0))
[1] "MayJuneJULY"
> paste("may","june","july")
[1] "may june july"
> str_c("may","june","july")
[1] "mayjunejuly"
```

 # String Length

The length of a string is extracted by the function str_length(), which acts like nchar(). In comparison to nchar(), str_length() is more robust and has NA values in a better and consistent manner. One major difference in char() and str_length() is that nchar() is not able to convert factors to characters, while str_length can convert factors to characters.

```
> str_length(month.name)
[1] 7 8 5 5 3 4 4 6 9 7 8 8
> month.name
[1]  "January"   "February"   "March"
[4]  "April"     "May"        "June"
[7]  "July"      "August"     "September"
[10] "October"   "November"   "December"
> str_length(factor("month.name"))
[1] 10
> a1="\u00fd"
> a1
[1] "ý"
```

```
> str_length(a1)
[1] 1
> str=c("may","june","july",NULL)
> str_length(str)
[1] 3 4 4
```

 # String Duplication

Text duplication is required for various data preprocessing tasks. R provides the function str_dup() to duplicate and concatenate data within string vector.

```
> cars<-c("Maruti","Honda","Hyndai","Toyota")
> str_dup(cars,2)
[1] "MarutiMaruti" "HondaHonda"
[3] "HyndaiHyndai" "ToyotaToyota"
> str_dup(cars,1:4)
[1] "Maruti"
[2] "HondaHonda"
[3] "HyndaiHyndaiHyndai"
[4] "ToyotaToyotaToyotaToyota"

> str_c("happy",str_dup("very",1:5))
[1] "happyvery"
[2] "happyveryvery"
[3] "happyveryveryvery"
[4] "happyveryveryveryvery"
[5] "happyveryveryveryveryvery"
```

 # String Trimming

One of the major functions in string manipulation is parsing. Parsing is of great importance in text analytics and its functionality is to parse text to words. After parsing, we receive string with white spaces. To remove white spaces at the end or beginning, 'stringr' has a function named str_trim().

```
> str_trim("  R Programming   ")
[1] "R Programming"
> text_data=c(" R"," Programming ","Workshop ")
> str_trim(text_data,side="left")
[1] "R"             "Programming "
[3] "Workshop "
> str_trim(text_data,side="right")
[1] "R"             "Programming"
[3] "Workshop"
> str_trim(text_data,side="both")
[1] "R"             "Programming" "Workshop"
```

 # Substring Extraction

For substring extraction in R, the sub_str() function is used. Sub_str() has an additional feature to work with negative start and end value. If a negative index is used to specify a substring position, then this function starts working and counts backward from the position of the last character .str_sub() also takes indices that will be recycled over the entire character vector.

```
> library(stringr)
> s="R Programming workshop"
> str_sub(s,2,7)
[1] " Progr"
> str_sub(s,c(2,4),c(7,11))
[1] " Progr"  "rogrammi"
> str_sub(s,-3)
[1] "hop"
> str_sub(s,end=-5)
[1] "R Programming work"
> str_sub("hello",2:3)
[1] "ello" "llo"
> s1= c("delhi", "noida", "agra", "hyderabad")
> str_sub(s1,start=-3,end=-1)
[1] "lhi" "ida" "gra" "bad"
> # Replace text with other data
> str_sub(s,1,2)<-"python "
> s
[1] "python Programming workshop"
```

String Padding

str_pad() is used for string padding in R language. It works by taking a character object as an input and then pads it with characters to the specified width. A space is used as a default padding string.

```
> s
[1] "R Programming workshop"
> str_pad(s,40,"left")
[1] "                  R Programming workshop"
> str_pad(s,40,"right")
[1] "R Programming workshop                  "
> str_pad(s,40,"both")
[1] "         R Programming workshop         "
> str_pad(s,40,"both",pad="%")
[1] "%%%%%%python Programming workshop%%%%%%%"
```

String Count

To extract the count of matches for a specified symbol/character, str_count() is used. str_count() is vectorized over a pattern and character object.

```
> d<-head(iris$Species)
> d
[1] setosa setosa setosa setosa setosa setosa
Levels: setosa versicolor virginica
> str_count(d,"s")
[1] 2 2 2 2 2 2
> str_count(d,"t")
[1] 1 1 1 1 1 1
> str_count(c("a.","...",".a.a"),".")
[1] 2 3 4
```

 # String Wrapping

The function str_wrap() is used to wrap text in R programming. str_wrap() is used to format a paragraph by first splitting the long text according to the width and then performing a line indentation.

```
> s=c("Capital of India is New Delhi", "Capital of Sri Lanka is
Colombo","Capital of USA Washington, D.C", "Capital of Australia
is Canberra")
> s
[1] "Capital of India is New Delhi"
[2] "Capital of SriLanka is Colombo"
[3] "Capital of USA Washington, D.C"
[4] "Capital of Australia is Canberra"
> data=paste(s,collapse = "")
> data
[1] "Capital of India is New DelhiCapital of SriLanka is
ColomboCapital of USA Washington, D.CCapital of Australia is
Canberra"
> str_wrap(data,width=40)
[1] "Capital of India is New DelhiCapital\nof SriLanka is
ColomboCapital of USA\nWashington, D.CCapital of Australia
is\nCanberra"
> str_wrap(data,width=40,indent=2,exdent=2)
[1] " Capital of India is New DelhiCapital\n of SriLanka is
ColomboCapital of USA\n Washington, D.CCapital of Australia
is\n Canberra"
```

 # String Extraction

The str_extract() and str_extracct_all() functions are used to fetch the exact specified pattern from the given string.

```
> car_list
[1] "maruti dzire"  "maruti ciaz"
[3] "hyundai creta" "hyundai verna"
[5] "honda city"    "honda civic"
[7] "toyota fortuner"
> str_extract(car_list,"\\d")
[1] NA NA NA NA NA NA NA
> str_extract(car_list,"[a-z]+")
[1] "maruti"  "maruti"  "hyundai"  "hyundai"
[5] "honda"   "honda"   "toyota"
> str_extract(car_list,"\\b[a-z]+\\b")
[1] "maruti"  "maruti"  "hyundai"  "hyundai"
[5] "honda"   "honda"   "toyota"
> str_extract_all(car_list,"\\b[a-z]+\\b")
[[1]]
[1] "maruti"   "dzire"

[[2]]
[1] "maruti"   "ciaz"

[[3]]
[1] "hyundai"  "creta"

[[4]]
[1] "hyundai" "verna"
```

```
[[5]]
[1] "honda"   "city"

[[6]]
[1] "honda"   "civic"

[[7]]
[1] "toyota"  "fortuner"
```

 # *Word Extraction*

Another function in R for extraction is word(), which is used to extract a word from a given sentence/paragraph.

```
> car_list
[1] "maruti dzire"    "maruti ciaz"
[3] "hyundai creta"   "hyundai verna"
[5] "honda city"      "honda civic"
[7] "toyota fortuner"
> word(car_list,3)
[1] NA NA NA NA NA NA NA
> word(car_list,2)
[1] "dzire"    "ciaz"    "creta"
[4] "verna"    "city"    "civic"
[7] "fortuner"
> word(car_list,-1)
[1] "dzire"    "ciaz"    "creta"
[4] "verna"    "city"    "civic"
[7] "fortuner"
> s="xyz.jhk..567.6666.999"
> word(s,1,sep=fixed(".."))
[1] "xyz.jhk"
> word(s,2,sep=fixed(".."))
[1] "567.6666.999"
```

 # *Case Conversion*

For case conversion, the str_to_upper() and str_to_lower() functions are used. The usages are shown below.

```
> car_list
[1] "maruti dzire"    "maruti ciaz"
[3] "hyundai creta"   "hyundai verna"
[5] "honda city"      "honda civic"
[7] "toyota fortuner"
> str_to_upper(car_list)
[1] "MARUTI DZIRE"    "MARUTI CIAZ"
[3] "HYUNDAI CRETA"   "HYUNDAI VERNA"
[5] "HONDA CITY"      "HONDA CIVIC"
[7] "TOYOTA FORTUNER"
> str_to_lower(car_list)
[1] "maruti dzire"    "maruti ciaz"
[3] "hyundai creta"   "hyundai verna"
[5] "honda city"      "honda civic"
[7] "toyota fortuner"
```

 # Truncate String

To truncate a string, the str_trunc() method is available in the 'stringr' package. The user needs to input the maximum desired width along with the position and content. The content is the part of the string that has been removed.

```
> s=" String Concepts in R"
> rbind(
+      str_trunc(s, 10, "right"),
+      str_trunc(s, 10, "left"),
+      str_trunc(s, 10, "center")
+ )
[,1]
[1,] " String..."
[2,] "...ts in R"
[3,] " Str...n R"
```

 # Split String

For splitting the string into subsequent pieces, the str_split() method is used in the 'stringr' package. The character vector is provided as an argument, and then the pattern is based on which string needs to be split into subparts. fixed() is for fixed pattern matching and boundary() is for matching line, word, and character boundaries. The total number of pieces is returned by 'n' (i.e., it is used to specify the total subsequent pieces after the split). For displaying output as a character vector, set the FALSE value in the simplify parameter, and for displaying output as a character matrix, set it to TRUE. The uses are shown below.

```
> book=c("Data mining & Machine Learning","R Programming & Python
Programming","Database & Datawarehouse","Networking & Cloud")
> str_split(book,"&")
[[1]]
[1] "Data mining "    " Machine Learning"

[[2]]
[1] "R Programming "
[2] " Python Programming"

[[3]]
[1] "Database "         " Datawarehouse"

[[4]]
[1] "Networking "       " Cloud"

> str_split(book,"&",simplify = TRUE)
        [,1]              [,2]
[1,] "Data mining "     " Machine Learning"
[2,] "R Programming "  " Python Programming"
[3,] "Database "        " Datawarehouse"
[4,] "Networking "      " Cloud"
> str_split(book,"&",n=3)
[[1]]
[1] "Data mining "     " Machine Learning"
```

```
[[2]]
[1] "R Programming "
[2] " Python Programming"

[[3]]
[1] "Database "" Datawarehouse"

[[4]]
[1] "Networking " " Cloud"
> str_split(book,"&",n=1)
[[1]]
[1] "Data mining & Machine Learning"

[[2]]
[1] "R Programming & Python Programming"

[[3]]
[1] "Database & Datawarehouse"

[[4]]
[1] "Networking & Cloud"
```

 # String Ordering

To arrange a data element in ascending/descending order, the str_order() function is used. Another function, str_sort(), is used to sort data elements. Following is the implementation of both the functions:

```
> str_order(book)
[1] 1 3 4 2
> str_sort(book)
[1] "Data mining & Machine Learning"
[2] "Database & Datawarehouse"
[3] "Networking & Cloud"
[4] "R Programming & Python Programming"
> str_order(month.name)
[1]  4  8 12  2  1  7  6  3  5 11 10  9
> str_sort(month.name)
[1] "April"     "August"    "December"
[4] "February"  "January"   "July"
[7] "June"      "March"     "May"
[10] "November" "October"   "September"
```

4.5 Pattern Matching

In the domain of data science and analysis (Sinkovics, 2018), pattern matching is the process of scanning a given segment of tokens for the existence of the component of some pattern (Hak & Dul, 2009).

Consider the given example:

String: X X Y X X Z X X T X X Y X X Y X
Pattern: X X Y X

```
X X Y X                      X X Y X
X X Y X X Z X X T X X  Y  X  X  Y  X
1 2 3 4 5 6 7 8 9 10 11 12 13 14 15 16
                  X  X  Y  X
```

So, the given pattern 'X X Y X' is found at the 1, 10, and 13 positions. Following is the list of functions provided in R for pattern matching.

 # *Searching Patterns*

The grep() method is used to find a particular pattern in the given string. grep (pattern, string) provides the list of indices for the supplied pattern. If the pattern matches a particular sequence in the given string, it returns the index (Bathie et al., 2024). Set the option value= 'TRUE' if you want confirmed matching element values.

```
> str<-c("xyz","zxx","xxyzx","yzxxy","z x xy")
> grep("xy",str)
[1] 1 3 4 5
> grep("xy",str,value = TRUE)
[1] "xyz"    "xxyzx"  "yzxxy"  "z x xy"
```

 # Searching and Replacing Patterns

The sub() and gsub() built-in functions are used to search and replace patterns (Navarro, 2004). sub(pattern, new_string, string) replaces the first pattern with the new_string while gsub(pattern, new_string, string) replaces all occurrences of a given pattern with the new_string.

Example:

```
> s<-c("information regarding Demography of Asia")
> sub("in","XY",s)
[1] "XYformation regarding Demography of Asia"
> gsub("in","XY",s)
[1] "XYformation regardXYg Demography of Asia"
```

4.6 Regular Expressions

We have covered some rudimentary and other functions for working with text and string data in R programming. All these built-in methods are very useful and permit us to perform tasks related to data and text analysis

(Jockers & Thalken, 2020). These capabilities are quite useful and allow us to do a range of interesting tasks. However, in order to fully realize the potential of string manipulation, we must elevate our technique and venture into the domain of regular expressions. Regular expression (regex) is a distinct text string that describes a set amount of text (Edition, 2021). R provides a large number of utilities for working with regex.

4.6.1 Regular Expression in R

Regular expressions are a succinct and adaptable method for describing patterns in strings. Regular expressions serve as the primary pattern engine in stringr (Hocking, 2019). This implies that when employing a pattern matching function with a plain string, it is akin to encapsulating it within a call to regex().

Example:

The str_extract() function in R's stringr package extracts the first occurrence of a pattern from a string. The regular call:

```
>str_extract(electronic_item, "sams")
```

is indeed shorthand for:

```
>str_extract(electronic_item, regex("sams"))
```

The regex() function creates a regular expression pattern object from the input string 'sams.' This pattern object is then supplied to str_extract(), which matches the input string electronic_item and extracts the first occurrence of the pattern. Using regex() explicitly gives you more control, such as specifying additional arguments or changing the regular expression pattern before passing it to str_extract(). However, for basic patterns such as 'sams,' shorthand notation is more concise and widely used. For further information on regular expressions in R, refer to the help pages: help(regex) and help(regexp).

4.6.2 Metacharacter

Metacharacters in regular expressions are characters with specific meanings that are utilized to generate text matching patterns. Extended regular expressions (EREs) include the following metacharacters: | . () \ { $ [+ * ? In most traditional programming languages, we escape any of the metacharacters in a regex pattern with\(backslash); however, in R programming, double backslashes are required to escape them. Table 4.5 presents the metacharacters used in R and how to escape them.

TABLE 4.5
R Metacharacters and Escape Sequence

Metacharacter	R Escape
.	\\.
$	\\$
*	*
+	\\+
?	\\?
|	\\\|
\	\\\\
^	\\^
[	\\[
]	\\]
{	\\{
}	\\}
(	\\(
)	\\)

In R, you can escape metacharacters and replace them with an empty " ". A few examples are shown below.

Example:

```
> # dollar
> sub("\\$", "", "$R-Programming")
[1] "R-Programming"
> # dot
> sub("\\.", "", "R.Programming")
[1] "RProgramming"
> # caret
> sub("\\^", "", "R^Programming")
[1] "RProgramming"
> # vertical bar
> sub("\\|", "", "R|Programming")
[1] "RProgramming"
```

4.6.3 Sequences

Sequences are a collection of elements arranged in a specified order that are identified or matched within a bigger set of data using specific criteria. This notion is widely employed in a variety of domains, including computer science, mathematics, and bioinformatics, and is frequently associated with algorithms capable of detecting these sequences effectively (Chapman et al., 2019). Commonly sequences used in R are detailed in Table 4.6.

TABLE 4.6
R Sequences

Sequence	Elucidation
\\d	Digit Match
\\D	Nondigit Match
\\w	Word Match
\\W	Nonword Match
\\b	Word Boundary Match
\\B	Nonword Boundary Match
\\s	Space Match
\\S	Nonspace Match
\\v	Vertical Space Match
\\V	Nonvertical Space Match
\\h	Horizontal Space Match
\\H	Nonhorizontal Space Match

Here are some instances of how anchor sequences can be used for replacement.

Example:

```
> # replace space with '+'
> sub("\\s", "+", "Statistical Programming with R")
[1] "Statistical+Programming with R"
> gsub("\\s", "+", "Statistical Programming with R")
[1] "Statistical+Programming+with+R"
> # replace word with '$'
> sub("\\b", "$", "Statistical Programming with R")
[1] "$Statistical Programming with R"
> gsub("\\b", "$", "Statistical Programming with R")
[1] "$S$t$a$t$i$s$t$i$c$a$l$ $P$r$o$g$r$a$m$m$i$n$g$
$w$i$t$h$ $R$"
```

4.6.4 Character Class

A character class or character set refers to a group of characters that are enclosed within square brackets [ ]. A character class or set specifies a collection of characters that a single character in the input string must match. They enable you to conduct complicated pattern matching operations efficiently on strings. A detailed look at the sorts of character classes available in R's implementation of regular expressions, which are typically Portable Operating System Interface (POSIX) standard regular expressions, are provided in Tables 4.7 and 4.8. In R, [[]] brackets are used to represent POSIX character classes.

TABLE 4.7
R Character Classes

Character Class	Description
[aeiou]	Vowel Match (Lowercase)
[AEIOU]	Vowel Match (Uppercase)
[^aeiou]	Nonvowel Match
[a-z]	ASCII character (Lowercase)
[A-Z]	ASCII character (Uppercase)
[0-9]	Digit Match
[^0-9]	Nondigit Match

TABLE 4.8
R POSIX Character Classes

POSIX Character Class	Description
[[:alpha:]]	Alphabetic characters
[[:digit:]]	Digits
[[:lower:]]	Lowercase letters
[[:upper:]]	Uppercase letters
[[:cntrl:]]	Control characters
[[:punct:]]	Punctuation characters
[[:xdigit:]]	Hexadecimal digits
[[:graph:]]	Graphical characters
[[:blank:]]	Blank spaces
[[:space:]]	Space characters

Below are several examples illustrating the use of character classes and POSIX character classes in R.

Example:

```
Languages = c("Hindi", "English", "German", "French")
> # look for 'a' or 'u'
> grep(pattern = "[au]", Languages, value = TRUE)
[1] "German"
> # look for 'e' or 'i'
> grep(pattern = "[ei]", Languages, value = TRUE)
[1] "Hindi" "English" "German" "French"
> data1 = c("4567", "25-August", "IX-X-XI", "R")
> # match any digit
> grep(pattern = "[0-9]", data1, value = TRUE)
[1] "4567"      "25-August"
> # negated digit
> grep(pattern = "[^0-9]", data1, value = TRUE)
```

```
[1] "25-August"  "IX-X-XI"  "R"
> s1 = "String 1  #AS#011C (rose);\nGood Language!
\nSentiment Analysis"
> # remove digits
> gsub(pattern = "[[:xdigit:]]", replacement = "", s1)
[1] "String   #S# (ros);\nGoo Lngug! \nSntimnt nlysis"
> # remove printable characters
> gsub(pattern = "[[:print:]]", replacement = "", s1)
[1] "\n\n"
> # remove graphical characters
> gsub(pattern = "[[:graph:]]", replacement = "", s1)
[1] "    \n  \n "
```

4.6.5 Quantifiers

Quantifiers in regular expressions describe how many times a character, group, or character class should be found in the input text (Forta, 2018). They offer flexibility and control over the pattern matching process. Table 4.9 illustrates commonly utilized quantifiers in R.

Examples below showcase the usage of quantifiers in various contexts.

Example:

```
sname = c("Tanya", "Atharv", "Neha", "mehmet", "John", "anna",
"Marie", "Rose", "Jane", "Rehana", "Flora", "Aditya")
> grep(pattern = "a?", sname, value = TRUE)
[1] "Tanya" "Atharv" "Neha" "mehmet" "John"
[6] "anna" "Marie" "Rose" "Jane" "Rehana"
[11] "Flora" "Aditya"
> # match 'a' exactly once
> grep(pattern = "a{1}", sname, value = TRUE, perl = FALSE)
[1] "Tanya" "Atharv" "Neha" "anna" "Marie" "Jane" "Rehana"
"Flora" "Aditya"
> # match 'a' zero or more times, and 'v'
> grep(pattern = "a*v", sname, value = TRUE)
[1] "Atharv"
```

TABLE 4.9

R Quantifiers

Quantifier	Interpretation
?	Data item is not mandatory and can occur at most once
*	Data item is matched zero or more times
{n}	Exactly 'n' times data is matched
{n,}	'n' or more times data is matched
{n,m}	At least 'n' times and maximum 'm' times data item is matched
+	Data item is matched at least one time

TABLE 4.10
R Regex Functions

Fucntion_Name	Description
grep()	Returns the elements matched; typically refers to the values within the data set that satisfy the specified regular expression pattern (index / value)
grep1()	Returns the elements that match the pattern that returns TRUE, while those that don't match return FALSE. This provides a logical vector indicating which elements fulfil the specified criteria.
regexpr()	Returns the 'first match position'; typically refers to the indices within the string or vector where the first occurrence of the pattern is found.
regexec()	Potentially combines the functionality of regexpr() and gregexpr(), providing information about the positions of the first match (like regexpr()) and also returning all matches (like gregexpr()).
sub()	Employed for replacing only the first occurrence of a pattern within a string.
gsub()	Replaces all occurrences of a pattern within a string with a specified replacement.
strsplit()	Splits a vector according to matches found in its elements.
help()	Seeks help on regular expression functions in R; you can use the built-in help system.

4.6.6 Regular Expression Functions

In R programming language, regular expressions serve as robust tools for detecting patterns and manipulating strings, offering versatile capabilities in text retrieval and extraction tasks. In the foundational package of R, there exists a collection of functions that facilitate the discovery of pattern matches. Table 4.10 presents the list of widely used regex functions in R (Hocking, 2019).

4.7 Word Cloud

A word cloud is a type of data visualization where each word's size represents its relevance or frequency in the text data (Feng et al, 2022). With a word cloud, important textual data points can be highlighted. Word clouds are frequently employed in social network data analysis (Xu et al., 2016). By altering the size of each word, a word cloud created in R emphasizes the frequency of word recurrence and is a visually appealing approach to describe text data. In order to construct a word cloud, you usually need to utilize a library that can generate the visualization after preprocessing the text to remove common stop words, punctuation, etc. (Kabir et al., 2018). Wordcloud and Wordcloud2 are two packages in R that let you generate word clouds. Major points emphasizing word cloud significance are listed below:

1. Clarity and simplicity enhanced with the help of a word cloud. In a word cloud, the most popular keywords are more prominently shown.
2. Word clouds are an effective means of communication. They have an impact, are simple to comprehend, and can be shared.
3. Word clouds provide more visual interest than table data.

Following are the steps to create a word cloud and analyse data (Fellows et al., 2018):

Step 1: Install and upload required packages

```
> # Install required packages
> install.packages(c("tm", "wordcloud","SnowballC","RColorBre
wer"))
> library("tm")
> library("SnowballC")
> library("wordcloud")
> library("RColorBrewer")
```

The 'tm' package in R is primarily used for text preparation, document-term matrix construction, and other textual data-related activities for text mining applications (Feinerer et al., 2015). Although 'tm' does not directly construct word clouds, it does include preprocessing capabilities that can be used to prepare text input for use in word cloud generation using other packages like 'wordcloud.'

The R 'wordcloud' package offers a straightforward yet effective method for visualizing text data (Fellows et al., 2018). To create a word cloud, you must first have text data to show. Any text data source, including character vectors, text files, and data frames with text columns, can be used. Assume for the purposes of this example that your text data is contained in a character vector text_data. Using the wordcloud function, you can generate a word cloud after you have your text data. Here's how to accomplish it:

```
wordcloud(words = text_data, min.freq = 1, random.order =
FALSE)
```

The minimum threshold for a word to appear in the word cloud is represented by 'min.freq,' where 'words' is the vector of words or a named vector where names represent words and values denote frequencies. You can change the words that are included by using this parameter. The 'random.order' option shows you if words should be plotted in decreasing frequency order or randomly. When set to FALSE, words are plotted in decreasing order of frequency.

The Snowball stemmers are word stemming algorithms that may be accessed via the 'snowballc' package in R. In order to normalize text data for natural language processing (NLP) and text mining jobs, stemming is the process of reducing words to their root or base form. A selection of colour schemes for use in data visualization may be found in the 'RColorBrewer' package in R. These colour schemes are intended to be aesthetically pleasing and appropriate for use with maps, graphs, charts, and other visualizations of data. A variety of colour schemes, including sequential, divergent, and qualitative palettes, are readily available and usable in R plots thanks to the 'RColorBrewer' package.

Step 2: Prepare the text data

We have used the text shown in Figure 4.1 for the creation of a cloud. You can use any data and save it as a .txt file. Also, you can directly browse data from any web resource directly.

```
And so even though we face the difficulties of today and tomorrow, I still have
a dream. It is a dream deeply rooted in the American dream.

I have a dream that one day this nation will rise up and live out the true
meaning of its creed:

We hold these truths to be self-evident, that all men are created equal.

I have a dream that one day on the red hills of Georgia, the sons of former
slaves and the sons of former slave owners will be able to sit down together at
the table of brotherhood.

I have a dream that one day even the state of Mississippi, a state sweltering
with the heat of injustice, sweltering with the heat of oppression, will be
transformed into an oasis of freedom and justice.
```

FIGURE 4.1
Example.txt data for word cloud creation.

Enter the following R code to import the locally saved file on your computer or directly from a web repository. It will prompt you to interactively select the text file:

```
> text <- readLines(file.choose())
# Read the text file from internet
filePath <- "http://www.example.co.in/worcloud-files/example.txt"
text <- readLines(filePath)
```

Step 3: Mining text document

The actions stated below must be followed when mining text documents. The section that follows goes into detail about each of these steps.

Step 3.1: Creation of data corpus and data inspection
Step 3.2: Data transformation
Step 3.3: Data cleaning

Step 3.4: Creation of term-document matrix
Step 3.5: Word cloud creation

Step 3.1: Using the Corpus() function from the text mining (tm) package, the text is loaded. A corpus is a list of documents; in this instance, there is just one text document. Also, inspect data to check the content of the uploaded text file.

Figure 4.2 presents investigated data along with the corpus of data.

```
> docs <- Corpus(VectorSource(text))
> inspect(docs)
<<SimpleCorpus>>
Metadata:  corpus specific: 1, document level (indexed): 0
Content:  documents: 45

 [1] And so even though we face the difficulties of today and tomorrow, I still have a dream. It is a drea
m deeply rooted in the American dream.
 [2]
 [3] I have a dream that one day this nation will rise up and live out the true meaning of its creed:
 [4]
 [5] We hold these truths to be self-evident, that all men are created equal.
 [6]
 [7] I have a dream that one day on the red hills of Georgia, the sons of former slaves and the sons of fo
rmer slave owners will be able to sit down together at the table of brotherhood.
```

FIGURE 4.2
Data corpus and inspected data.

Step 3.2: Data transformation

The tm_map() function is used to conduct transformation, such as replacing special characters in the text. Swapping out "/", "@", and "|" for spaces is outlined in Figure 4.3.

```
# Transforming Loaded Data
  > toSpace <- content_transformer(function (x , pattern ) gsub(pattern, " ", x))
  > docs <- tm_map(docs, toSpace, "/")
  >docs <- tm_map(docs, toSpace, "@")
  >docs <- tm_map(docs, toSpace, "\\|")
```

FIGURE 4.3
Data transformation.

Step 3.3: Data cleaning

The tm_map() function is used to eliminate common stopwords like 'them' and 'i', remove extraneous white space, and convert the text to lowercase. Because 'stopwords' are so prevalent in a language, their information value is almost nil. Before doing additional analysis, it is helpful to remove words of this type. Given below is the snapshot of R commands used to clean data. The entire text has been changed to lowercase, and all stopwords and numerals have been eliminated, along with any superfluous punctuation and spaces. Additionally, text stemming is done. The entire cleaning of R commands is highlighted in Figure 4.4.

```
# Convert the text to lower case
docs <- tm_map(docs, content_transformer(tolower))
# Remove numbers
docs <- tm_map(docs, removeNumbers)
# Remove english common stopwords
docs <- tm_map(docs, removeWords, stopwords("english"))
# Remove your own stop word
# specify your stopwords as a character vector
docs <- tm_map(docs, removeWords, c("blabla1", "blabla2"))
# Remove punctuations
docs <- tm_map(docs, removePunctuation)
# Eliminate extra white spaces
docs <- tm_map(docs, stripWhitespace)
# Text stemming
# docs <- tm_map(docs, stemDocument)
```

FIGURE 4.4
Data cleaning.

Step 3.4: Term-document matrix creation

A document matrix is a table that lists the terms' frequencies (Anandarajan et al., 2019). Row names are documents and column names are words. The text mining package's TermDocumentMatrix() method can be used and is described below in Figure 4.5.

```
> # Text stemming
> # docs <- tm_map(docs, stemDocument)
> dtm <- TermDocumentMatrix(docs)
> m <- as.matrix(dtm)
> v <- sort(rowSums(m),decreasing=TRUE)
> d <- data.frame(word = names(v),freq=v)
> head(d, 10)
              word freq
will          will   17
freedom    freedom   13
ring          ring   12
dream        dream   11
day            day   11
let            let   11
every        every    9
one            one    8
able          able    8
together together    7
```

FIGURE 4.5
Word matrix creation with frequencies.

Step 3.5: Generation of word cloud

Method 'wordcloud' is used to generate a word cloud, where d$word contains all the terms that will be plotted, along with their frequencies. Words with frequencies less than minimum frequency will not be plotted. Most words that can be plotted are described by max.words. The text plot is created in random order using random.order. If untrue, they will be plotted in rot.per's decreasing frequency: proportional terms rotated by 90 degrees (vertical text). Colours is used to manage colour terms. For a single colour, use colours = "red", for instance. Word cloud generation with R is shown in Figure 4.6.

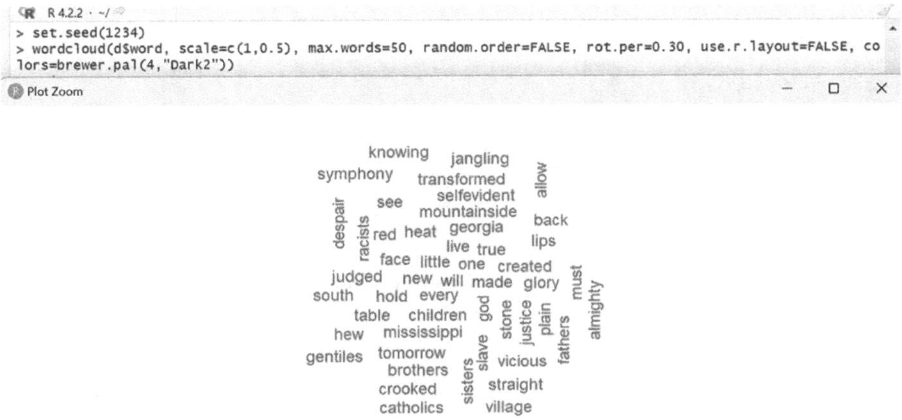

```
> set.seed(1234)
> wordcloud(d$word, scale=c(1,0.5), max.words=50, random.order=FALSE, rot.per=0.30, use.r.layout=FALSE, co
lors=brewer.pal(4,"Dark2"))
```

FIGURE 4.6
Word cloud.

Step 4: Explore frequent terminologies and their meanings

A list of frequently occurring terms in the term-document matrix is manifested in Figure 4.7. We are looking for words that appear at least four times in the example below.

```
> findFreqTerms(dtm, lowfreq = 4)
 [1] "dream"    "day"      "nation"   "one"      "will"     "able"     "together" "freedom"  "every"
[10] "mountain" "shall"    "faith"    "free"     "let"      "ring"
> findAssocs(dtm, terms = "freedom", corlimit = 0.3)
$freedom
         let         ring  mississippi        stone mountainside        state     mountain
        0.89         0.85         0.34         0.34         0.34         0.32         0.32
       every
        0.31
```

FIGURE 4.7
Association of frequent terms.

Step 5: Word frequency in table format and visualization

Figures 4.8 and 4.9 show the frequency of the first ten frequently occurring words as a bar plot and as a table.

```
> # Frequency table of word
> head(d, 10)
             word freq
will         will   17
freedom   freedom   13
ring         ring   12
dream       dream   11
day           day   11
let           let   11
every       every    9
one           one    8
able         able    8
together together    7
> barplot(d[1:10,]$freq, las = 2, names.arg = d[1:10,]$word,col ="lightblue", main ="Most frequent words",
ylab = "Word frequencies")
```

FIGURE 4.8
Frequency of terms in tabular format.

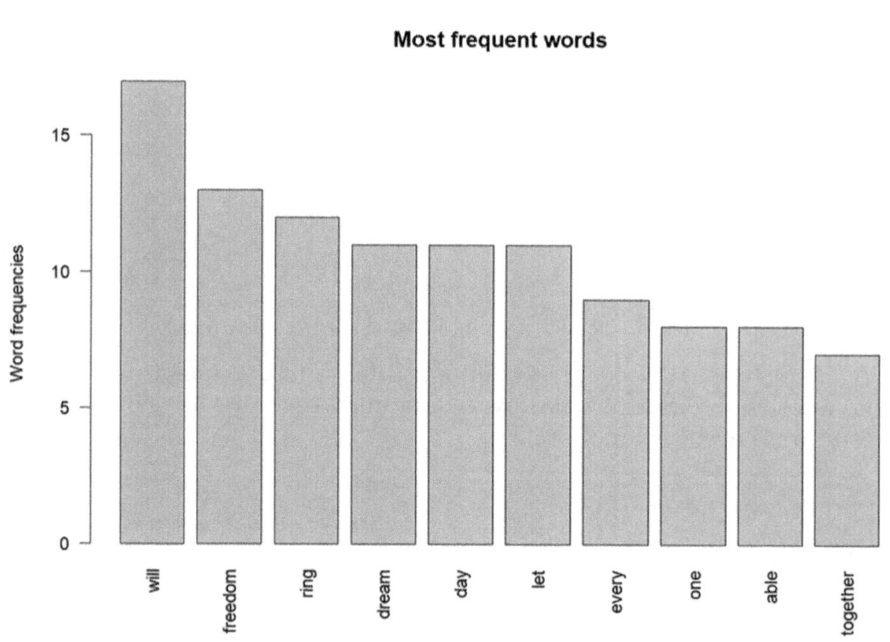

FIGURE 4.9
Bar plot of words' frequency.

4.8 Case Study

4.8.1 Journal Data Analysis

You are doing an analysis on a data set made up of different journal entries. Different forms for dates may or may not be present in each entry. The task at hand is to create an R function that can recognize and extract every unique date from a supplied text. The dates may appear in any of the subsequent formats:

- dd-mm-yyyy
- mm/dd/yyyy
- yyyy/mm/dd

Objective:
Create a function in R that scans each entry for dates matching any of the above formats and extracts all unique dates present in the text.

Solution:

Step 1: Create Sample Data

```
> journal_entries <- c("The project started on 15-08-2022
and was supposed to end by 10-06-2023, but it got delayed.
","Check-in dates are 15/11/2022 and check-out is on
15/12/2023.","Our next meeting is scheduled for 2024/01/01
at the new office location.","No dates are mentioned in
this entry.")
```

Step 2: Define a Function to Extract Dates

```
> extract_dates <- function(text_entries) {
+    pattern <- "\\b(?:\\d{2}-\\d{2}-
\\d{4}|\\d{2}/\\d{2}/\\d{4}|\\d{4}/\\d{2}/\\d{2})\\b"
+    lapply(text_entries, function(entry) {
+      dates_found <- regmatches(entry, gregexpr(pattern, entry))
+      unique(unlist(dates_found))
+    })
+ }
```

Step 3: Apply the Function and Print Results

```
> extracted_dates <- extract_dates(journal_entries)
> print(extracted_dates)
[[1]]
[1] "15-08-2022" "10-06-2023"

[[2]]
[1] "15/11/2022" "15/12/2023"

[[3]]
[1] "2024/01/01"

[[4]]
character(0)
```

Explanation:

A list with each element corresponding to a journal entry will be produced by this script. Every item in the list has a vector of distinct dates that were retrieved from that specific entry using the predefined patterns. This case study and its solution show how to use regular expressions in R for text processing, with an emphasis on pattern recognition–based information extraction. This is a typical preparation and cleaning activity for data, particularly for unstructured textual data.

4.8.2 Sentiment Analysis

You are tasked with preparing a customer review data set for sentiment analysis. The reviews come in a variety of formats and could include numerous date formats, email addresses, URLs, and nonstandard text characters.

Objective:
Clean and preprocess the text data to:

1. Remove any URLs, email addresses, and nonstandard punctuation or special characters.
2. Extract and standardize date formats.
3. Tokenize the text into individual words.
4. Convert all text to lowercase.

Solution:

Step 1: Create Sample Data

```
reviews <- c("Great service, really appreciated it! Although,
it was pricey. Email us at feedback@exampl.com or visit
http://example.com. Next appointment: 12-05-2021.","Terrible
experience:( Never coming back! Contact: badservice@example.
net. See our terrible reviews at https://badexample.com","An
average experience, nothing special. Could be better! But not
the worst. Date of visit was 05/12/2021.","Loved it! Excellent
food and staff. Visit on 2022/01/15 was memorable. Email:
goodtimes@resto.com")
```

Step 2: Clean and Extract Data

```
> library(stringr)
> clean_reviews <- function(text) {
+ text <- str_remove_all(text, "http[s]?://\\S+")
+ text <- str_remove_all(text, "\\b[\\w.-]+@[\\w.-]+\\.
[A-Za-z]{2,6}\\b")
+ dates <- str_extract_all(text, "\\b(?:\\d{2}-\\d{2}-\\
d{4}|\\d{2}/\\d{2}/\\d{4}|\\d{4}/\\d{2}/\\d{2})\\b")
```

```
+ dates <- unlist(lapply(dates, function(date) {
+     ifelse(grepl("-", date), as.Date(date,
format="%d-%m-%Y"),
as.Date(date, format="%m/%d/%Y"))
+ }))
+ text <- str_replace_all(text, "\\b(?:\\d{2}-\\d{2}-
\\d{4}|\\d{2}/\\d{2}/\\d{4}|\\d{4}/\\d{2}/\\d{2})\\b",
as.character(dates))
+ text <- str_replace_all(text, "[^\\w\\s]", "")
+ text <- tolower(text)
+ words <- unlist(str_split(text, "\\s+"))
+ words[words != ""]
+ }
> # Apply cleaning function
> cleaned_reviews <- lapply(reviews, clean_reviews)
```

Step 3: Print Results

```
> print(cleaned_reviews)
[[1]]
 [1] "great"   "service"    "really"    "appreciated"    "it"
"although"  "it"
 [8] "was"     "pricey"     "email"     "us"             "at"
"or"            "visit"
[15] "next"     "appointment" "18759"

[[2]]
NULL

[[3]]
 [1] "an" "average" "experience" "nothing" "special" "could"
"be"
 [8] "better" "but"  "not"       "the"     "worst"   "date"
"of"
[15] "visit"  "was"     "18759"

[[4]]
[1] NA
```

Explanation:

URL and Email Removal: To concentrate solely on the significant text for sentiment analysis, they are eliminated.

Date Extraction and Standardization: To ensure uniformity, dates are recognized and transformed into a consistent format.

Tokenization and Lowercasing: Text is divided into individual words and changed to lowercase in order to get ready for tasks involving frequency analysis or additional NLP.

4.8.3 Email ID Matching for a Marketing Campaign

As a data analyst for a marketing company, your job is to examine client information in order to create a customized email marketing campaign. In order to provide customized promotional offers, the marketing team wishes to divide the client database according to email addresses. It is your responsibility to recognize and extract the valid email addresses from the database, which has a mixture of valid and invalid email addresses. Develop a script in R to efficiently identify and extract valid email IDs from a data set.

Here are the suggested actions for matching email addresses:

1. Data Preprocessing: Load the data set into R. Clean and preprocess the email address column to remove any leading or trailing white spaces.
2. Email ID Validation: Implement regular expressions to match and validate email addresses based on common patterns. Consider factors such as the presence of the '@' symbol, valid domain names, and appropriate email structure.
3. Matching and Extraction: Utilize string handling functions to extract valid email IDs from the data set. Store the extracted email IDs for further analysis and segmentation.
4. Analysis and Reporting: Conduct exploratory data analysis to understand the distribution of valid email IDs across different customer segments. Generate reports and visualizations to present insights to the marketing team.

Solution:

Step 1: Load Data

```
customer_data <- read.csv("customer_data.csv")
```

Step 2: Data Preprocessing and Email Validation

```
# Data preprocessing
customer_data$email <- trimws(customer_data$email)
# Email ID validation function
validate_email <- function(email)
{
pattern <- "^\\b[A-Za-z0-9._%+-]+@[A-Za-z0-9.-]  +\\.
[A-Z|a-z]{2,}\\b$"
return(grepl(pattern, email))
}
```

Step 3: Analysing and Reporting

```
# Apply validation function to data set
valid_emails <- customer_data[validate_email(customer_
data$email), "email"]
```

```
# Analysis and Reporting (Example: Counting valid email IDs)
valid_email_count <- length(valid_emails)
print(paste("Total Valid Email IDs:", valid_email_count))
# End of Script
```

We will focus on processing typical email addresses that adhere to one (or a similar variation) of the following formats:

rbookauthor@mail.com

rbookauthor99@mail.com

rbook.author@mail.com

rbook.author@an-mail.com

rbook.author@an.mail.com

4.9 Summary

- Character vectors are used in R to represent strings, and each element in the vector can hold a string of characters.
- Square brackets [] can be used to subset and index strings, which makes it possible to extract particular characters or substrings.
- Strings can be divided into substrings using the strsplit() function depending on a given delimiter.
- Regular expressions are useful for manipulating strings and matching patterns. Pattern matching tasks are made easier by functions like grep(), grepl(), gsub(), and sub().
- 'stringr' offers functions such as str_to_lower() and str_to_upper() for converting strings to lowercase or uppercase, respectively.
- Strings can have leading and trailing white space removed using functions like trimws().
- Regular expression matches can be found within strings using the regexpr() and gregexpr() methods, which can also provide the locations or substrings of the matches.
- With functions like str_detect(), str_extract(), and str_replace(), package 'stringr' offers a standardized interface for string manipulation tasks.
- The package 'stringi' provides functions for pattern matching, manipulation, and collation, along with strong string processing capabilities with support for several encodings and locales.
- 'stringr' allows you to manipulate strings in a variety of languages and character sets while handling Unicode characters with ease.

- R package 'stringr' allows you to manipulate strings in a variety of languages and character sets while handling unicode characters with ease.
- A word cloud is a graphic representation of text data in which the magnitude of a word in a particular piece of text represents how frequently or how important it is.
- The following procedures are usually involved in constructing a word cloud in R. Text data can be imported from a file or text source initially, then eliminate punctuation, digits, and frequently occurring stopwords from the text data and convert it to lowercase. After that, determine how often each word appears in the text and create and display the word cloud.

EXERCISES

I. Multiple-Choice Questions

1. Which of the given functions is used to split the given string, bname="Introduction, to R programming"
   a. strsplit(bname,"[,]")
   b. strsplit(bname,"[.]")
   c. strsplit(bname,"[]")
   d. strsplit(bname,"[“”]")
2. What values need to be substituted for?? to achieve the expected result?

```
> str <- paste(c(??), "3", sep = ":")
> print (str)
[1] "2:3" "3:3" "4:3" "5:3" "6:3"
```

   a. 3:6
   b. 2:6
   c. 4:6
   d. 1:6
3. Which R command extracts 'Japan' from the given string: >d=`"India Japan Germany Australia France"`?
   a. substr(d[7:11])
   b. substr[d(7,11)]
   c. substr(d,7,11)
   d. None of the above

4. What will be the output of following R command?

```
> u <- "ASMPR2493S"
> print(nchar(u))
```

   a. "ASMPR2493S"
   b. 10
   c. NULL
   d. Error

5. For the given data, = c("New Year","New Delhi","New York","Car New","Happy New Year"), which of the given commands produces a different output?
   a. grep (pattern = "New*", x = data)
   b. grep (pattern = "^New", x = data)
   c. grep (pattern = "New", x = data)
   d. grep (pattern = "New^", x = data)

6. Predict the outcome of the following: `> s2 <- paste(c(2:6), c(4:7), sep = "$$") > print (s2)`
   a. "23456" "4567" "$$$$$" "$$$$"
   b. "2$$4" "3$$5" "4$$6" "5$$7" "6$$4"
   c. "2$$3" "3$$4" "4$$5" "5$$6" "6$$7"
   d. "2$4$" "3$5$" "4$6$" "5$7$" "6$4$"

7. Which syntax is correct to generate the following output: "#ython #@" "Java #@" from the string: "Python PL", "Java PL"?
   a. chartr("PL", "#@", c("Python PL", "Java PL"))
   b. chartr("PL", "#@", "Python PL" "Java PL")
   c. Both of the above
   d. None of the above

8. What is the output of the following syntax: > chartr("ram", "#@", "Programming Language")?
   a. Prog#@ ing Language
   b. Progr#@mming Language
   c. Prog#@#@ing Language
   d. Error

9. Given the string s3: 'R programming book', along with additional strings 'with', 'easy', 'learning', and 'examples', choose the appropriate command to replace the third character of the first string with '$', the next string with '#', and so forth.
   a. substr(s3, 3, 3) <- c("$", "#")
   b. substr(s3, 3, 3) <- c("s3[1]$", "s3[2]#")

   c. substr(s3, 3[1]($), 3[2](#)) <- c("$", "#")
   d. None of the above
10. What is the correct output of the following R command: print (casefold(c("R Programming", "Language")))?
   a. R PROGRAMMING LANGUAGE
   b. r pROGRAMMING lANGUAGE
   c. r programming language
   d. R programming language
11. Which of the ensuing R packages is not usually utilized for word cloud production and visualization?
   a. tm
   b. ggplot2
   c. RColorBrewer
   d. wordcloud
12. Which TM package function converts every word in a text corpus to lowercase?
   a. content_transformer()
   b. Corpus()
   c. tm_map()
   d. tolower()

II. Brief Questions

1. Explain the differences between the paste() and paste0() functions and how you might concatenate numerous strings in R.
2. Describe the operation of the R substr() function. Describe how to separate the substring "ndi" from the string "India" in an example.
3. Describe how to use the methods grep() and grepl(). How would you apply them to determine whether the vector c("programming", "data", "statistics", "science") contains the string "statistics"?
4. Explain the use of regular expressions in R's string handling. Give an example of how to find matches in a vector of strings and retrieve email addresses.
5. Compare the functions offered by the stringr and stringi packages with the standard R functions for handling strings. What are the benefits of utilizing base R functions over stringr or stringi?

6. Describe the features of the stringi package that allow for more complex string manipulations. Using stri_sub(), stri_replace_all_fixed(), and stri_split_regex(), give an example.
7. What is the handling of regular expressions by stringr and stringi? Give an example of how to use functions from both packages to find and replace every digit in a string using a regular expression.
8. Discuss the performance issues that come with handling big amounts of text data in R. How do base R functions perform in comparison to the stringr package in R?
9. How can special characters be handled in R regular expressions for string matching? Provide an example of matching a string with special characters like $.+*?[^].
10. Discuss the steps needed to create a word cloud in R. Describe each R package's function in the procedure and the importance of text preprocessing.

III. Coding Exercise

1. Write R code to extract email addresses from a string using base R functions and functions from the stringr package.
2. Write R code to generate a vector containing 10-character alphanumeric strings. Then, you will clean these strings to extract only the numeric digits. Finally, you will format the cleaned digits into the pattern "XXX-XXX-XXXX".
3. Use the stringi package to perform the following tasks:
   a. Extract the substring "istical" from the string "Statistical Programming Language".
   b. Replace all occurrences of the string "python" with "R" in the given string: "I like R and Python Programming".
   c. Split the string "Data, analytics; science" using a regular expression function provided by the stringi package.
4. A data set of product reviews, each with several sentences, is provided to you. It is your responsibility to locate and extract every product ID from the reviews. The product IDs all begin with the letters "PROD" and consist of exactly six numbers (e.g., "PROD123456"). To further anonymize the data, you must locate all instances of these product IDs and substitute them with the string "[PRODUCT ID]".

5. You are given a data set of customer feedback comments stored in a character vector. Perform the following tasks using required packages in R:
   a. Identify and extract any email addresses from the comments.
   b. Count the number of occurrences of specific keywords (e.g., "service", "price", "quality") in each comment.
   c. Clean the comments by removing any URLs.
   d. Normalize the comments by converting them to lowercase and removing any leading or trailing white space.
6. To find the value "a" in a vector of strings, write a R command. Additionally, indicate the position of the corresponding values.
7. Write an R function named reverse_words that takes a string and returns a new string where each word in the original string is reversed, but the order of words remains the same. Sample Input/Output:

   Input: "Data Science"
   Output: "ataD ecneicS"
8. Develop an R function named extract_numbers that takes a string and extracts all numeric values, returning them as a vector of numbers. Sample Input/Output:

   Input: "There are 12 apples for $15"
   Output: c(12, 15)
9. Create a function named is_palindrome that checks if a given string is a palindrome (reads the same forward and backward, ignoring spaces, punctuation, and case). Sample Input/Output:

   Input: "A man, a plan, a canal, Panama"
   Output: FALSE
10. You are assisting a client who has given you a sizable collection of customer reviews as part of a text analysis project. In order to learn more about the opinions and input of their customers, the client want to determine which words are most often spoken. Explain your R-based strategy to this task. Provide the procedures you would use to prepare the text data, create a word cloud, and analyse the outcome.

References

Anandarajan, M., Hill, C., Nolan, T., Anandarajan, M., Hill, C., & Nolan, T. (2019). Term-document representation. In Practical Text Analytics: Maximizing the Value of Text Data (pp. 61–73). Springer.

Bathie, G., Charalampopoulos, P., & Starikovskaya, T. (2024). Pattern matching with mismatches and wildcards. arXiv preprint arXiv:2402.07732.
Boehmke, B. C. (2016). Data wrangling with R. Switzerland: Springer International Publishing.
Chapman, C., Wang, P., & Stolee, K. T. (2017, October). Exploring regular expression comprehension. In 2017 32nd IEEE/ACM International Conference on Automated Software Engineering (ASE) (pp. 405–416). IEEE.
Edition, T. (2021). Mastering regular expressions. O'Reilly Media.
Feinerer, I., Hornik, K., & Feinerer, M. I. (2015). Package 'tm'. Corpus, 10(1) 1–64.
Fellows, I., Fellows, M. I., Rcpp, L., & Rcpp, L. (2018). Package 'wordcloud'. R package version, 2, 331.
Feng, K. K., Gao, A., & Karras, J. S. (2022, October). Towards semantically aware word cloud shape generation. In Adjunct Proceedings of the 35th Annual ACM Symposium on User Interface Software and Technology (pp. 1–5).
Forta, B. (2018). Learning regular expressions. Addison-Wesley Professional.
Gagolewski, M. (2022). stringi: Fast and portable character string processing in R. Journal of statistical software, 103, 1–59.
Hak, T. & Dul, J. (2009). Pattern matching. ERIM Report Series Reference No. ERS-2009-034-ORG. https://ssrn.com/abstract=1433934
Hocking, T. D. (2019). Comparing namedCapture with other R packages for regular expressions. R journal, 9(2) 328–346.
Ionescu, R. T., Popescu, M., Ionescu, R. T., & Popescu, M. (2016). State-of-the-art approaches for string and text analysis. In Knowledge Transfer Between Computer Vision and Text Mining: Similarity-Based Learning Approaches (pp. 135–147). Springer.
Jockers, M. L., & Thalken, R. (2020). Text analysis with R. Springer International Publishing.
Kabir, A. I., Karim, R., Newaz, S., & Hossain, M. I. (2018). The Power of Social Media Analytics: Text Analytics Based on Sentiment Analysis and Word Clouds on R. Informatica Economica, 22(1).
Navarro, G. (2004). Pattern matching. Journal of applied statistics, 31(8), 925–949.
Sanchez, G. (2013). Handling and processing strings in R. Trowchez Editions, Berkeley.
Sinkovics, N. (2018). Pattern matching in qualitative analysis. In The Sage Handbook of Qualitative Business and Management Research Methods (pp. 468–485). Sage Journals.
Welbers, K., Van Atteveldt, W., & Benoit, K. (2017). Text analysis in R. Communication methods and measures, 11(4), 245–265.
Wickham, H., & Bryan, J. (2023). R packages. O'Reilly Media, Inc.
Xu, J., Tao, Y., & Lin, H. (2016, April). Semantic word cloud generation based on word embeddings. In 2016 IEEE Pacific Visualization Symposium (PacificVis) (pp. 239–243). IEEE.
Zhang, X. Z., Gong, Y. Z., Wang, Y. W., Xing, Y., & Zhang, M. Z. (2017). Automated string constraints solving for programs containing string manipulation functions. Journal of computer science and technology, 32, 1125–1135.

5

Data Import and Export

Learning Objectives:

After studying this chapter, students and researchers will be able to:

- Load data in various file formats in R
- Use data manipulation
- Use packages and functions for file handling operations in R
- Use XML data format import and export
- Use CSV and Excel data import and export
- Use structured data format import and export
- Use interfacing of R with backend language like MySQL
- Use import/export of other types of data
- Use real-time case studies

5.1 Introduction

The procedures and functions that let users work with files that are stored locally on their systems or remotely are referred to as file handling. This covers managing directories and writing to and reading from files (Villalobos Alva, 2024). File handling is primarily used to make data import and export easier, allowing users to load data sets for analysis and store the outcomes of their calculations. R is a flexible programming language for manipulating data since it has built-in functions for handling a variety of file formats, including CSV, text, and Excel files. R has methods for more complex file handling activities like listing files in a directory, creating and removing directories, and interactively selecting files in addition to fundamental read and write operations. We work with massive volumes of data in the R programming language by storing data in files. To access data in this way, we

DOI: 10.1201/9781003489382-5

can execute certain commands like creating, reading, renaming, and so on. Several file operations in R include (Visne et al., 2012):

- Creating files
- Adding text to the files
- Obtaining information from a file
- Verifying the current state of a file
- Renaming the current files

Numerous advantages of file management with R are listed below.

1. Flexibility in Data Structures
    - Numerous data types, including as CSV, text, Excel, JSON, and more, are supported by R. Because of its adaptability, users can deal with a variety of data sets without requiring a lot of extra software or conversions.
    - Data import into R for analysis is made simple by built-in functions, which expedites the process and cuts down on the amount of time needed for data preparation.
2. Ease of Data Manipulation
    - Once data is brought into R, it may be effectively cleaned, transformed, and analysed using robust data manipulation packages.
    - Simple functions make it easy to write processed data back to files so that it can be shared with others, used for additional analysis, and saved.
3. Smooth Integration with Tools for Analysis
    - R offers a wide range of analytical and statistical tools that work in perfect harmony with its file handling capabilities. Without having to jump between separate software environments, users can now run sophisticated analysis directly on imported data thanks to this connectivity.
    - The reliability and scalability of data analysis projects are improved by the ability to automate file handling operations through scripting.
4. Assistance with Big Data Sets
    - Large data set handling is optimized by packages like 'data.table' and 'readr', which offer quick and memory-efficient methods for reading and writing data. R is therefore appropriate for big data applications where efficiency is essential.
    - R's capacity to efficiently manage and analyse massive volumes of data is further expanded by its ability to link to databases and handle data directly from these sources.

5.2 Import and Export of Data in CSV Format

The import and export of data in CSV (comma-separated values) format is a fundamental aspect of data management in R, facilitating seamless interaction with external data sets. Tabular data is frequently stored in CSV files, which are platform-independent and easily accessible by both computers and humans. CSV files are plain-text files. Functions like 'read.csv' or 'read.csv2', which parse the CSV file and load its contents into R data structures like data frames, are used in the R data import process for CSV data (Team, 2000). Conversely, exporting data to CSV format is achieved by the use of functions such as 'write.csv' or 'write.csv2', which allow users to store their processed data sets or analytical findings in a format that is readily shared, edited, and opened in other software programmes. With this flexible and standardized method for importing and exporting data in CSV format, R users may effectively manage a wide range of data sets for statistical modelling, exploratory data analysis, and data-driven decision-making.

5.2.1 Reading Data from a CSV File

R's read.csv() function can be used to read a CSV file's contents as a data frame. Either the CSV file to be read is already in the current working directory, or the directory has been adjusted with the help of the R setwd() command. The read.csv() function can also be used to read a CSV file from a URL (Jones et al., 2022).

Enter the following command to find the current working directory:

```
> getwd()
[1] "C:/Users/user1/OneDrive/Documents"
```

The working directory's new location is set using the setwd() method, as demonstrated below:

```
> setwd('C:/Users/megha/Downloads')
> getwd()

[1] "C:/Users/user1/Downloads"
```

The total number of files in the current working directory is also visible with the help of "length" method. List.files() is the function used to display all files that are stored in the current working directory.

```
> length(list.files())
[1] 8
> list.files()
[1] "Custom Office Templates" "desktop.ini" "IAJIT - latest.docx"
[4] "My Tableau Repository"   "plot.pdf"    "plot.png"
[7] "Rplot.png"               "Zoom"
```

	A	B	C	D	E	F	G	H	I
1	Pregnancies	Glucose	BloodPressure	SkinThickness	Insulin	BMI	DiabetesPedigre	Age	Outcome
2	6	148	72	35	0	33.6	0.627	50	1
3	1	85	66	29	0	26.6	0.351	31	0
4	8	183	64	0	0	23.3	0.672	32	1
5	1	89	66	23	94	28.1	0.167	21	0
6	0	137	40	35	168	43.1	2.288	33	1
7	5	116	74	0	0	25.6	0.201	30	0
8	3	78	50	32	88	31	0.248	26	1
9	10	115	0	0	0	35.3	0.134	29	0
10	2	197	70	45	543	30.5	0.158	53	1

FIGURE 5.1
'Diabetes.csv' data.

Figure 5.1 illustrates the process of reading CSV files in R. Let us assume that we have the following data in a CSV file called 'diabetes.csv' as shown in Figure 5.1

The read.csv() function in R is used to read a CSV file that is located in our current directory. The example in Figure 5.2 uses the head() method to read 'diabetes.csv' and displays the first few records. As shown in Figure 5.3, we can supply the whole file path to extract data if it is not saved in the current working directory.

The result of the read.csv() function is typically provided as a data frame. A detailed description and full syntax of the read.csv() method is described below:

```
read.csv(file, header = TRUE, sep = ",", quote = "\"",dec = ".",
fill = TRUE, comment.char = "", stringsAsFactors = FALSE, ...)
```

where 'file' is the CSV file that has to be read; 'header' denotes if column names are present in the CSV file's first row with the default value = TRUE; 'sep' is the CSV file's field separator character; 'quote' indicates the character that will be used to enclose fields that contain special characters, such as newline or commas; 'dec' specifies the character used for decimal points; and 'fill' denotes if fields that have no value should be filled in. It has a default value of TRUE. The CSV file's comment.char specifies the character that will be used to denote comments, and 'stringsAsFactors' defines if

```
> data <- read.csv(file = 'diabetes.csv')
> head(data)
  Pregnancies Glucose BloodPressure SkinThickness Insulin  BMI DiabetesPedigreeFunction Age Outcome
1           6     148            72            35       0 33.6                    0.627  50       1
2           1      85            66            29       0 26.6                    0.351  31       0
3           8     183            64             0       0 23.3                    0.672  32       1
4           1      89            66            23      94 28.1                    0.167  21       0
5           0     137            40            35     168 43.1                    2.288  33       1
6           5     116            74             0       0 25.6                    0.201  30       0
```

FIGURE 5.2
Reading 'diabetes.csv' data.

```
> data1<- read.csv(file = 'C:/Users/megha/Documents/diabetes.csv')
> tail(data1)
    Pregnancies Glucose BloodPressure SkinThickness Insulin  BMI DiabetesPedigreeFunction Age Outcome
763           9      89            62             0       0 22.5                    0.142  33       0
764          10     101            76            48     180 32.9                    0.171  63       0
765           2     122            70            27       0 36.8                    0.340  27       0
766           5     121            72            23     112 26.2                    0.245  30       0
767           1     126            60             0       0 30.1                    0.349  47       1
768           1      93            70            31       0 30.4                    0.315  23       0
```

FIGURE 5.3
Reading 'diabetes.csv' data not placed in current working directory.

character vectors ought to be converted to factors. Here's a demonstration of loading diabetes data with the header = TRUE and header = FALSE, illustrated in Figure 5.4. When header = TRUE, the feature names are displayed at the top of the data set, while when header = FALSE, the headers V1 to V9 appear as column names.

5.2.2 Dimensions of a CSV File

With R, it's simple to find the rows and columns of a CSV file by using the 'dim()' function. The number of rows and the number of columns are the two elements of the numeric vector that this method returns. The dimensions of the data set can be easily retrieved by users by applying dim() to the data frame that is formed after reading the CSV file. For instance, dim(data) would yield a vector representing the total number of rows and columns in the data set if data were the data frame holding the CSV data. Understanding the size and organization of the data set is critical for activities involving data exploration, manipulation, and analysis in R. Furthermore, users can determine the memory and processing resources required for efficient data operations by knowing the dimensions of the data set. To find the total number of rows

```
> data <- read.csv(file = 'diabetes.csv',header = TRUE, stringsAsFactors = FALSE)
> tail(data)
    Pregnancies Glucose BloodPressure SkinThickness Insulin  BMI DiabetesPedigreeFunction Age Outcome
763           9      89            62             0       0 22.5                    0.142  33       0
764          10     101            76            48     180 32.9                    0.171  63       0
765           2     122            70            27       0 36.8                    0.340  27       0
766           5     121            72            23     112 26.2                    0.245  30       0
767           1     126            60             0       0 30.1                    0.349  47       1
768           1      93            70            31       0 30.4                    0.315  23       0
> data <- read.csv(file = 'diabetes.csv',header = FALSE, stringsAsFactors = FALSE)
> tail(data)
    V1  V2 V3 V4  V5   V6    V7 V8 V9
764  9  89 62  0   0 22.5 0.142 33  0
765 10 101 76 48 180 32.9 0.171 63  0
766  2 122 70 27   0 36.8  0.34 27  0
767  5 121 72 23 112 26.2 0.245 30  0
768  1 126 60  0   0 30.1 0.349 47  1
769  1  93 70 31   0 30.4 0.315 23  0
> |
```

FIGURE 5.4
Demonstration of loading diabetes data with different header values.

and columns in the data set, we can alternatively utilize the ncol() and nrow() functions. The example is shown below:

```
> dim(data)
[1] 768   9
> # print number of columns
> print (ncol(data))
[1] 9
> # print number of rows
> print(nrow(data))
[1] 768
```

5.2.3 Analysing CSV Data

In today's data-driven world, data analysis is essential since it enables businesses and individuals to make well-informed decisions. It converts unprocessed data into useful insights by methodically using statistical and logical approaches to describe, show, condense, summarize, and assess data. Through trend detection, pattern recognition, and actionable intelligence extraction, this approach aids in making strategic decisions, streamlining processes, and forecasting future results. Data analysis with R is especially important due to its powerful capabilities and flexibility in handling complex data sets. R is a feature-rich statistical software environment that is frequently utilized in business, education, and research. It provides a wide range of tools and packages for advanced statistical modelling, data processing, and visualization (Pathak, 2014). This makes it simple for analysts to carry out complex analysis and produce excellent, customizable visualizations. The result of the read.csv() function is typically provided as a data frame.

On CSV data, you can apply aggregator functions such as min, max, and count, etc. To extract columns, you can use the $sign. For example, if you want to extract the 'Glucose' and 'BloodPressure' attributes under the condition that the 'Glucose' value is greater than 190, you can use this approach. We have loaded the 'diabetes' data set for analysis and extracted the 'Glucose' and 'BloodPressure' columns based on the specified condition, as shown below:

```
> data<-read.csv(file = 'diabetes.csv')
> #Selecting 'Glucose' and 'Blood Pressure' columns for patients
with Glucose value greater than 190
> result<-data[data$Glucose > 190, c("Glucose", "BloodPressure")]
> result
    Glucose BloodPressure
9       197            70
23      196            90
186     194            68
207     196            76
229     197            70
259     193            50
261     191            68
320     194            78
360     196            76
```

The max and min functions can be used to determine the maximum and least value from a CSV file or data frame. Below is an example of how both functions operate:

```
> # Get the max value for Insulin column from diabetes data.
> max_value<-max(data$Insulin)
> print(max_value)
[1] 846
> min_value<-min(data$Insulin)
> print(min_value)
[1] 0
```

Similar to a SQL (Structured Query Language) where clause, we can retrieve rows that satisfy particular filter criteria. The person's details with the highest insulin value will be presented, just like in the example below. The 'subset' method is employed to extract data for additional examination.

```
> res<-subset(data, Insulin == max(Insulin))
> print(res)
Pregnanicies  Glucose  BloodPressure  SkinThickness  Insulin  BMI
1             189      60             23             846      30.1
DiabetesPedigreeFunction  Age  Outcome
0.398                     59   1
```

In a different instance, we are asking people who are older than 40 and have diabetes. The result is displayed below:

```
># Patient details who are diabetic and Age>40
>details<-subset(data,Outcome=='1' & Age >40)
> print(details)
Pregnanicies  Glucose  BloodPressure  SkinThickness  Insulin  BMI
6             148      72             35               0      33.6
2             197      70             45             543      30.5
8             125      96              0               0       0.0
1             189      60             23             846      30.1
5             166      72             19             175      25.8
DiabetesPedigreeFunction  Age  Outcome
0.627                     50   1
0.158                     53   1
0.232                     54   1
0.398                     59   1
0.587                     51   1
```

The mean function can also be used to calculate the average of any attribute. Figure 5.5 provides a working instance of the mean method:

```
> result <- tapply(data$BloodPressure, data$Age, mean)
> result
      21       22       23       24       25       26       27       28       29       30       31
65.93651 63.72222 64.31579 64.95652 59.66667 64.18182 73.50000 68.31429 68.24138 64.85714 64.37500
      32       33       34       35       36       37       38       39       40       41       42
70.06250 65.64706 74.00000 75.60000 69.12500 75.94737 71.12500 72.66667 69.23077 67.59091 73.38889
      43       44       45       46       47       48       49       50       51       52       53
78.46154 61.75000 83.06667 76.00000 78.33333 78.40000 81.40000 78.25000 84.50000 81.50000 79.00000
      54       55       56       57       58       59       60       61       62       63       64
89.33333 70.25000 76.33333 76.80000 78.28571 74.00000 80.00000 76.00000 71.50000 78.00000 78.00000
      65       66       67       68       69       70       72       81
78.66667 86.00000 72.66667 82.00000 81.00000 82.00000  0.00000 74.00000
>
```

FIGURE 5.5
Demonstration of mean and apply methods.

5.2.4 Writing Data to a CSV File

Writing processed or analysed data to a CSV file in R is necessary to save it for distribution or future use. Commonly used for this purpose are the write_csv() function from the readr package and the write.csv() function from base R (Pathak, 2014). You can save matrices or data frames as CSV files using these functions. R data structures are transformed into a tabular format and then written to a text file with a. csv extension as part of the process. Writing data to CSV files makes data sharing and interchange simple and guarantees compatibility with a wide range of software programmes. The syntax for writing data into a new file is shown below:

```
> # Example: Writing data frame 'my_data' to a CSV file using
base R
> write.csv(my_data, file = "new_file.csv", row.names = FALSE)
# Example: Writing data frame 'my_data' to a CSV file using
readr package
>library(readr)
>write_csv(my_data, path = "new_file.csv")
```

where 'my_data' is the data frame or matrix that you wish to send to a CSV file. The 'row.names' = FALSE option in the write.csv() function prevents row names from being put as the first column in the CSV file, and the file directory where the CSV file will be saved is specified by the path parameter when using the 'write_csv()' function from the 'readr' package.

R can generate a CSV file from an existing data frame using the write.csv() function, with the working directory as the file's location. Figure 5.6 provided demonstration on how to create a new CSV file named 'new_data.csv' from a data set. By employing the write.csv() function, we produce the desired file containing the supplied data. To verify its content, you can use the read.csv() method to read the newly created file.

```
> write_data
  Pregnancies Glucose BloodPressure SkinThickness Insulin  BMI DiabetesPedigreeFunction Age Outcome
1           6     148            72            35       0 33.6                    0.627  50       1
2           1      85            66            29       0 26.6                    0.351  31       0
3           8     183            64             0       0 23.3                    0.672  32       1
4           1      89            66            23      94 28.1                    0.167  21       0
5           0     137            40            35     168 43.1                    2.288  33       1
6           5     116            74             0       0 25.6                    0.201  30       0
> write.csv(write_data,"new_data.csv")
> n<- read.csv("new_data.csv")
> n
  X Pregnancies Glucose BloodPressure SkinThickness Insulin  BMI DiabetesPedigreeFunction Age Outcome
1 1           6     148            72            35       0 33.6                    0.627  50       1
2 2           1      85            66            29       0 26.6                    0.351  31       0
3 3           8     183            64             0       0 23.3                    0.672  32       1
4 4           1      89            66            23      94 28.1                    0.167  21       0
5 5           0     137            40            35     168 43.1                    2.288  33       1
6 6           5     116            74             0       0 25.6                    0.201  30       0
>
```

FIGURE 5.6
Demonstration of writing data to a CSV File.

5.2.5 Packages for Data Analysis in R

R boasts an extensive package ecosystem catering to many facets of data research (Maindonald & Braun, 2010). The following list of frequently used packages is arranged according to their main purposes.

1. Data Manipulation: 'dplyr' and 'data.table' are two widely used packages available in R for data manipulation (Spector, 2008). 'dplyr' is a package that provides a consistent set of verbs that will assist you in resolving the majority of data manipulation problems is known as the grammar of data manipulation. 'data.table' is a 'data.frame' extension that offers effective data manipulation capabilities. The commands for installation are listed below.

```
> install.packages("dplyr")
> library(dplyr)
> install.packages("data.table")
> library(data.table)
```

2. Data Import and Export: Commonly utilized packages for importing and exporting data are (Chan & Chan, 2018): 'readr', 'readxl', and 'haven'. 'readr' provides functions to read text files, 'readxl' provide data import and export functions to handle Excel data, and 'haven' provides various methods for import and export data from Statistical Package for Social Sciences (SPSS), Statistics and Data (Stata), and Statistical Analysis System (SAS). Installation commands are mentioned below.

```
>install.packages("readr")
>library(readr)
>install.packages("readxl")
>library(readxl)
>install.packages("haven")
>library(haven)
```

3. Data Cleaning: Data cleaning in R involves removing or correcting errors and inconsistencies in data sets to ensure accuracy and reliability. This process typically includes handling missing values, correcting data types, removing duplicates, and standardizing formats. It aims to prepare the data for analysis by enhancing its quality and integrity. Popular packages like 'dplyr,' 'janitor,' and 'tidyr' offer streamlined functions for efficient data cleaning in R (Van der Loo & De Jonge, 2018). For reshaping and tidying data, the 'tidyr' package is used and 'janitor' provides methods for data inspection and cleaning. Below are instructions for downloading and installing data cleaning packages:

```
>install.packages("tidyr")
>library(tidyr)
>install.packages("janitor")
>library(janitor)
```

4. Statistical Modelling: Statistical modelling with R involves using mathematical techniques to describe and analyse relationships between variables in data. It includes a range of techniques, including machine learning algorithms, time series analysis, logistic regression, and linear regression (Golding, 2019). With packages like 'caret,' 'randomForest,' and 'glmnet,' R offers a comprehensive environment for statistical modelling that helps analysts and researchers create prediction models and make data-driven choices. The steps for downloading and setting up statistical modelling packages are provided below.

```
>install.packages("caret")
>library(caret)
>install.packages("randomForest")
>library(randomForest)
>install.packages("glmnet")
>library(glmnet)
```

5. Information Reporting: Using R, data reporting entails producing dynamic, interactive reports that share insights gained from data analysis. Reproducible reports that include code, analysis findings, and visualizations in one document are frequently made with 'RMarkdown'. Web-based dashboards for interactive data exploration and presentation can be made with packages such as 'Shiny' (Sievert, 2020). Below are the instructions for obtaining and installing reporting packages.

```
>install.packages("shiny")
>library(shiny)
>install.packages("rmarkdown")
>library(rmarkdown)
```

5.3 Import and Export of Data in Excel Format

Import and export data from Excel into R enables users to take advantage of the smooth connection between R and Excel, maximizing their ability to analyse and visualize data. Functions to import/export Excel files into R data frames are provided by packages such as 'readxl', 'writexl', and 'openxlsx', which enable users to access and work with Excel data using R's robust data analysis features. Data analysts, researchers, and business users can collaborate more easily, improve reproducibility, and streamline workflows via bidirectional data sharing between R and Excel (Gardener, 2017).

5.3.1 Reading Data from an Excel File

Reading Excel data in R involves several steps, including installing necessary packages, loading the data, understanding the structure of the data, and performing any necessary data preprocessing. The steps to read data from an Excel file are listed below.

Step I: Import and load package 'readxl' or 'openxlsx'.

Step II: Specify the Excel file path in case the data is not stored in a current working directory.

Step III: Reading an Excel File with the read_excel() method.

Step IV: Data view loads data to check the content of the loaded data.

Let us take into account the provided Excel data set 'Rgrades' containing both students' grades and corresponding marks presented in Figure 5.7.

To read Excel data, the 'readxl' package is required. You can use the following method to install and load it. Once the package has been successfully installed, we can load Excel data, as seen in Figure 5.7. To read the

S no	Roll No	Name	Batch	T1+T2	T3	TA	Total	Grade
1	16103004	Neha Rastc	CSE(B6)	25	30.5	25	80.5	A+
2	16103006	Tarushi Agi	CSE(B10)	20	20.5	25	65.5	B+
3	16103007	Kanishk	CSE(B5)	24	13.5	23	60.5	B+
4	16103008	Rishabh Jai	CSE(B10)	18	16.5	21.5	56	B
5	16103009	Rajat Junej	CSE(B15)	20	20.5	21	61.5	B+
6	16103011	Yashwant \	CSE(B8)	12	10	23	45	C+
7	16103015	Anjali Shar	CSE(B15)	12	21.5	19	52.5	B
8	16103018	Naman Dat	CSE(B5)	19	17	24	60	B+
9	16103019	Rishab Ary	CSE(B4)	24	17	21	62	B+

Grade Scal	Grade	Count
0	F	0
30	D	1
37	C	0
44	C+	6
52	B	16
60	B+	23
70	A	15
80	A+	2

FIGURE 5.7
Sample Excel data for analysis and exploration.

Excel file, use the read_excel() method. As the first argument, give the file's path. Additional arguments, such as the sheet name, range, etc., can also be specified.

```
> install.packages("readxl")
> library('readxl')
> d1<- read_excel("Rgrades.xlsx")
```

You can use the name or index number of the sheet in R when reading an Excel file that has many sheets. When specifying the sheet by number, it reads the sheet corresponding to that index. Sheet indexing starts from 1. The sheet can also be identified by name. This is especially helpful if the sheets have names that make sense.

```
> # Specify sheet by its number
> d2 <- read_excel("Rgrades.xlsx", sheet = 2)
New names:
• `Grade` -> `Grade...9`
• `` -> `...10`
• `Grade` -> `Grade...12`
```

When reading Excel files in R that have missing values represented by characters other than blank cells, you can define how to interpret these data as missing values by using the na option of the read_excel() function, as illustrated below:

```
> d3<- read_excel("Rgrades.xlsx", sheet = 2, na = "---")
```

This guarantees that "—" occurrences in the Excel file are treated by R as missing values, allowing for the proper handling of missing data during pre-processing or analysis.

5.3.2 Exploring Data from an Excel File

Understanding the structure, contents, and features of the data set is necessary in order to explore Excel data in R. A detailed description on using R programming to examine Excel data is given below.

1. View Data: You can use the print(), head(), or tail() methods to see the contents of an Excel file format. An overview of how to load and output data on a console is shown in Figure 5.8.
2. Structure of Data: The str() function in R is used to show an object's structure. Applying str() on data frames, including those made by reading Excel files, reveals important details about the data set's dimensions, structures, and data types. When you read Excel data into R using packages like readxl, you will typically get a data frame object. Using str() on this data frame reveals details about Excel data's structure, as depicted in Figure 5.9.

```
> d1<- read_excel("Rgrades.xlsx")
New names:
• `` -> `...10`
• `` -> `...11`
• `` -> `...12`
• `` -> `...13`
> head(d1)
# A tibble: 6 × 13
  `S no` `Roll No` Name              Batch     `T1+T2`    T3    TA Total Grade ...10 ...11     ...12 ...13
   <dbl>     <dbl> <chr>             <chr>       <dbl> <dbl> <dbl> <dbl> <chr> <lgl> <chr>     <chr> <chr>
1      1  16103004 Neha Rastogi      CSE(B6)        25  30.5  25    80.5 A+    NA    Grade ... Grade Count
2      2  16103006 Tarushi Agrawal   CSE(B10)       20  20.5  25    65.5 B+    NA    0         F     0
3      3  16103007 Kanishk           CSE(B5)        24  13.5  23    60.5 B+    NA    30        D     1
4      4  16103008 Rishabh Jaiswal   CSE(B10)       18  16.5  21.5  56   B     NA    37        C     0
5      5  16103009 Rajat Juneja      CSE(B15)       20  20.5  21    61.5 B+    NA    44        C+    6
6      6  16103011 Yashwant Wardhan  CSE(B8)        12  10    23    45   C+    NA    52        B     16
> print(d1)
# A tibble: 68 × 13
  `S no` `Roll No` Name              Batch     `T1+T2`    T3    TA Total Grade ...10 ...11     ...12 ...13
   <dbl>     <dbl> <chr>             <chr>       <dbl> <dbl> <dbl> <dbl> <chr> <lgl> <chr>     <chr> <chr>
1      1  16103004 Neha Rastogi      CSE(B6)        25  30.5  25    80.5 A+    NA    Grade...  Grade Count
2      2  16103006 Tarushi Agrawal   CSE(B10)       20  20.5  25    65.5 B+    NA    0         F     0
3      3  16103007 Kanishk           CSE(B5)        24  13.5  23    60.5 B+    NA    30        D     1
4      4  16103008 Rishabh Jaiswal   CSE(B10)       18  16.5  21.5  56   B     NA    37        C     0
5      5  16103009 Rajat Juneja      CSE(B15)       20  20.5  21    61.5 B+    NA    44        C+    6
```

FIGURE 5.8
Excel data loading and printing.

```
> str(d1)
tibble [68 × 13] (S3: tbl_df/tbl/data.frame)
 $ S no   : num [1:68] 1 2 3 4 5 6 7 8 9 10 ...
 $ Roll No: num [1:68] 16103004 16103006 16103007 16103008 16103009 ...
 $ Name   : chr [1:68] "Neha Rastogi" "Tarushi Agrawal" "Kanishk" "Rishabh Jaiswal" ...
 $ Batch  : chr [1:68] "CSE(B6)" "CSE(B10)" "CSE(B5)" "CSE(B10)" ...
 $ T1+T2  : num [1:68] 25 20 24 18 20 12 12 19 24 20 ...
 $ T3     : num [1:68] 30.5 20.5 13.5 16.5 20.5 10 21.5 17 17 23 ...
 $ TA     : num [1:68] 25 25 23 21.5 21 23 19 24 21 21 ...
 $ Total  : num [1:68] 80.5 65.5 60.5 56 61.5 45 52.5 60 62 64 ...
 $ Grade  : chr [1:68] "A+" "B+" "B+" "B" ...
 $ ...10  : logi [1:68] NA NA NA NA NA NA ...
 $ ...11  : chr [1:68] "Grade Scale" "0" "30" "37" ...
 $ ...12  : chr [1:68] "Grade" "F" "D" "C" ...
 $ ...13  : chr [1:68] "Count" "0" "1" "0" ...
```

FIGURE 5.9
Structure of data.

3. Data Dimensions: To determine the size (both the number of rows and columns) of an Excel file in R, you can utilize the dim() function after importing the Excel data into a data frame. Additionally, for extracting column and row names, you can employ colnames() and rownames(), respectively, as demonstrated in Figure 5.10.

```
> # Get dimensions of the data frame
> dimensions <- dim(d1)
> dimensions
[1] 68 13
> colnames(d2)
 [1] "Sr.No."       "Roll No."     "Name"         "Group"         "Labtest(30)" "T3(40)"       "TA(30)"
 [8] "Total"        "Grade...9"    "...10"        "Grade Scale"   "Grade...12"  "Count"
> rownames(d2)
 [1] "1"  "2"  "3"  "4"  "5"  "6"  "7"  "8"  "9"  "10" "11" "12" "13"
> |
```

FIGURE 5.10
Data dimensions.

```
> # Summary Statistics
> summary(d2)
    Sr.No.        Roll No.              Name               Group             Labtest(30)
 Min.   :74   Min.   :16104007   Length:13          Length:13          Min.   :18.00
 1st Qu.:77   1st Qu.:16104018   Class :character   Class :character   1st Qu.:24.00
 Median :80   Median :16104049   Mode  :character   Mode  :character   Median :25.00
 Mean   :80   Mean   :16104037                                         Mean   :24.38
 3rd Qu.:83   3rd Qu.:16104053                                         3rd Qu.:27.00
 Max.   :86   Max.   :16104058                                         Max.   :28.00

     T3(40)          TA(30)          Total         Grade...9            ...10           Grade Scale
 Min.   :10.00   Min.   :20.00   Min.   :60.00   Length:13          Mode:logical   Min.   : 0.00
 1st Qu.:17.50   1st Qu.:21.00   1st Qu.:60.00   Class :character   NA's:13        1st Qu.:35.25
 Median :21.50   Median :24.00   Median :71.00   Mode  :character                  Median :48.00
 Mean   :21.42   Mean   :23.73   Mean   :69.54                                     Mean   :46.62
 3rd Qu.:25.00   3rd Qu.:26.00   3rd Qu.:80.00                                     3rd Qu.:62.50
 Max.   :31.00   Max.   :28.00   Max.   :80.00                                     Max.   :80.00
                                                                                   NA's   :5
  Grade...12            Count
 Length:13          Min.   :0.00
 Class :character   1st Qu.:0.00
 Mode  :character   Median :0.00
                    Mean   :1.50
                    3rd Qu.:3.25
                    Max.   :5.00
                    NA's   :5
```

FIGURE 5.11
Summarized statistics of loaded data.

4. Condensed Statistics: Let's now use our data to apply some summary statistics. Quick summaries of data are provided by summary statistics, which are especially helpful when comparing one statistical method/project to another. Figure 5.11 shows the condensed statistics of all columns, which include the minimum, median, first, third, and highest values of each column.

Statistical measures of central tendency play a crucial role in statistical analysis. With R, it's feasible to compute the minimum, maximum, standard deviation, mean, and median from Excel data, facilitating comprehensive statistical analysis (Walesiak et al, 2024). The dollar sign ($) in the example case shown below for the computation of standard deviation, min, and max designates the data collection and column that we will utilize in the analysis. When performing calculations, we're determining the standard deviation of the 'Total' column, representing the summation of all acquired marks. Likewise, we've also calculated the minimum and maximum marks in the data set.

```
> sd(d2$Total)
[1] 9.136606
> min(d2$Total)
[1] 60
> max(d2$Total)
[1] 80
```

5.3.3 Data Manipulation

Data manipulation encompasses various processes aimed at cleaning, transforming, and analysing data (Lanzetta et al., 2018). In R, several packages facilitate these tasks by enabling easy reading from and writing to Excel files, as well as performing comprehensive data manipulations. R users can manipulate Excel data by reading it into the language, utilizing dplyr and tidyr to do various transformations and analysis, and then writing the results back to an Excel file. In many data-driven projects, this workflow is crucial for data transformation, cleansing, and analysis. The 'readxl,' 'writexl,' 'dplyr,' and 'tidyr' packages are essential for data manipulation in R. A brief description is as follows:

readxl: For reading Excel files

vctrs: The R package vctrs is used to define and work with structured data types. It offers capabilities for building and verifying unique vector classes that behave consistently across operations.

dplyr: Offers a data manipulation grammar with a regular collection of verbs for typical activities.

tidyr: Used for data organisation, which converts data into a format in which variables are organized into columns, observations into rows, and different kinds of observational units into tables.

magirttr: A robust and well-liked R package called 'magrittr' offers a syntax for creating code that is more expressive and understandable by chaining processes together. Introduced as a carryover from the Unix shell concept, it offers the pipe operator %>%, which lets you transmit the parameter passed from one function call's result to the next.

Some data modification operations are listed below:

1. Filtering Rows: Using dplyr's function filter(), you can subset the rows according to criteria, for example, choosing rows where a variable satisfies specific requirements (e.g., values greater than a threshold).
2. Attribute/Column selection: The rows can be subset according to requirements using the sing filter() function from dplyr, for example, choosing rows when a variable satisfies specific requirements (values over a threshold, for example).
3. Data Mutation: Using dplyr's function filter(), you can subset the rows according to criteria, for example choosing rows where a variable satisfies specific requirements (e.g., values greater than a threshold).

4. Grouping and Summarizing Data: Utilize group_by() to organize data according to one or more variables, and then use summarize() to compile it. For aggregation activities such as group counts, sums, or means calculations, this is crucial.
5. Merging Data: Combining data frames based on a common key using functions like left_join(), right_join(), inner_join(), and full_join() from dplyr. This is crucial for integrating data from multiple sources.
6. Reshaping Data: The tidyr package's spread() and gather() functions can be used to convert data between wide and long formats, which can help organize data for different kinds of analysis and visualization.

This is a comprehensive example showing how to read data from an Excel file and applying various manipulation on it:

```
> install.packages("readxl")
> install.packages("vctrs")
> install.packages("dplyr")
> install.packages("magrittr")
> install.packages("tidyr")
> library(vctrs)
> library(dplyr)
> library(magrittr)
> library(tidyr)
> data <- read_excel("Rgrades.xlsx")
> head(data)
# A tibble: 6 × 13
  `S no` `Roll No` Name      Batch `T1+T2`    T3    TA Total Grade ...10
   <dbl>  <dbl>    <chr>     <chr>   <dbl> <dbl> <dbl> <dbl> <chr> <lgl>
1      1 16103004  Neha Ras… CSE(…   25     30.5  25    80.5  A+    NA
2      2 16103006  Tarushi … CSE(…   20     20.5  25    65.5  B+    NA
3      3 16103007  Kanishk   CSE(…   24     13.5  23    60.5  B+    NA
4      4 16103008  Rishabh … CSE(…   18     16.5  21.5  56    B     NA
5      5 16103009  Rajat Ju… CSE(…   20     20.5  21    61.5  B+    NA
6      6 16103011  Yashwant… CSE(…   12     10    23    45    C+    NA

> # Filter data based on condition with Total Value greater than 80
> filtered_data <- data %>% filter(Total > 80)
> filtered_data
S no` `Roll No` Name          Batch   `T1+T2` T3   TA Total Grade
1      16103004 Neha Rastogi CSE(B6)   25    30.5  25  80.5    A+
```

In the example provided, we imported student grades data containing attributes such as S.No, Roll number, Name, Batch, Test 1 + Test 2 (T1+T2), Test 3 (T3), Total (total marks out of 100), and Grades (based on total marks). We filtered the data to extract rows where the Total value exceeds 80, resulting in a single tuple as an output.

```
> # Select specific columns
> selected_data <- data %>% select('T1+T2', 'T3', 'Total')
> selected_data
# A tibble: 68 × 3
   `T1+T2`    T3  Total
     <dbl> <dbl>  <dbl>
 1      25  30.5   80.5
 2      20  20.5   65.5
 3      24  13.5   60.5
 4      18  16.5   56
 5      20  20.5   61.5
 6      12  10     45
 7      12  21.5   52.5
 8      19  17     60
 9      24  17     62
10      20  23     64
```

You can also choose particular columns, as shown earlier. The select() function is used for this purpose. In this instance, we've opted for three columns: T1+T2, T3, and total. It is also possible to modify existing data using the mutate() function. In the case illustrated below, we've created a new column that results from multiplying two existing columns: T3 and TA.

```
> # Create a new column based on existing columns
> mutated_data <- data %>% mutate(NewColumn = T3 * TA)
> mutated_data
> filtered_data <- data %>% filter(T3 > 15) %>% select(T3, TA,
Total) %>% mutate(NewColumn = T3 * TA)
> filetered_data
> filtered_data
# A tibble: 46 × 4
      T3    TA  Total  NewColumn
   <dbl> <dbl>  <dbl>      <dbl>
 1  30.5  25     80.5       762.
 2  20.5  25     65.5       512.
 3  16.5  21.5   56         355.
 4  20.5  21     61.5       430.
 5  21.5  19     52.5       408.
 6  17    24     60         408
 7  17    21     62         357
 8  23    21     64         483
 9  17.5  21     62.5       368.
10  15.5  24     60.5       372
```

In this code snippet, we have combined filter, select, and mutate operations. First, we filtered records where T3 exceeds a value of 15. Next, we selected columns T3, TA, and Total, and finally, we created a new column based on these selections.

```
> #Group by a column and summarize
> grouped_data <- data %>% group_by(Grade) %>% summarize
(MeanValue = mean(Total, na.rm = TRUE))
> grouped_data
# A tibble: 6 × 2
  Grade MeanValue
<chr>        <dbl>
1 A           71.4
2 A+          80.2
3 B           53.8
4 B+          61.6
5 C+          45.5
6 D           33
```

We can also perform grouping of data and summarize it. In this example, we grouped the data by grades and then calculated the mean within each group based on grades. Another working instance of grouping and summarization is presented below.

```
> grouped_data <- filtered_data %>% group_by(TA) %>%
summarize(MeanValue = mean(NewColumn, na.rm = TRUE),TotalValue
= sum(NewColumn, na.rm = TRUE))
> grouped_data
# A tibble: 13 × 3
      TA MeanValue TotalValue
   <dbl>     <dbl>      <dbl>
 1  17.5      341.       341.
 2  19        408.       408.
 3  20        370        740
 4  21        424.      2541
 5  21.5      480.      1440.
 6  22        448.      1793
 7  22.5      360        360
 8  23        405.      2024
 9  24        478.      5256
10  25        638.      1275
11  26        611       3055
12  27        557.      2228.
13  28        574        574
```

Data reshaping from one format to another, such as from long to wide format or vice versa, is a common activity in data analysis. In many cases, this transformation is required to get data ready for reporting, analysis, or visualization. Here's an illustration of how to restructure data in R using the 'tidyr' package, complete with step-by-step instructions.

Imagine you have a data frame named df1 that stores student test scores in a long format, with each row representing a student and their corresponding test score:

```
> df1 <- data.frame(Student = c("Student A1", "Student A2",
"Student A1", "Student A2"),Test = c("Test 2", "Test 2",
"Test 3", "Test 3"),Score = c(60, 75, 45, 80))
> df1
     Student   Test Score
1 Student A1 Test 2    60
2 Student A2 Test 2    75
3 Student A1 Test 3    45
4 Student A2 Test 3    80
> # Reshaping Data
> wide_data <- df1 %>% spread(Test, Score)
> wide_data
     Student Test 2 Test 3
1 Student A1     60     45
2 Student A2     75     80
> long_data <- df1 %>% gather(Score, Test, -Student)
> long_data
     Student Score   Test
1 Student A1  Test Test 2
2 Student A2  Test Test 2
3 Student A1  Test Test 3
4 Student A2  Test Test 3
4 Student A1 Score     60
6 Student A2 Score     75
5 Student A2 Score     80
```

5.3.4 Writing Data to an Excel File

There are various packages available in R for writing data to an Excel file, but the most widely used ones are 'writexl' and 'openxlsx'. I've included example code, theoretical explanations, and a step-by-step guide for both packages below. Detailed instructions for writing data with the 'writexl' package are manifested as follows:

Step I: Install and Load the Package

```
> # Writing data with wrtitexl package
> install.packages("writexl")
> library("writexl")
```

Step II: Data Frame Creation for Writing Data into an Excel File

```
> # Preparing data for writing into dataframe
> employee <- data.frame(Name = c("Amit", "Jaya", "Tanya",
"Atharv"),Age = c(38, 42, 26,27), Department = c("CSE",
"Finance", "IT","R&D"),Salary=c(100000,50000,65000,300000))
```

```
> employee
    Name Age Department Salary
1   Amit  38        CSE 100000
2   Jaya  42    Finance  50000
3  Tanya  26         IT  65000
4 Atharv  27        R&D 300000
```

Step III: Writing Data to an Excel File

```
> # Writing Data to Excel File
> write_xlsx(employee, "Employee.xlsx")
```

Step IV: Read Data with the read_excel Method

```
> read_excel('Employee.xlsx')
# A tibble: 4 × 4
  Name     Age Department Salary
  <chr>  <dbl> <chr>       <dbl>
1 Amit      38 CSE        100000
2 Jaya      42 Finance     50000
3 Tanya     26 IT          65000
4 Atharv    27 R&D        300000
```

Detailed instructions for using the 'openxlsx' package to write data are as follows:

Step I: Install and Load the Package

```
> # Writing data with openxlsx package
> install.packages("openxlsx")
> library("openxlsx")
```

Step II: Data Preparation (Creating a Data Frame)

```
> # Data Preparation
> org <- data.frame(Organization = c("Comp1", "Comp2",
"Comp3"),Established_date = c(1928, 2022, 1935),
Product = c("Car", "Bus", "Truck"))
> org
  Organization Established_date Product
1        Comp1             1928     Car
2        Comp2             2022     Bus
3        Comp3             1935   Truck
```

Step III: Creating Workbook and Adding a Worksheet

```
> # Create Workbook
> wb <- createWorkbook()
> # Add worksheet
> addWorksheet(wb, "o1")
```

Step IV: Writing Data

```
> writeData(wb, sheet = "o1", x = org)
> saveWorkbook(wb, "Organization.xlsx", overwrite = TRUE)
```

Step V: Reading Data

```
> read.xlsx("Organization.xlsx")
  Organization Established_date Product
1        Comp1             1928     Car
2        Comp2             2022     Bus
3        Comp3             1935   Truck
```

5.3.5 Data Deletion and Addition

Data addition and deletion in Excel files are handled by loading the workbook, modifying the data in R, and then saving the updated workbook back to Excel. Simplified data processing workflows are made possible by these operations, which offer flexibility and automation in managing Excel data straight from your R environment. For efficient data management, modify the code snippets in accordance with your unique Excel file structure and data processing needs.

The 'openxlsx' utility in R can be used to delete data from an Excel file. This may include eliminating some cells, rows, or columns. Code snippets of data deletion in an Excel format file are as follows.

Step I: Install and Load the Package

```
> install.packages("openxlsx")
> library("openxlsx")
```

Step II: Load Data

```
> # Load Excel Data
> emp<-read.xlsx("Employee.xlsx")
> emp
    Name Age Department Salary
1   Amit  38        CSE 100000
2   Jaya  42    Finance  50000
3  Tanya  26         IT  65000
4 Atharv  27        R&D 300000
```

Step III: Row Deletion

```
> # Remove rows 3 and 4
> data <- emp[-c(3, 4), ]
> data
  Name Age Department Salary
1 Amit  38        CSE  1e+05
2 Jaya  42    Finance  5e+04
```

In a different example, we have demonstrated how to delete rows based on a specific condition. For instance, in the example below, we retain the rows where the age value is 30 or greater.

```
> d1 <- emp[emp$Age >= 30, ]
> d1
  Name Age Department Salary
1 Amit  38        CSE  1e+05
2 Jaya  42    Finance  5e+04
```

We can combine two conditions to remove data from an Excel file. As shown in the code snippet below, we filter to remove rows where the department is "IT" or the age is greater than 30.

```
> emp
    Name Age Department Salary
1   Amit  38        CSE 100000
2   Jaya  42    Finance  50000
3  Tanya  26         IT  65000
4 Atharv  27    R&amp;D 300000
> d11 <- emp[!(emp$Age > 30 | emp$Department == "IT"), ]
> d11
     Name Age Department Salary
4  Atharv  27        R&D  3e+05
```

Step IV: Column Deletion

```
> # Remove columns 2 and 3
> data <- emp[, -c(2, 3)]
> data
     Name Salary
1    Amit 100000
2    Jaya  50000
3   Tanya  65000
2  Atharv 300000

 > # Remove columns 2 and 3
> data <- emp[, -c(1, 3)]
> data
  Age Salary
1  38 100000
2  42  50000
3  26  65000
3  27 300000

> # Remove columns in a range
> data <- emp[, -c(1:3)]
> data
[1] 100000  50000  65000 300000
```

Step V: Add the Updated Data to the Worksheet

```
> emp
    Name Age Department Salary
1   Amit  38        CSE 100000
2   Jaya  42    Finance  50000
3  Tanya  26         IT  65000
4 Atharv  27    R&amp;D 300000
> res<-emp[-c(1,4),]
> res
   Name Age Department Salary
2  Jaya  42    Finance  50000
3 Tanya  26         IT  65000
> # Overwrite the worksheet with the modified data
> writeData(wb, sheet = "Sheet1", x = res, withFilter = FALSE)
```

Step VI: Save the Updated Workbook

```
> #Save the updated workbook
> saveWorkbook(wb, "new_excel.xlsx", overwrite = TRUE)
> read.xlsx('new_excel.xlsx')
    Name Age Department Salary
2   Jaya  42    Finance  50000
3  Tanya  26         IT  65000
```

Using the openxlsx package to load an existing workbook, create a new one, write data to a designated sheet, and save the workbook are fundamental steps involved in adding data to an Excel file using R. The step-by-step procedure to append new data to an existing file is as follows:

Step I: Install and Load Libraries

```
> install.packages("openxlsx")
> library("openxlsx")
```

Step II: Load Existing Workbook

```
> # Appending Data
> # Load workbook
> wb <- loadWorkbook("Rgrades.xlsx")
```

Step III: Generating new_data for Appending Data into an Existing File

```
> # Data Preparation
> Icecream_data <- data.frame(Flavour = c("Chocolate",
"Vanilla", "strawberry"),Manufacturer = c("Quality Walls",
"Amul", "Vadilal"),Price = c(70, 30, 40))
> Icecream_data
```

```
      Flavour  Manufacturer Price
1  Chocolate Quality Walls    70
2    Vanilla          Amul    30
3 strawberry       Vadilal    40
```

Step IV: Create a New Worksheet and Add Data

```
> #Add or Write Data to a Worksheet
> # Create a new worksheet in workbook
> addWorksheet(wb, "Icecream")
> writeData(wb, sheet = "Icecream", x = Icecream_data)
```

Step V: Save Data

```
> # Save data
> saveWorkbook(wb, "Icecream.xlsx", overwrite = TRUE)
> read.xlsx("Icecream.xlsx", sheet='Icecream')
      Flavour  Manufacturer Price
1  Chocolate Quality Walls    70
2    Vanilla          Amul    30
3 strawberry       Vadilal    40
```

In the example demonstrated earlier, we created a new worksheet and populated it with data. Additionally, we can append data to the end of an existing file. The process for adding data to an existing file is outlined below:

Step I: Install and Load the Package

```
> install.packages("openxlsx")
> library("openxlsx")
```

Step II: Load the Existing Workbook

```
> #Load an existing workbook
> wb <- loadWorkbook("Employee.xlsx")
> # Read existing data from the specified sheet
> current_data <- read.xlsx(wb, sheet = "Sheet1")
> current_data
    Name Age Department Salary
1   Amit  38        CSE 100000
2   Jaya  42    Finance  50000
3  Tanya  26         IT  65000
4 Atharv  27        R&D 300000
```

Step III: Data Preparation for Adding It to the Existing File

```
> # Prepare new data to append
> emp1 <- data.frame(Name = c("Neha", "Arjun"),Age = c(29,
28),Department = c("CSE", "ECE"), Salary=c(100000,78000))
> emp1
```

```
   Name Age Department Salary
1  Neha  29        CSE 100000
2 Arjun  28        ECE  78000
```

Step IV: Append New Data to the Currently Loaded Existing File

```
> # Append new data to existing data
> appended_data <- rbind(current_data, emp1)
```

Step V: Write and Save the Data

```
> # Write combined data back to the existing worksheet
> writeData(wb, sheet = "Sheet1", x = appended_data,
startCol = 1, startRow = 1, colNames = TRUE)
> # Save the workbook with changes
> saveWorkbook(wb, "Employee_New.xlsx", overwrite = TRUE)
```

Step V: Read Updated Data

```
> read.xlsx(wb, sheet = "Sheet1")
    Name Age Department Salary
1   Amit  38        CSE 100000
2   Jaya  42    Finance  50000
3  Tanya  26         IT  65000
4 Atharv  27        R&D 300000
5   Neha  29        CSE 100000
6  Arjun  28        ECE  78000
```

5.3.6 Querying Data

Effectively extracting, modifying, and analysing data to produce insightful conclusions is the main goal of querying Excel data in R from a data analysis standpoint. Understanding the structure and organization of the data within the Excel sheets is essential before beginning any analysis. This comprises of recognizing sheets and examining the data types and column names. It takes a methodical strategy to load data into R and give insights for efficient Excel data querying for data analysis. To extract valuable insights and make wise decisions, every stage of the process—from loading data to exploring, cleaning, analysing, and reporting—contributes. To make these chores easier, the 'openxlsx' package offers necessary functions that allow Excel data to be easily integrated into R's analytical capabilities (Daróczi, 2015).

Let's delve more closely into some specific instances of querying Excel data. In this context, querying includes actions such as collecting data, filtering rows, choosing particular columns, and generating summaries. In data analysis workflows, these operations are essential for obtaining actionable insights from data sets. These querying strategies are demonstrated in depth in the examples below.

```
> # Load Data
> wb <- loadWorkbook("Employee_New.xlsx")
> # Read Data
> r <- read.xlsx(wb, sheet = "Sheet1")
> r
    Name Age Department Salary
1   Amit  38        CSE 100000
2   Jaya  42    Finance  50000
3  Tanya  26         IT  65000
4 Atharv  27    R&amp;D 300000
5   Neha  29        CSE 100000
6  Arjun  28        ECE  78000

> # Querying Data
> #Row Extraction based on condition
> extract_data <- r[r$Salary > 60000 & r$Department == "CSE", ]
> extract_data
  Name Age Department Salary
1 Amit  38        CSE  1e+05
5 Neha  29        CSE  1e+05
```

In the R code above, we loaded an existing data file and subsequently filtered rows based on specific conditions.

Moreover, we can sort data values using the order() function. Below, we have sorted the Name column in descending order.

```
> # Data Sorting
> sorted_data <- r[order(data$Name, decreasing = TRUE), ]
> sorted_data <- r[order(r$Name, decreasing = TRUE), ]
> sorted_data
    Name Age Department Salary
3  Tanya  26         IT  65000
5   Neha  29        CSE 100000
2   Jaya  42    Finance  50000
4 Atharv  27    R&amp;D 300000
6  Arjun  28        ECE  78000
1   Amit  38        CSE 100000
```

Performing calculations on Excel data is straightforward using R programming. In this instance, we updated the salaries of all employees by multiplying their age by a constant value of 1000.

```
> # Performing Calculations
> r$Salary <- r$Age * 1000
> r$Salary
[1] 38000 42000 26000 27000 29000 28000
```

Summarization is an essential stage in data analysis that helps to understand the distribution and features of data; it functions similarly to a pivot table in spreadsheet software, as depicted below.

```
> # Table like Summarization
> summary_table <- with(r, table(Department), Age = tapply(Age,
Department, mean))
> summary_table
Department
CSE ECE Finance IT R&D
1   1   1       1  1
```

5.4 Import and Export of XML Data

XML, short for Extensible Markup Language, consists of markup tags, where each tag represents the information conveyed by a specific attribute within the XML file (Nolan & Lang, 2014). The main reasons why XML is used in data analysis are its structured format and adaptability when it comes to expressing hierarchical data. Here are some important contexts in which XML is used in data analysis:

1. Integration of Data: With XML, structured data can be exchanged between systems in a standard format, facilitating the integration of data from diverse sources.
2. Representation of Data: Because of its hierarchical structure, XML may be used to describe layered data structures and complex data linkages, which is helpful for a variety of analytical activities.
3. Transformation of Data: XSLT (Extensible Stylesheet Language Transformations) can be used to convert XML documents into other formats that are appropriate for display or analysis.
4. Management for Metadata: Metadata is the information about the content, structure, and relationships of data; is frequently managed using XML; and is essential for efficient data analysis.
5. Querying Data: To help with data analysis activities, XML may be queried to extract certain data items or run complicated searches across XML documents.
6. Visualization of Data: Data visualization capabilities can be improved by transforming and rendering XML data into various visual forms for analysis and reporting needs.

5.4.1 Reading XML Data with the xml2 Package

Using packages 'xml' or 'xml2,' which offer functions to parse XML documents into R data structures that can be edited and examined, is necessary to read XML data in R. Below are the detailed steps on how to read XML data in R using the 'xml' package, known for its modern and efficient XML parsing capabilities. To read XML data, the xml/xml2 package is required. You can import and parse the XML file into an R object by using the read_xml() function from xml2, as demonstrated.

```
> # Install and load xml2 package
> install.packages("xml2")
> library('xml2')
> # Read Data
> x1 <- read_xml("employee.xml")
> x1
{xml_document}
<records>
[1] <employee_info>\n  <id>1</id>\n  <name>Atharv</name>\n
<salary>340000</salary>\n  <date>14/1/20 ...
[2] <employee_info>\n  <id>2</id>\n  <name>Misha</name>\n
<salary>450000</salary>\n  <date>6/7/1999 ...
[3] <employee_info>\n  <id>1</id>\n  <name>Nisha</name>\n
<salary>90000</salary>\n  <date>4/7/2020< ...
[4] <employee_info>\n  <id>1</id>\n  <name>Sheetal</name>\n
<salary>78000</salary>\n  <date>10/5/19 ...
[5] <employee_info>\n  <id>1</id>\n  <name>Arjun</name>\n
<salary>97000</salary>\n  <date>9/4/2021< ...
[6] <employee_info>\n  <id>1</id>\n  <name>Ayush</name>\n
<salary>56000</salary>\n  <date>9/1/2022< ...
[7] <employee_info>\n  <id>1</id>\n  <name>Ananya</name>\n
<salary>87900</salary>\n  <date>5/1/2019 ...
[8] <employee_info>\n  <id>1</id>\n  <name>Sahil</name>\n
<salary>150000</salary>\n  <date>1/1/2018 ..
```

In the above example, we have used the "Employee.xml" file and the structure of the file is shown in Figure 5.12.

5.4.2 Accessing XML Data with the xml2 Package

When accessing specific data, one must navigate the hierarchical structure of an XML document in order to access XML nodes and elements. The 'xml2' package in R is usually used for this purpose, as it offers functions like xml_find_all() and xml_find_first() that allow you to use XPath expressions to search for nodes. The properties, text content, and child elements of every node in the XML tree can all be accessed, retrieved, and manipulated. This approach makes it possible to extract and process data efficiently, giving users the ability to work with complex XML documents in an organized and methodical way (Brahmia et al., 2020).

```
<records>
▼<employee_info>
   <id>1</id>
   <name>Atharv</name>
   <salary>340000</salary>
   <date>14/1/2019</date>
   <dept>CSE</dept>
 </employee_info>
▼<employee_info>
   <id>2</id>
   <name>Misha</name>
   <salary>450000</salary>
   <date>6/7/1999</date>
   <dept>Finance</dept>
 </employee_info>
▼<employee_info>
   <id>1</id>
   <name>Nisha</name>
   <salary>90000</salary>
   <date>4/7/2020</date>
   <dept>ECE</dept>
 </employee_info>
```

FIGURE 5.12
Employee.xml file structure.

Once parsed, you can use the functions given by xml2 to traverse through the XML structure. Below are several examples of extracting different structural elements from XML data.

1. Root Node Extraction: To access the XML document's root node, use xml_root(). Figure 5.13 shows the commands and output of root extraction from XML data.

```
> # Extract root node
> root_node <- xml_root(x1)
> root_node
{xml_document}
<records>
[1] <employee_info>\n  <id>1</id>\n  <name>Atharv</name>\n  <salary>340000</salary>\n  <date>14/1/20 ...
[2] <employee_info>\n  <id>2</id>\n  <name>Misha</name>\n  <salary>450000</salary>\n  <date>6/7/1999 ...
[3] <employee_info>\n  <id>1</id>\n  <name>Nisha</name>\n  <salary>90000</salary>\n  <date>4/7/2020< ...
[4] <employee_info>\n  <id>1</id>\n  <name>Sheetal</name>\n  <salary>78000</salary>\n  <date>10/5/19 ...
[5] <employee_info>\n  <id>1</id>\n  <name>Arjun</name>\n  <salary>97000</salary>\n  <date>9/4/2021< ...
[6] <employee_info>\n  <id>1</id>\n  <name>Ayush</name>\n  <salary>56000</salary>\n  <date>9/1/2022< ...
[7] <employee_info>\n  <id>1</id>\n  <name>Ananya</name>\n  <salary>87900</salary>\n  <date>5/1/2019 ...
[8] <employee_info>\n  <id>1</id>\n  <name>Sahil</name>\n  <salary>150000</salary>\n  <date>1/1/2018 ...
```

FIGURE 5.13
Root node extraction.

2. Child Node Extraction: To navigate through child nodes, use functions such as xml_children(), xml_find_all(), or xml_find_first(). In the code below, all elements with the tag "name" are extracted, and then we use the xml_text() method to retrieve only their text contents.

```
> #Find all elements with "name" tag
> nodes <- xml_find_all(x1, "//name")
> nodes
{xml_nodeset (8)}
[1] <name>Atharv</name>
[2] <name>Misha</name>
[3] <name>Nisha</name>
[4] <name>Sheetal</name>
[5] <name>Arjun</name>
[6] <name>Ayush</name>
[7] <name>Ananya</name>
[8] <name>Sahil</name>
> # Get text content of a node
> tc <- xml_text(nodes)
> tc
[1] "Atharv" "Misha" "Nisha" "Sheetal" "Arjun" "Ayush"
"Ananya" "Sahil"
```

5.4.3 Reading XML Data with the 'XML' Package

A strong and popular package for parsing and extracting data from XML documents is the R XML package. It has methods for reading XML files, navigating their hierarchy, and extracting the necessary information. Given below is the code to read and parse XML data with the 'XML' package.

```
> install.packages('XML')
> library('XML')
> r <- xmlParse(file = "Employee.xml")
> r
<?xml version="1.0"?>
<records>
  <employee_info>
    <id>1</id>
    <name>Atharv</name>
    <salary>340000</salary>
    <date>14/1/2019</date>
    <dept>CSE</dept>
  </employee_info>
  <employee_info>
    <id>2</id>
    <name>Misha</name>
    <salary>450000</salary>
    <date>6/7/1999</date>
    <dept>Finance</dept>
  </employee_info> .........</records>
```

5.4.4 Accessing XML Data with the 'XML' Package

The XML package in R is a comprehensive package for parsing and manipulating XML data. This package provides functions to read XML files, navigate the XML tree structure, and extract specific information using XPath expressions. Below is a detailed guide on accessing XML data using the XML package.

1. Extracting Rootnode: The XML document's root node can be retrieved using the 'xmlRoot' function.

```
> rnode <- xmlRoot(r)
> rnode
<records>
  <employee_info>
    <id>1</id>
    <name>Atharv</name>
    <salary>340000</salary>
    <date>14/1/2019</date>
    <dept>CSE</dept>
  </employee_info>
  <employee_info>
    <id>2</id>
    <name>Misha</name>
    <salary>450000</salary>
    <date>6/7/1999</date>
    <dept>Finance</dept>
  </employee_info>
  <employee_info>
    <id>1</id>
    <name>Nisha</name>
    <salary>90000</salary>
    <date>4/7/2020</date>
    <dept>ECE</dept>
  </employee_info>
  <employee_info>
    <id>1</id>
    <name>Sheetal</name>
    <salary>78000</salary>
    <date>10/5/1998</date>
    <dept>IT</dept>
  </employee_info>
  ...............

</records>
```

2. Extracting Rootnode Size: To determine the size of an XML document, use the xmlSize() function.

```
> rsize <- xmlSize(rnode)
> rsize
[1] 8
```

3. Extracting Specific Node: Demonstrated below are several instances of extracting certain nodes from an XML document. In this example, the first and seventh nodes have been taken.

```
> print(rnode[1])
$employee_info
<employee_info>
  <id>1</id>
  <name>Atharv</name>
  <salary>340000</salary>
  <date>14/1/2019</date>
  <dept>CSE</dept>
</employee_info>

attr(,"class")
[1] "XMLInternalNodeList" "XMLNodeList"

> print(rnode[7])
$employee_info
<employee_info>
  <id>1</id>
  <name>Ananya</name>
  <salary>87900</salary>
  <date>5/1/2019</date>
  <dept>IT</dept>
</employee_info>

attr(,"class")
[1] "XMLInternalNodeList" "XMLNodeList"
```

4. Extracting a Specific Element of a Specific Node: To extract a specific element from a specific node, you need to include the node in square brackets followed by the element in another set of square brackets, as demonstrated in the following example.

```
> # Getting the first element of the first node.
> print(rnode[[1]][[1]])
<id>1</id>

> # Extracting Fourth element of First node:

> print(rnode[[1]][[4]])
<date>14/1/2019</date>
```

5.4.5 Converting XML Data to Another Data Type with the 'XML' Package

Data analysis and manipulation sometimes need converting XML data to different data types in R. Strong facilities for parsing XML and converting it to data frames or lists or more suitable formats for analysis are included in the XML package. An XML document named 'Employee' is loaded for conversion using the xmlParse() function.

```
> r <- xmlParse(file = "Employee.xml")
```

1. Converting XML to List Type: In R, a list is an object that has the ability to hold elements of various types, such as strings, numbers, and other lists. The xmlToList() function can be used to turn the XML data into a list.

```
> # Converting XML to List type
> x1 <- xmlToList(r)
> x1
$employee_info
$employee_info$id
[1] "1"

$employee_info$name
[1] "Atharv"

$employee_info$salary
[1] "340000"

$employee_info$date
[1] "14/1/2019"

$employee_info$dept
[1] "CSE"


$employee_info
$employee_info$id
[1] "2"

$employee_info$name
[1] "Misha"

$employee_info$salary
[1] "450000"

$employee_info$date
[1] "6/7/1999"

$employee_info$dept
[1] "Finance"


$employee_info
$employee_info$id
[1] "1"
```

```
$employee_info$name
[1] "Nisha"

$employee_info$salary
[1] "90000"

$employee_info$date
[1] "4/7/2020"

$employee_info$dept
[1] "ECE"


$employee_info
$employee_info$id
[1] "1"

$employee_info$name
[1] "Sheetal"

$employee_info$salary
[1] "78000"

$employee_info$date
[1] "10/5/1998"

$employee_info$dept
[1] "IT"


$employee_info
$employee_info$id
[1] "1"

$employee_info$name
[1] "Arjun"

$employee_info$salary
[1] "97000"

$employee_info$date
[1] "9/4/2021"

$employee_info$dept
[1] "HR"


$employee_info
$employee_info$id
[1] "1"
```

```
$employee_info$name
[1] "Ayush"

$employee_info$salary
[1] "56000"

$employee_info$date
[1] "9/1/2022"

$employee_info$dept
[1] "HR"


$employee_info
$employee_info$id
[1] "1"

$employee_info$name
[1] "Ananya"

$employee_info$salary
[1] "87900"

$employee_info$date
[1] "5/1/2019"

$employee_info$dept
[1] "IT"


$employee_info
$employee_info$id
[1] "1"

$employee_info$name
[1] "Sahil"

$employee_info$salary
[1] "150000"

$employee_info$date
[1] "1/1/2018"

$employee_info$dept
[1] "CSE"
```

2. Converting XML to a Data Frame: A data frame is a table or two-dimensional array-like structure where each row has a collection of values from each column and each column includes the values of a

single variable. The xmlToDataFrame() function can be used to turn the XML data into a list.

```
> x1 <- xmlToDataFrame(r)
> x1
  id    name salary      date    dept
1  1  Atharv 340000 14/1/2019     CSE
2  2   Misha 450000  6/7/1999 Finance
3  1   Nisha  90000  4/7/2020     ECE
4  1 Sheetal  78000 10/5/1998      IT
5  1   Arjun  97000  9/4/2021      HR
6  1   Ayush  56000  9/1/2022      HR
7  1  Ananya  87900  5/1/2019      IT
8  1   Sahil 150000  1/1/2018     CSE
```

5.4.6 Converting a List to XML Type

You can use the XML or xml2 packages in R to convert a list to an XML format. The detailed instructions for converting a list to XML type with the 'XML' package are provided below.

1. Install and Load the XML Package

```
> install.packages("XML")
> library(XML)
```

2. Prepare List Data

```
> # List Creation
> l1 <- list(item1 = list(name = "Item 1", price =
10.00),item2 = list(name = "Item 2", price = 20.00))
> l1
$item1
$item1$name
[1] "Item 1"

$item1$price
[1] 10


$item2
$item2$name
[1] "Item 2"

$item2$price
[1] 20
```

3. List to XML Conversion: Initially, create a root node, then add elements to this root node using a user-defined function, and finally convert the XML structure into a string, as depicted further down.

```
> #Convert List to XML
> ## Create the root node
> root <- newXMLNode("root")
> # Add elements to the root node
> for (item in names(l1)) {
+     node <- newXMLNode("item", parent = root)
+     newXMLNode("name", l1[[item]]$name, parent = node)
+     newXMLNode("price", l1[[item]]$price, parent = node)
+ }
> # Convert the XML document to a string
> xml_string <- saveXML(root)
>
> print(xml_string)
[1] "<root>\n  <item>\n    <name>Item 1</name>\n
<price>10</price>\n  </item>\n  <item>\n
<name>Item 2</name>\n    <price>20</price>\n
</item>\n</root>"
```

5.4.7 Converting a Data Frame to XML Type

In R, to convert a data frame to XML with the 'XML' package, follow the steps mentioned below:

1. Install and Load the XML Package

```
> install.packages("XML")
> library(XML)
```

2. Prepare a Data Frame

```
> product_df <- data.frame(product = c("Item 1", "Item 2"),
product_price = c(10.00, 20.00),stringsAsFactors = FALSE)
> product_df
  product product_price
1  Item 1            10
2  Item 2            20
```

3. Data Frame to XML Conversion: Initially, create a root node, then add elements to this root node using a user-defined function, and finally convert the XML structure into a string.

```
> # Create the root node
> root <- newXMLNode("root")
> # Add items to the root
> for (i in 1:nrow(product_df)) {
+     item <- newXMLNode("item", parent = root)
```

```
+      newXMLNode("product name", product_df$product[i],
parent = item)
+      newXMLNode("product price",product_df$product_
price[i], parent = item)
+ }
> xml_string <- saveXML(root)
> print(xml_string)
[1] "<root>\n  <item>\n    <product name>Item 1</product
name>\n    <product price>10</product price>\n  </item>\n
<item>\n    <product name>Item 2</product name>\n
<product price>20</product price>\n  </item>\n</root>"
```

5.5 Import and Export of JSON Data

JavaScript Object Notation, or JSON, is a data-interchange format that is simple for computers to understand and produce as well as for people to read and write (Marrs, 2017). It is frequently used to send text data between a web application and server. R has the capability to read JSON files using packages like 'rjson' and 'jsonlite' (Ooms, 2014). R objects (such data frames or lists) can be converted into JSON format and vice versa when importing and exporting JSON data. JSON files or strings can be read into R objects. R has several packages that can be used for this operation.

5.5.1 Importing JSON Data

JSON data is arranged using curly brackets {} to organize key-value pairs. Every key is a string that has the value after it, followed by a colon (:). In JSON, arrays are surrounded by square brackets []. Copy the information below into a text editor, such as Notepad, to create a JSON file. Choose the file type as all files (*.*) and save the file with a .json extension.

```
{
  "EID":["1","2","3","4","5","6","7","8" ],
  "Emp_Name":["Reyansh","Sneha","Mayank","Ryan","Gargi","Neeta",
"Shelly","Garvit"],
  "Emp_Salary":["623.3","515.2","611","729","843.25","578",
"632.8","722.5"]"Job_start_Date":["4/1/2018","7/25/1998",
"10/16/2024","12/12/2020","4/14/2015","8/25/2010",
"11/13/1999","2/21/2017"],
"Dept":[ "HR","Accounts","Finance","IT","Logistic","HR",
"Accounts","Logistic"]
}
```

The 'jsonlite' package facilitates the importation of data from diverse sources such as files, URLs, or strings in order to load data. Below are examples of R code that demonstrate how to load data from various sources.

1. Import JSON from a File:

```
> install.packages('jsonlite')
> library('jsonlite')
> # Reading Json file
> jfile <- "empdata.json"
> # Import JSON data from file
> jdata <- fromJSON(jfile)
> jdata
$EID
[1] "1" "2" "3" "4" "5" "6" "7" "8"

$Emp_Name
[1] " Reyansh"  "Sneha"  "Mayank" "Ryan"  "Gargi"  "Neeta"
"Shelly"  "Garvit"

$Emp_Salary
[1]  "623.3"   "515.2"   "611"   "729"  "843.25"  "578"
 "632.8"  "722.5"

$Job_start_Date
[1] "4/1/2018"   "7/25/1998"  "10/16/2024" "12/12/2020"
"4/14/2015"  "8/25/2010"  "11/13/1999"
[8] "2/21/2017"

$Dept
[1] "HR" "Accounts" "Finance" "IT" "Logistic" "HR"
"Accounts" "Logistic"
```

2. Import JSON from a URL:

```
> # Import JSON data from URL
>jurl <- "https://microsoftedge.github.io/Demos/json-dummy-
data/64KB.json"
> jdata <- fromJSON(jurl)
> jdata
                  name    language                id
1        Adeel solangi      Sindhi  V59OF92YF627HFY0
2        Afzal Ghaffar      Sindhi  ENTOCR13RSCLZ6KU
3        Aamir Solangi      Sindhi  IAKPO3R4761JDRVG
4        Abla Dilmurat      Uyghur  5ZVOEPMJUI4MB4EN
5            Adil Eli      Uyghur  6VTI8X6LL0MMPJCC
6          Adile Qadir      Uyghur  F2KEU5L7EHYSYFTT
7   Abdukerim Ibrahim      Uyghur  LO6DVTZLRK68528I
8           Adil Abro      Sindhi  LJRIULRNJFCNZJAJ
9    Afonso Vilarchán    Galician  JMCL0CXNXHPL1GBC
10       Mark Schembri     Maltese  KU4T500C830697CW
```

3. Import JSON from a String:

```
> # Import JSON data from String
> jstring <- '{"name": "Atharv", "age": 24, "city":
"Noida", "Education":"B.Tech IIT"}'
> jdata <- fromJSON(jstring)
> jdata
$name
[1] "Atharv"

$age
[1] 24

$city
[1] "Noida"

$Education
[1] "B.Tech IIT"
```

5.5.2 Exporting JSON Data

R objects (such as lists, data frames, or vectors) must be converted into JSON format in order to export JSON data from R.

```
> # Export Data
> sample_data <- data.frame(ename = c("Atharv", "Aryan"),
age = c(24, 25),city = c("Noida", "Dehradun"))
> sample_data
   ename age     city
1 Atharv  24    Noida
2  Aryan  25 Dehradun
> # Convert data frame to JSON and write to file
> jfile <- "Jdata.json"
> writeLines(toJSON(sample_data), jfile)
> cat("Exported JSON content:\n")
Exported JSON content:
> cat(readLines(jfile), sep = "\n")
[{"ename":"Atharv","age":24,"city":"Noida"},{"ename":"Aryan",
"age":25,"city":"Dehradun"}]
```

Using functions from packages like 'jsonlite,' R objects must be converted into JSON format in order to export JSON data from R. Data can be saved and transferred in a standardized, human-readable text format by understanding the JSON format and utilizing the relevant functionalities. This supports data interchange and storage requirements in a variety of applications.

5.5.3 Querying JSON Data

When querying JSON data with R, this usually means importing JSON data as R objects (such lists or data frames) and utilizing R's functions to subset, filter, or otherwise work with the data as needed (Liu, 2019). After the JSON data is loaded into R, you can query it by utilizing R's manipulation and subsetting features. After Emp.json data have been loaded, here are some instances of JSON data queries:

1. Accessing Specific Elements:

```
> # Querying JSON data
> jdata <- fromJSON("empdata.json")
> # Print the entire JSON data
> jdata
$EID
[1] "1" "2" "3" "4" "5" "6" "7" "8"

$Emp_Name
[1] "Rick"    "Dan"     "Michelle" "Ryan"    "Gary"     "Nina"
"Simon"    "Guru"

$Emp_Salary
[1] "623.3"  "515.2"  "611"     "729"     "843.25"    "578"
"632.8"  "722.5"

$Job_start_Date
[1] "4/1/2018"   "7/25/1998"  "10/16/2024" "12/12/2020"
"4/14/2015"  "8/25/2010"  "11/13/1999"
[8] "2/21/2017"

$Dept
[1] "HR"      "Accounts" "Finance"  "IT"      "Logistic" "HR"
"Accounts" "Logistic"

> # Access specific elements
> print(jdata$Emp_name)
NULL
> print(jdata$Dept)
[1] "HR"      "Accounts" "Finance"  "IT"      "Logistic" "HR"
"Accounts" "Logistic"
> print(jdata$Job_start_Date)
[1] "4/1/2018"   "7/25/1998"  "10/16/2024" "12/12/2020"
"4/14/2015"  "8/25/2010"  "11/13/1999"
[8] "2/21/2017"
> print(jdata$Emp_Name)
[1] "Rick"    "Dan"     "Michelle" "Ryan"    "Gary"     "Nina"
"Simon"    "Guru"
```

2. Data Subsetting: The approach to accessing specific elements in your JSON data depends on its structure. For straightforward JSON objects where json_data consists of a simple top-level array or list of objects, you can directly apply filters based on conditions. However, if json_data includes nested structures, accessing specific elements within these nested lists may require using indexing or the lapply() function.

```
> jdata$EID <- as.numeric(jdata$EID)
> # Filter based on condition (EID > 3)
> filtered_data <- lapply(jdata, function(x) x[jdata$EID > 3])
> print(filtered_data)
$EID
[1] 4 5 6 7 8

$Emp_Name
[1] "Ryan"  "Gary"  "Nina"  "Simon" "Guru"

$Emp_Salary
[1] 729.00 843.25 578.00 632.80 722.50

$Job_start_Date
[1] "12/12/2020" "4/14/2015"  "8/25/2010"  "11/13/1999"
"2/21/2017"

$Dept
[1] "IT"       "Logistic" "HR"       "Accounts" "Logistic"
```

To ensure accurate numerical comparisons, make sure the Employee ID(EID) component is transformed to numeric using as.numeric(), as shown above. Since json_data is a list, subset elements based on the criterion json_data$EID > 3 by iterating over each component (EID, Emp_Name, etc.) using lapply().The list filtered_data will only include elements that meet the requirement that EID > 3 for each component (EID, Emp_Name, etc.).

3. Use of Aggregate Functions: In R, numerical values expressed as characters (such as salary "623.3", "515.2") that are imported from JSON data are handled as strings (character class). These character representations must be converted into numeric values (numeric class) in order to perform numerical operations, such as salary averaging. Ensuring that numerical operations (such as averaging) produce accurate results requires converting textual representations to numeric. Data analysis and reporting tasks are made easier by the compatibility of numerical data formats with statistical and numerical functions in R.

```
> jdata$Emp_Salary <- as.numeric(jdata$Emp_Salary)
> print(jdata$Emp_Salary)
[1] 623.30 515.20 611.00 729.00 843.25 578.00 632.80 722.50
```

```
> avg_salary <- mean(jdata$Emp_Salary)
> avg_salary
[1] 656.8813
```

5.6 Import and Export of Relational Database (MySQL)

Data is stored in a structured and normalized format in a relational database management system. Relational databases are a type of database that can be used to store, create, retrieve, and perform other actions on data in the form of tables. We require extremely sophisticated and intricate SQL queries to complete statistical computing (Rockoff, 2021). SQL queries are needed to retrieve data from the enormous and massive amounts of data that are contained in tables. Data frames are the normal format in which the information is extracted when it comes from these databases. The data will become a typical R data set as soon as it is imported into the R environment from the database. With the aid of all the strong packages and functionalities, the data analyst may quickly evaluate or work with the data. In the context of R programming for database interaction, dbSendQuery() and dbGetQuery() are important functions provided in database interface packages.

Structure query language (SQL) is a widely used database language in relational database management systems (RDBMS) such as MySQL, Oracle, SQLite, and others (Liu et al., 2024). R has the capability to establish connections with various relational databases to retrieve records from them. The 'RMySQL' package facilitates establishing a connection between the R environment and a MySQL server (Boggaram Gopinath, 2009). Once the connection is established, the data set can be manipulated and analysed using a variety of functions. We have created two tables, as depicted in Figure 5.14. The 'stu' table includes attributes such as student ID (sid), student name (sname), address, and age, where sid serves as the primary key. The 'Course' table contains attributes course ID (cid), course name (cname), credit, and student ID (sid), with sid acting as a foreign key referencing the 'stu' table, and cid as the primary key.

5.6.1 dbSendQuery() and dbGetQuery()

Database interface packages like RMySQL or DBI offer two main functions, dbSendQuery() and dbGetQuery(), which are crucial when using R programming for database interactions. A SQL query can be sent to a database server using the dbSendQuery() function. It initiates the execution of the query and prepares to fetch results if applicable. The syntax of dbSendQuery() is mentioned below. You provide a connection object inside the parentheses and then write a valid SQL statement. However, dbSendQuery() itself does not

```
mysql> select * from stu;
+-----+--------+------------------+------+
| sid | sname  | address          | age  |
+-----+--------+------------------+------+
| 101 | megha  | noida 62         |   23 |
| 102 | Atharv | Noida sector 52  |   24 |
| 103 | Arjun  | Gurgaon          |   28 |
| 104 | Deepti | Dehradun         |   26 |
| 105 | Adi    | Goa              |   23 |
+-----+--------+------------------+------+
5 rows in set (0.00 sec)

mysql> select * from Course;
+-----+--------------------+--------+------+
| cid | cname              | credit | sid  |
+-----+--------------------+--------+------+
|  11 | R Programming      |      2 |  101 |
|  12 | Python Programming |      2 |  102 |
|  13 | Data Science       |      3 |  102 |
|  14 | Data Mining        |      3 |  103 |
|  15 | Cloud Computing    |      3 |  104 |
|  16 | Java Programming   |      3 |  105 |
|  17 | Data Analytics     |      3 |  104 |
+-----+--------------------+--------+------+
7 rows in set (0.00 sec)
```

FIGURE 5.14
Student, course relations in MySQL.

directly return results; instead, you need to use the fetch() method to retrieve records based on the SQL statements executed.

```
dbSendQuery(connection_statement, "SQL Statement")
```

The syntactical structure of dbGetQuery is shown below and returns a data frame as the query's outcome. However, the advantage of using the dbGetQuery() method is that it has a default implementation that calls dbSendQuery(), dbFetch(), and dbClearResult() to guarantee that the result is always freed. This implementation should work with the majority of backends.

```
dbGetQuery(connection_statement, "SQL Statement")
```

dbGetQuery() integrates the submission of the query and the retrieval of results into a single phase, in contrast to dbSendQuery(), which merely submits the query and gets ready for result retrieval. In R's database interface, the functions dbSendQuery() and dbGetQuery() have different uses. A comparison to show how they differ from one another is in Table 5.1.

5.6.2 Establishing Connection: R with MySQL

To interact with a MySQL database, you need to establish a connection between R and the database. This connection is created using the dbConnect() function provided by R. This function accepts input parameters

TABLE 5.1

Comparative Analysis of dbSendQuery() and dbGetQuery() Methods

Comparison Points	dbSendQuery()	dbGetQuery()
Purpose	A SQL query can be sent to a database server via dbSendQuery() to be executed there.	A SQL query can be sent to a database server using dbGetQuery(), which then retrieves the results and puts them into a R data frame.
Usage	When you wish to run SQL queries that retrieve data (SELECT) or alter data (INSERT, UPDATE, DELETE) without immediately fetching the results into R, you can use dbSendQuery().	When you anticipate a tiny result set that can be easily loaded into memory in R, you utilize dbGetQuery(). The query results are immediately returned as a R data frame, which may be utilized right away for additional processing or analysis.
Work Flow	To get the actual data rows from the database, you usually use fetch () or dbFetch() after running dbSendQuery().	If you want to quickly retrieve and work with data in R without having to deal with DBI result objects and manual fetching, use dbGetQuery().
Example	`> q<-dbSendQuery (mysqlconnection, "select * from stu")` `> d = fetch(q,n=-1)`	`q<-dbGetQuery (mysqlconnection, "select * from stu")`

such as username, password, database name, and host name. Let's explore an example to illustrate how dbConnect() is utilized to establish a database connection.

```
> install.packages("RMySQL") # Installing package 'RMySQL'
> library('RMySQL')
> # Connect MySQL to R
> mysqlconnection = dbConnect(RMySQL::MySQL(),dbname='db1',
host='localhost',port=3306,user='root',password='test123')
> #List the tables
> dbListTables(mysqlconnection)
[1] "course" "stu"
```

The code above uses the RMySQL package to connect from R to a MySQL database, RMySQL::MySQL() indicates the MySQL driver to connect to MySQL databases from within the RMySQL package (Ooms et al., 2024), 'dbname='db1" indicates the name of the database (db1) to which you wish to establish a connection, the hostname of the MySQL server is specified using host='localhost', 'port=3306' indicates the port number that the MySQL server is running on, user='root' indicates the username (root) that is utilized for MySQL database authentication and connection, and password='test123'

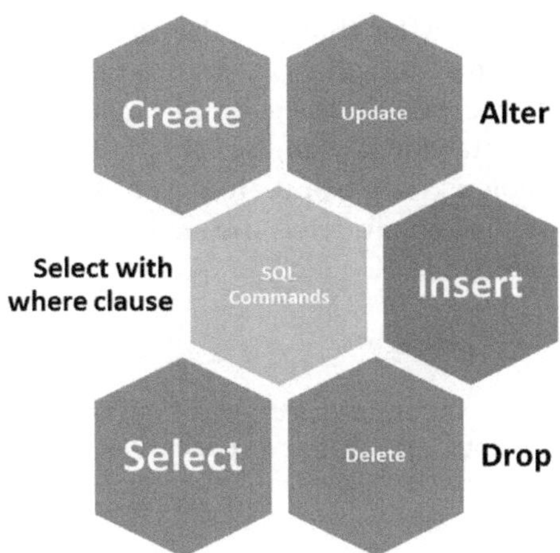

FIGURE 5.15
Basic SQL commands.

indicates the password (test123) linked to the username in order to facilitate authentication.

5.6.3 MySQL Commands

In R, you can execute SQL commands such as insert, delete, update, and more. To perform queries on a database, R offers the dbSendQuery() function. This function sends the query to MySQL for execution, and the resulting data set can be retrieved using the fetch() function in R. Figure 5.15 provides an overview of SQL commands.

Relational databases are managed and manipulated by SQL. A new database or table can be created with the CREATE command; to change the structure of an existing database or table, use the ALTER command; and to change already-existing records in a table, use the UPDATE command. A database's contents can be retrieved using the SELECT command. A table can have new records added to it using the INSERT command, To delete records from a table, use the DELETE command. To remove a whole database, table, or other database item, use the SQL DROP command. The selected object, along with all of its data and structure, is totally deleted from the database when you use the DROP command. This action cannot be undone.

5.6.4 Create, Populate, and Select Data with R Interface

You can create a new table, populate it, and retrieve data by interfacing between R and MySQL. First, you need to load the necessary packages to establish a connection between MySQL and R. Then, create the connection, as shown in the example. After that, use the 'dbSendQuery' method in R to create a new table. This method allows you to write SQL queries just like you would in the MySQL interface. In the example, a new table named 'faculty' is created with the attributes faculty_id, faculty_name, department, and experience. The faculty table is then populated, and data retrieval is performed using the SELECT command. To fetch the saved details from the tables, you can use the fetch() method in R.

```
> install.packages("RMySQL")
> library("RMySQL)
> mysqlconnection = dbConnect(RMySQL::MySQL(),dbname='db1',
host='localhost',port=3306,user='root',password='test123')
> dbListTables(mysqlconnection)
[1] "course" "stu"

> # Create Table with R Interface
> q1 = dbSendQuery(mysqlconnection, "create table faculty
(fid varchar(25) primary key, fname char(10),department
varchar(20),experience integer)")
> dbListTables(mysqlconnection)
[1] "course"  "faculty" "stu"

> # Populate Faculty Relation
> q2 = dbSendQuery(mysqlconnection,"insert into faculty
values('F101', 'Shreya','CSE',10)")
> q3 = dbSendQuery(mysqlconnection,"insert into faculty
values('F102', 'Aastha','CSE',2)")
> q4 = dbSendQuery(mysqlconnection,"insert into faculty
values('F103', 'Shikha','IT',16)")
> q5 = dbSendQuery(mysqlconnection,"insert into faculty
values('F104', 'Anamika','IT',9)")

> # Data Retrieval
> res = dbSendQuery(mysqlconnection,"select * from faculty")
> q6 = fetch(res)
> q6
   fid   fname department experience
1 F101  Shreya        CSE         10
2 F102  Aastha        CSE          2
3 F103  Shikha         IT         16
4 F104 Anamika         IT          9
```

You can also verify the newly created table using the MySQL interface. Figure 5.16 displays all the tables in the existing database 'db1', and Figure 5.17 shows the content stored in all the tables.

```
mysql> use db1;
Database changed
mysql> show tables;
+---------------+
| Tables_in_db1 |
+---------------+
| course        |
| faculty       |
| stu           |
+---------------+
3 rows in set (0.00 sec)
```

FIGURE 5.16
Tables in database 'db1'.

```
mysql> select * from stu;
+-----+--------+------------------+-----+
| sid | sname  | address          | age |
+-----+--------+------------------+-----+
| 101 | megha  | noida 62         |  23 |
| 102 | Atharv | Noida sector 52  |  24 |
| 103 | Arjun  | Gurgaon          |  28 |
| 104 | Deepti | Dehradun         |  26 |
| 105 | Adi    | Goa              |  23 |
+-----+--------+------------------+-----+
5 rows in set (0.05 sec)
mysql> select * from course;
+-----+--------------------+--------+-----+
| cid | cname              | credit | sid |
+-----+--------------------+--------+-----+
|  11 | R Programming      |      2 | 101 |
|  12 | Python Programming |      2 | 102 |
|  13 | Data Science       |      3 | 102 |
|  14 | Data Mining        |      3 | 103 |
|  15 | Cloud Computing    |      3 | 104 |
|  16 | Java Programming   |      3 | 105 |
|  17 | Data Analytics     |      3 | 104 |
+-----+--------------------+--------+-----+
7 rows in set (0.05 sec)
mysql> select * from faculty;
+------+---------+------------+------------+
| fid  | fname   | department | experience |
+------+---------+------------+------------+
| F101 | Shreya  | CSE        |         10 |
| F102 | Aastha  | CSE        |          2 |
| F103 | Shikha  | IT         |         16 |
| F104 | Anamika | IT         |          9 |
+------+---------+------------+------------+
```

FIGURE 5.17
Relations content stored in database 'db1'.

5.6.5 Data Retrieval Based on Condition

Using R to get data from a MySQL database depends on predetermined criteria, establishing a connection, submitting a query, and retrieving the results. To obtain the result, any legitimate select query can be passed. In the processed query, we retrieved data from the faculty table based on the condition that the department is 'IT'.

```
> # Data Selection based on condition
> q7 = dbSendQuery(mysqlconnection, "select * from faculty
where department='IT'")
> dfm = fetch(q7, n = -1)
> dfm
   fid   fname department experience
1 F103  Shikha         IT         16
2 F104 Anamika         IT          9
```

In a different instance, we have taken data extraction based on the requirement that student names be shown in descending order.

```
> # Data Sorting
> q8<-dbSendQuery(mysqlconnection, "select * from stu order by
sname desc")
> dfm = fetch(q8,n=-1)
> dfm
  sid  sname          address age
1 101  megha        noida 62  23
2 104 Deepti        Dehradun  26
3 102 Atharv Noida sector 52  24
4 103  Arjun         Gurgaon  28
5 105    Adi             Goa  23
```

The following code shows an example of using an aggregate function, such as max, min, count, sum, etc. Aggregate functions calculate a single value after applying a set of values.

```
> # Example of Aggregate Function Count
> q9<-dbSendQuery(mysqlconnection, "select count(experience)
from faculty where department='CSE'")
> dfm = fetch(q9,n=-1)
> dfm
  count(experience)
1                 2

> # Example of Aggregate Function Max,  Min
> q10<-dbSendQuery(mysqlconnection, "select max(age) from stu")
> dfm = fetch(q10,n=-1)
> dfm
  max(age)
1       28
```

```
> q10<-dbSendQuery(mysqlconnection, "select min(age) from stu")
> dfm = fetch(q10,n=-1)
> dfm
  min(age)
1       23

> # Example of Aggregate Function Sum
> q11<-dbSendQuery(mysqlconnection, "select sum(age) from stu")
> dfm = fetch(q11,n=-1)
> dfm
  sum(age)
1      124
```

In a WHERE clause, the LIKE operator is used to look for a given pattern in a column. We are searching for a pattern where the student name begins with "a," like in this example.

```
> # Example of LIKE operator
> q12<-dbSendQuery(mysqlconnection, "select * from stu where
sname like 'a%'")
> dfm = fetch(q12,n=-1)
> dfm
  sid  sname         address age
1 102 Atharv Noida sector 52  24
2 103  Arjun         Gurgaon  28
3 105    Adi             Goa  23
```

SQL joins are used to extract data from two or more relations. Combining rows from two or more tables based on a shared column between them is done with a JOIN clause. SQL joins are essential in relational database management because they enable you to merge rows from many tables by using a common column between them. Joins facilitate normalization, which divides data into similar tables to eliminate redundancy and preserve data integrity. Additionally, joins can improve performance by quickly retrieving related data in a single query as opposed to several.

When a primary key from one table is referenced in another table, it is called a foreign key. In our case, we have established two tables, 'stu' and 'course,' where sid serves as the primary key in the student table and as a foreign key in the course table. The following R code demonstrates how we can extract data from two relations.

```
> # Join Example
> q13<-dbSendQuery(mysqlconnection, "select * from stu,course
where age>23 and stu.sid=course.sid")
> dfm = fetch(q13,n=-1)
> dfm
  sid  sname         address age cid              cname  credit sid
1 102 Atharv Noida sector 52  24  12 Python Programming       2 102
2 102 Atharv Noida sector 52  24  13       Data Science       3 102
```

```
3 103  Arjun     Gurgaon 28  14        Data Mining     3 103
4 104 Deepti    Dehradun 26  15    Cloud Computing     3 104
5 104 Deepti    Dehradun 26  17     Data Analytics     3 104
```

As seen in the example above, a JOIN clause is used to retrieve data where the age is greater than 23, displaying content from both tables.

5.6.6 Data Updation

In SQL, updating data entails making changes to already-existing records in a table. The UPDATE statement is used to modify existing records in a table based on specified conditions. We have changed the new address = "New Delhi" where address = "Goa", as the code indicates.

```
> # Data Updation
> #Data before updation
> q14<-dbSendQuery(mysqlconnection, "select * from stu")
> dfm = fetch(q14,n=-1)
> dfm
  sid  sname          address age
1 101  megha         noida 62  23
2 102 Atharv Noida sector 52  24
3 103  Arjun          Gurgaon  28
4 104 Deepti         Dehradun  26
5 105    Adi              Goa  23

> # Update Command to update Data
> q15<-dbSendQuery(mysqlconnection, "update stu set address =
'New Delhi' where address = 'Goa'")
> q14<-dbSendQuery(mysqlconnection, "select * from stu")
> dfm = fetch(q14,n=-1)
> dfm
  sid  sname          address age
1 101  megha         noida 62  23
2 102 Atharv Noida sector 52  24
3 103  Arjun          Gurgaon  28
4 104 Deepti         Dehradun  26
4 105    Adi        New Delhi  23
```

5.6.7 Data Deletion

Deleting data in SQL involves removing records from a table based on specified conditions. Here's an example of how to use the dbSendQuery() function to pass the delete query and remove a particular row from the table.

```
> # student data before deletion
> q14<-dbSendQuery(mysqlconnection, "select * from stu")
> dfm = fetch(q14,n=-1)
```

```
> dfm
  sid  sname          address age
1 101  megha         noida 62  23
2 102 Atharv Noida sector 52  24
3 103  Arjun          Gurgaon  28
4 104 Deepti         Dehradun  26
5 105    Adi       New Delhi  23

> # Delete Command
> q15<-dbSendQuery(mysqlconnection, "delete from stu where
sid=105")
> dfm = fetch(q15,n=-1)

> # student data after deletion
> q14<-dbSendQuery(mysqlconnection, "select * from stu")
> dfm = fetch(q14,n=-1)
> dfm
  sid  sname          address age
1 101  megha         noida 62  23
2 102 Atharv Noida sector 52  24
3 103  Arjun          Gurgaon  28
4 104 Deepti         Dehradun  26
```

We can also delete any relation with its schematic structure with the help of the 'drop' command. Here is the structure of the 'test1' table (Figure 5.18, and we need to delete this table along with all its inserted data.

Below is the syntax for dropping a table:

```
> # Drop Table
> q16<-dbSendQuery(mysqlconnection, "drop table test1")
> q16<-dbSendQuery(mysqlconnection, "select * from test1")
Error in dbSendQuery(mysqlconnection, "select * from test1") :
  could not run statement: Table 'db1.test1' doesn't exist
```

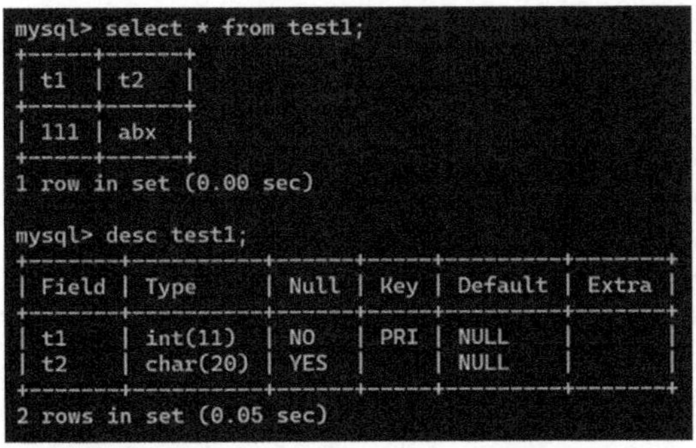

```
mysql> select * from test1;
+-----+-----+
| t1  | t2  |
+-----+-----+
| 111 | abx |
+-----+-----+
1 row in set (0.00 sec)

mysql> desc test1;
+-------+----------+------+-----+---------+-------+
| Field | Type     | Null | Key | Default | Extra |
+-------+----------+------+-----+---------+-------+
| t1    | int(11)  | NO   | PRI | NULL    |       |
| t2    | char(20) | YES  |     | NULL    |       |
+-------+----------+------+-----+---------+-------+
2 rows in set (0.05 sec)
```

FIGURE 5.18
Schematic structure of table 'test1'.

As exemplified earlier, after dropping the table, if we attempt to execute a SELECT statement to retrieve its contents, an error message will appear indicating that the table does not exist.

5.6.8 SQL Commands with dbgetQuery()

The method dbgetQuery() is especially helpful for searches that yield a manageable amount of results that fit in memory. With this method, you can easily run SQL queries against databases from within R and get the results right into the R environment. For data analysis tasks involving database interactions, its efficiency and simplicity make it the go-to option. Below are several examples illustrating how to connect R with MySQL and execute various SQL commands using this function.

```
> # Table Creation
> q15<-dbGetQuery(mysqlconnection,"create table Exam(examname
varchar(20) primary key,examdate date,Stream
varchar(20),Course varchar(20))")

> # Insert Data
> q16<-dbGetQuery(mysqlconnection,"insert into Exam
values('T1','2024-01-01','CSE','BTech')")
> q16<-dbGetQuery(mysqlconnection,"insert into Exam
values('T2','2024-03-10','CSE','MTech')")

> # Display Data
> q17<-dbGetQuery(mysqlconnection,"select * from Exam")
> q17
  examname   examdate Stream Course
1       T1 2024-01-01    CSE  BTech
2       T2 2024-03-10    CSE  MTech

> # Update Data
> q18<-dbGetQuery(mysqlconnection,"update Exam set Stream='IT'
where Course='MTech'")
> q<-dbGetQuery(mysqlconnection,"select * from Exam")
> q
  examname   examdate Stream Course
1       T1 2024-01-01    CSE  BTech
2       T2 2024-03-10     IT  MTech

> # Delete Data
> q19<-dbGetQuery(mysqlconnection,"delete from Exam where
examname='T1'")
> q<-dbGetQuery(mysqlconnection,"select * from Exam")
> q
  examname   examdate Stream Course
1       T2 2024-03-10     IT  MTech
```

5.6.9 Data Analytics on SQL Data with R

R can be used to connect to a database, retrieve data, conduct analysis, and visualize the findings. Integrating SQL data with R for data analytics allows you to leverage R's powerful statistical and visualization capabilities alongside SQL's robust data storage and querying capabilities. This combination allows for thorough data exploration, analysis, and display, which makes it appropriate for a variety of analytical tasks ranging from intricate modelling and reporting to basic data summaries. For data analytics to be performed on SQL data, this is a general rundown of how this procedure normally operates:

Step I: Loading the Packages

```
> # Data Analytics Example on SQL Data
> library(DBI)
> library(dplyr)
```

Step II: Connection between R and MySQL

```
>con=dbConnect(RMySQL::MySQL(),dbname='db1',host='localhost',
port=3306,user='root',password='test123')
```

Step III: Data Loading for Manipulation and Analysis

```
> q<-dbGetQuery(con,"select * from workman")
> q
  wid  wname dept salary
1  20  vikas  CSE  50000
2  21 shreya  CSE  78000
3  22 shipra   IT  87000
4  23   neha   IT  95000
5  24  sneha  ECE  95000
6  25   amit  ECE  87000
```

Step IV: Filtering and Summarizing Data

```
> q<-dbGetQuery(con,"select * from workman where
salary>50000")
> q
  wid  wname dept salary
1  21 shreya  CSE  78000
2  22 shipra   IT  87000
3  23   neha   IT  95000
4  24  sneha  ECE  95000
5  25   amit  ECE  87000

> # Calculate average age
> sal_avg <-q %>% summarise(avg_age = mean(salary,
na.rm = TRUE))
> sal_avg
  avg_age
1   88400
```

```
> # Count number of students
> workman_count <- nrow(q)
> workman_count
[1] 5
```

Step IV: Aggregating Data by Groups

```
> #Aggregate workman data department wise
> q<-dbGetQuery(con,"select dept,avg(salary),count(*)
from workman group by dept")
> q
  dept avg(salary) count(*)
1  CSE       64000        2
2  ECE       91000        2
3   IT       91000        2
```

Step V: Complex Query and Data Manipulation

This executes a complex SQL query to summarize department-specific salary statistics for employees aged over 30 and with average salaries above $50,000.

```
> #Data summarization and manipultion
> q<-dbGetQuery(con,"SELECT dept, COUNT(*),AVG(salary) AS
avg_salary,MIN(salary) AS min_salary,MAX(salary) AS max_
salary FROM workman WHERE salary > 50000 GROUP BY dept
HAVING AVG(salary) > 50000")
> q
  dept COUNT(*) avg_salary min_salary max_salary
1  CSE        1      78000      78000      78000
2  ECE        2      91000      87000      95000
2   IT        2      91000      87000      95000
```

5.7 Case Study 1—Analysis of Sales of Electronic Products

You are employed for a retail business that deals in electronics. The business keeps a CSV file with sales information from the previous year. It is your responsibility to examine this data and draw conclusions that can guide company choices. Figure 5.19 shows the data arrangement, which is in CSV format.

	Order ID	Product	Quantity (	Price Each	Order Date	Purchase Address	Month	Sales	City	Hour
0	295665	Macbook	1	1700	30-12-2019 00:01	136 Church St, New	12	1700	New York	0
1	295666	LG Washir	1	600	29-12-2019 07:03	562 2nd St, New Yc	12	600	New York	7
2	295667	USB-C Cha	1	11.95	12-12-2019 18:21	277 Main St, New Y	12	11.95	New York	18
3	295668	27in FHD I	1	149.99	22-12-2019 15:13	410 6th St, San Frai	12	149.99	San Franc	15
4	295669	USB-C Cha	1	11.95	18-12-2019 12:38	43 Hill St, Atlanta, (	12	11.95	Atlanta	12
5	295670	AA Batteri	1	3.84	31-12-2019 22:58	200 Jefferson St, N	12	3.84	New York	22
6	295671	USB-C Cha	1	11.95	16-12-2019 15:10	928 12th St, Portlar	12	11.95	Portland	15
7	295672	USB-C Cha	2	11.95	13-12-2019 09:29	813 Hickory St, Dall	12	23.9	Dallas	9
8	295673	Bose Soun	1	99.99	15-12-2019 23:26	718 Wilson St, Dall:	12	99.99	Dallas	23
9	295674	AAA Batte	4	2.99	28-12-2019 11:51	77 7th St, Dallas, T)	12	11.96	Dallas	11
10	295675	USB-C Cha	2	11.95	13-12-2019 13:52	594 1st St, San Frar	12	23.9	San Franc	13

FIGURE 5.19
Electronic_Product sales data used for analysis.

Objective: Conduct a detailed analysis on the CSV data to address important business queries and give stakeholders useful information.

Solution:

Step I: Data Loading

In this step, data is loaded and first few records of Sales data are visualized in Figure 5.20.

```
> # Case Study 1
> # Load Data
> sdata <- read.csv("sales.csv", header = TRUE)
> head(sdata)
  X Order.ID                Product Quantity.Ordered Price.Each
1 0   295665      Macbook Pro Laptop                1    1700.00
2 1   295666      LG Washing Machine                1     600.00
3 2   295667    USB-C Charging Cable                1      11.95
4 3   295668        27in FHD Monitor                1     149.99
5 4   295669    USB-C Charging Cable                1      11.95
6 5   295670  AA Batteries (4-pack)                 1       3.84
           Order.Date                           Purchase.Address Month
1 2019-12-30 00:01:00    136 Church St, New York City, NY 10001    12
2 2019-12-29 07:03:00       562 2nd St, New York City, NY 10001    12
3 2019-12-12 18:21:00      277 Main St, New York City, NY 10001    12
4 2019-12-22 15:13:00       410 6th St, San Francisco, CA 94016    12
5 2019-12-18 12:38:00             43 Hill St, Atlanta, GA 30301    12
6 2019-12-31 22:58:00 200 Jefferson St, New York City, NY 10001    12
    Sales            City Hour
```

FIGURE 5.20
Sales data loading demonstration.

Step II: Data Exploration

Examine the unique values in categorical variables, the data set's structure (str()), and summary statistics (summary()). Gain insights into how sales are distributed throughout different goods, nations, and clientele groups (Figures 5.21–5.23).

```
> # Data Exploration
> # Data Summary
> summary(sdata)
       X            Order.ID          Product          Quantity.Ordered   Price.Each        Order.Date
 Min.   :    0   Min.   :141234   Length:185950      Min.   :1.000    Min.   :   2.99   Length:185950
 1st Qu.: 3894   1st Qu.:185831   Class :character   1st Qu.:1.000    1st Qu.:  11.95   Class :character
 Median : 7786   Median :230368   Mode  :character   Median :1.000    Median :  14.95   Mode  :character
 Mean   : 8340   Mean   :230418                      Mean   :1.124    Mean   : 184.40
 3rd Qu.:11872   3rd Qu.:275036                      3rd Qu.:1.000    3rd Qu.: 150.00
 Max.   :25116   Max.   :319670                      Max.   :9.000    Max.   :1700.00
     Month            Sales              City               Hour
 Min.   : 1.000   Min.   :   2.99   Length:185950      Min.   : 0.00
 1st Qu.: 4.000   1st Qu.:  11.95   Class :character   1st Qu.:11.00
 Median : 7.000   Median :  14.95   Mode  :character   Median :15.00
 Mean   : 7.059   Mean   : 185.49                      Mean   :14.41
 3rd Qu.:10.000   3rd Qu.: 150.00                      3rd Qu.:19.00
 Max.   :12.000   Max.   :3400.00                      Max.   :23.00
```

FIGURE 5.21
Sales data summary.

```
> # Structure of Data
> str(sdata)
'data.frame':   185950 obs. of  11 variables:
 $ X               : int  0 1 2 3 4 5 6 7 8 9 ...
 $ Order.ID        : int  295665 295666 295667 295668 295669 295670 29567
 $ Product         : chr  "Macbook Pro Laptop" "LG Washing Machine" "USB-
 $ Quantity.Ordered: int  1 1 1 1 1 1 1 2 1 4 ...
 $ Price.Each      : num  1700 600 11.9 150 11.9 ...
 $ Order.Date      : chr  "2019-12-30 00:01:00" "2019-12-29 07:03:00" "20
 $ Purchase.Address: chr  "136 Church St, New York City, NY 10001" "562 2
y, NY 10001" "410 6th St, San Francisco, CA 94016" ...
 $ Month           : int  12 12 12 12 12 12 12 12 12 12 ...
 $ Sales           : num  1700 600 11.9 150 11.9 ...
 $ City            : chr  " New York City" " New York City" " New York Ci
 $ Hour            : int  0 7 18 15 12 22 15 9 23 11 ...
```

FIGURE 5.22
Sales data structure.

```
> # Unique values in categorical variables
> unique(sdata$Product)
 [1] "Macbook Pro Laptop"         "LG Washing Machine"
 [3] "USB-C Charging Cable"       "27in FHD Monitor"
 [5] "AA Batteries (4-pack)"      "Bose SoundSport Headphones"
 [7] "AAA Batteries (4-pack)"     "ThinkPad Laptop"
 [9] "Lightning Charging Cable"   "Google Phone"
[11] "Wired Headphones"           "Apple Airpods Headphones"
[13] "Vareebadd Phone"            "iPhone"
[15] "20in Monitor"               "34in Ultrawide Monitor"
[17] "Flatscreen TV"              "27in 4K Gaming Monitor"
[19] "LG Dryer"
> unique(sdata$Country)
NULL
> unique(sdata$City)
[1] " New York City" " San Francisco" " Atlanta"       " Portland"
[5] " Dallas"        " Los Angeles"   " Boston"        " Austin"
[9] " Seattle"
```

FIGURE 5.23
Unique categorical values.

Step III: Business Insights

Determine the total income and units sold from sales. List the top-selling goods and cities. Examine historical sales trends (such as quarterly or monthly data) (Figures 5.24 and 5.25).

```
> #Business Insights:
> # Total revenue and units sold
> total_revenue <- sum(sdata$Quantity.Ordered * sdata$Price.Each)
> total_revenue
[1] 34492036
> total_units_sold <- sum(sdata$Quantity.Ordered)
> total_units_sold
[1] 209079
```

FIGURE 5.24
Hidden business insights from data.

```
> # Sales trends over time
> sdata$Order.Date <- as.Date(sdata$Order.Date, format = "%Y-%m-%d")
> sdata$Month <- format(sdata$Order.Date, "%Y-%m")
> monthly_sales <- aggregate(Quantity.Ordered ~ Month, data = sdata, sum)
> monthly_sales
     Month Quantity.Ordered
1  2019-01            10862
2  2019-02            13449
3  2019-03            17005
4  2019-04            20558
5  2019-05            18667
6  2019-06            15253
7  2019-07            16072
8  2019-08            13448
9  2019-09            13109
10 2019-10            22703
11 2019-11            19798
12 2019-12            28114
13 2020-01               41
```

FIGURE 5.25
Sales trend over time.

Step IV: Order Segmentation
Segment your orders according to their spending patterns (e.g., frequency, total spending). Determine prospective orders and high-value orders for focused marketing initiatives (Figure 5.26).

```
> # Order Segmentation
> Order_summary <- aggregate(Quantity.Ordered ~Order.ID, data = sdata, sum)
> high_value_orders <- head(sort(Order_summary$Quantity.Ordered, decreasing = TRUE), 10)
> high_value_orders
 [1] 9 9 9 8 8 8 8 8 8 8
```

FIGURE 5.26
Sales order segmentation.

Step V: Decision Support
Provide suggestions and actionable insights that can be put into practice based on the analysis. Plots and succinct summaries can be used to visually represent the main findings (Figure 5.27).

This case study shows how to use R to analyse CSV data for a retail company's sales statistics. You can discern patterns, derive valuable insights, and efficiently assist company decisions by implementing these procedures. Expand and modify these studies in accordance with particular company needs, and investigate further using extra statistical techniques or visualizations to improve comprehension and reporting.

```
> ## Recommendations
> cat("Total Revenue: $", total_revenue, "\n")
Total Revenue: $ 34492036
> cat("Total Units Sold: ", total_units_sold, "\n\n")
Total Units Sold:  209079

> cat("Top Selling Products:\n")
Top Selling Products:
> print(top_products)

    USB-C Charging Cable Lightning Charging Cable    AAA Batteries (4-pack)
                   21903                    21658                     20641
> cat("\nMonthly Sales Trend:\n")

Monthly Sales Trend:
> print(monthly_sales)
    Month Quantity.Ordered
1 2019-01            10862
2 2019-02            13449
3 2019-03            17005
4 2019-04            20558
5 2019-05            18667
6 2019-06            15253
7 2019-07            16072
8 2019-08            13448
```

FIGURE 5.27
Decisive insights from sales data.

5.8 Case Study 2—Data Analysis on Structured Data of Research and Development (R&D) Division

You are employed by a tech corporation as a data analyst in the research and development (R&D) division. The following tables are kept up to date in the company's MySQL database:

1. Project includes information about R&D projects
   - pid: unique number assigned to every project.
   - pname: Project name.
   - start_date: Project start date.
   - end_date: Project end date.
   - budget: Budget allocated to the project.
2. Team includes information about the teams working on the projects
   - tid: Distinct identity for every team.
   - pid: Identifier that connects the team to a project.
   - ename: The employee name who is part of the team.
3. Employee (emp) includes employee details
   - eid: A distinct identity for every employee.
   - team_id: Identifier linking the member to a team
   - ename: Employee name.
   - role: The member's role within the team.

Objective: Get all of the project's specifics, including the members of the team and their details. Write out a report that summarizes the results for the following mentioned tasks:

1. Create all relations with primary and foreign key constraints in MySQL.
2. Connect to the MySQL database.
3. To merge data from the three tables, use join operations.
4. Get every project's full details, including the members of the involved team and their details.
5. Find out the total budget by team.
6. Determine the typical team size for each project.
7. Determine which projects have more teams or members than the predetermined amount.
8. Determine how each team's roles are distributed.
9. Write out a report that summarizes the results.

Solution:

1. Project, team, and employee relations were created in MySQL, and the tables are shown in Figures 5.28, 5.29, and 5.30 following data insertion.

```
mysql> select * from project;
+-----+-------------------+------------+------------+----------+
| pid | pname             | start_date | end_date   | budget   |
+-----+-------------------+------------+------------+----------+
| P11 | Airline           | 2018-04-06 | 2019-09-07 | 2.5e+006 |
| P12 | Bank              | 2020-10-10 | 2021-09-12 | 1.5e+006 |
| P13 | Stock Analysis    | 2020-05-09 | 2020-12-12 |   1e+006 |
| P14 | Drone Application | 2017-10-09 | 2019-12-12 |   4e+006 |
+-----+-------------------+------------+------------+----------+
```

FIGURE 5.28
Project table.

```
mysql> select * from team;
+---------+---------+-----+
| team_id | ename   | pid |
+---------+---------+-----+
| T101    | Arjun   | P13 |
| T102    | Atharv  | P14 |
| T103    | Rehan   | P14 |
| T104    | Shobhit | P12 |
| T105    | Arpit   | P11 |
+---------+---------+-----+
```

FIGURE 5.29
Team table.

```
mysql> select * from emp;
+-----+---------+--------------------+---------+
| eid | ename   | role               | team_id |
+-----+---------+--------------------+---------+
| 100 | Arpit   | Team Lead          | T105    |
| 101 | Shobhit | Senior Programmer  | T104    |
| 102 | Rehan   | Business Analyst   | T103    |
| 103 | Arjun   | Project Manager    | T101    |
| 104 | Atharv  | Project Manager    | T102    |
+-----+---------+--------------------+---------+
```

FIGURE 5.30
Employee table.

2. Load and install the required R packages and establish the MySQL database and R Connection. List all tables stored in database 'rd'.

```
> # Set up the MySQL connection
> library(RMySQL)
> library(dplyr)
> mysqlconnection = dbConnect(RMySQL::MySQL(),dbname='rd',
host='localhost',port=3306,user='root',password='test123')
> # List all tables
> dbListTables(mysqlconnection)
[1] "emp"     "project" "team"
```

3. To join the three tables according to their relationships, create SQL queries. Moreover, retrieve the amalgamated data into a R data frame.

```
> #Retrieve and join data from the MySQL database
> # Write SQL queries to join the three tables
> join_data<-dbGetQuery(mysqlconnection, "SELECT p.pid,
p.pname, p.start_date, p.end_date, p.budget, t.team_id,
t.ename, e.eid, e.ename, e.role FROM project p JOIN team t
ON p.pid = t.pid JOIN emp e ON t.team_id = e.team_id")
> join_data
```

Figure 5.31 presents the outcome of join command.

```
> join_data
  pid             pname start_date   end_date  budget team_id   ename eid   ename              role
1 P11           Airline 2018-04-06 2019-09-07 2500000    T105   Arpit 100   Arpit         Team Lead
2 P12              Bank 2020-10-10 2021-09-12 1500000    T104 Shobhit 101 Shobhit Senior Programmer
3 P14 Drone Application 2017-10-09 2019-12-12 4000000    T103   Rehan 102   Rehan  Business Analyst
4 P13    Stock Analysis 2020-05-09 2020-12-12 1000000    T101   Arjun 103   Arjun   Project Manager
5 P14 Drone Application 2017-10-09 2019-12-12 4000000    T102  Atharv 104  Atharv   Project Manager
```

FIGURE 5.31
Join operation on three relations.

4. Analyse and compile the information by project. The project summary should be printed. Furthermore, locating and printing specific team and member details for every project (Figure 5.32).

```
> #  Perform the analysis
> join_data<-dbGetQuery(mysqlconnection, "SELECT p.pid,
p.pname, p.start_date, p.end_date, p.budget, t.team_id,
t.ename, e.eid, e.role FROM project p JOIN team t ON
p.pid = t.pid JOIN emp e ON t.team_id = e.team_id")
> num_members = n_distinct(eid)
> project_summary <- join_data %>% group_by(pid, pname,
start_date, end_date, budget) %>%
+     summarise(
+         num_teams = n_distinct(team_id),
+         num_members = n_distinct(eid)
+     )
`
> # Print the project summary
> cat("Project Summary:\n")
Project Summary:
> print(project_summary)
# A tibble: 4 × 7
# Groups: pid, pname, start_date, end_date  [4]
  pid   pname            start_date end_date    budget num_teams num_members
  <chr> <chr>            <chr>      <chr>        <dbl>     <int>       <int>
1 P11   Airline          2018-04-06 2019-09-07 2500000         1           1
2 P12   Bank             2020-10-10 2021-09-12 1500000         1           1
3 P13   Stock Analysis   2020-05-09 2020-12-12 1000000         1           1
4 P14   Drone Application 2017-10-09 2019-12-12 4000000        2           2

> # Detailed team and member information for each project
> detailed_info <- join_data %>% arrange(pid, team_id, eid)
> detailed_info
```

```
> detailed_info
  pid             pname start_date   end_date  budget team_id   ename eid              role
1 P11           Airline 2018-04-06 2019-09-07 2500000    T105   Arpit 100         Team Lead
2 P12              Bank 2020-10-10 2021-09-12 1500000    T104 Shobhit 101 Senior Programmer
3 P13    Stock Analysis 2020-05-09 2020-12-12 1000000    T101   Arjun 103   Project Manager
4 P14 Drone Application 2017-10-09 2019-12-12 4000000    T102  Atharv 104   Project Manager
5 P14 Drone Application 2017-10-09 2019-12-12 4000000    T103  Rehan 102  Business Analyst
```

FIGURE 5.32
Detailed Information for each project.

5. Combine the tables for projects and teams, group by team ID, and add up the budgets.

```
> #Calculate the total budget by team
> total_budget_by_team <- join_data %>% group_by
(team_id, ename) %>% summarise(total_budget = sum
(budget, na.rm = TRUE))
> print(total_budget_by_team)
# A tibble: 5 × 3
# Groups:   team_id [5]
  team_id  ename   total_budget
     <chr>  <chr>          <dbl>
1    T101   Arjun        1000000
2    T102  Atharv        4000000
3    T103   Rehan        4000000
4    T104 Shobhit        1500000
5    T105   Arpit        2500000
```

6. Add a join to all three tables, group by team and project IDs, and determine the average number of members.

```
> #Find the average number of members per team per project
> avg_members_per_team <- join_data %>% group_by
(pid, pname, team_id, ename) %>% summarise
(num_members = n_distinct(eid)) %>% summarise
(avg_members = mean(num_members))
> print(avg_members_per_team)
# A tibble: 5 × 4
# Groups:   pid, pname [4]
  pid   pname             team_id avg_members
  <chr> <chr>             <chr>     <dbl>
1 P11   Airline           T105      1
2 P12   Bank              T104      1
3 P13   Stock Analysis    T101      1
4 P14   Drone Application T102      1
5 P14   Drone Application T103      1
```

7. Connect the three tables together and sort projects by pid and filter them according to the number of teams or participants.

```
> #Identify projects with more than a specified number of
teams or members
> threshold_teams <- 2
> threshold_members <- 2
> projects_with_many_teams <- join_data %>% group_by
(pid, pname) %>% summarise(num_teams = n_distinct
(team_id)) %>% filter(num_teams > threshold_teams)
```

```
> print(projects_with_many_teams)
# A tibble: 0 × 3
# Groups:   pid [0]
# i 3 variables: pid <chr>, pname <chr>, num_teams <int>

> projects_with_many_members <- join_data %>% group_by
(pid, pname) %>% summarise(num_members = n_distinct(eid))
%>% filter(num_members > threshold_members)
`summarise()` has grouped output by 'pid'. You can
override using the `.groups` argument.
> cat("\nProjects with More Than", threshold_teams,
"Teams:\n")

Projects with More Than 2 Teams:
> print(projects_with_many_teams)
# A tibble: 0 × 3
# Groups:   pid [0]
# i 3 variables: pid <chr>, pname <chr>, num_teams <int>
>
> cat("\nProjects with More Than", threshold_members,
"Members:\n")

Projects with More Than 2 Members:
> print(projects_with_many_members)
# A tibble: 0 × 3
# Groups:   pid [0]
# i 3 variables: pid <chr>, pname <chr>, num_members <int>
```

8. Join the tables for members and teams. Count the number of people in each role after grouping by team_id and role.

```
> #Calculate the distribution of roles within each team
> role_distribution <- join_data %>% group_by(team_id,
ename, role) %>% summarise(num_members = n())
> cat("\nRole Distribution within Each Team:\n")

Role Distribution within Each Team:
> print(role_distribution)
# A tibble: 5 × 4
# Groups:   team_id, ename [5]
  team_id ename   role               num_members
  <chr>   <chr>   <chr>                    <int>
1 T101    Arjun   Project Manager              1
2 T102    Atharv  Project Manager              1
3 T103    Rehan   Business Analyst             1
4 T104    Shobhit Senior Programmer            1
5 T105    Arpit   Team Lead                    1
```

9. Write all the summaries to a CSV file and close the connection.

```
> # Write the enhanced analysis results to CSV files
> write.csv(total_budget_by_team, "total_budget_by_team.
csv", row.names = FALSE)
> write.csv(avg_members_per_team, "avg_members_per_team.
csv", row.names = FALSE)
> write.csv(projects_with_many_teams, "projects_with_many_
teams.csv", row.names = FALSE)
> write.csv(projects_with_many_members, "projects_with_
many_members.csv", row.names = FALSE)
> write.csv(role_distribution, "role_distribution.csv",
row.names = FALSE)
> r1<-read.csv("total_budget_by_team.csv")
> r1
  team_id   ename total_budget
1    T101   Arjun      1000000
2    T102  Atharv      4000000
3    T103   Rehan      4000000
4    T104 Shobhit      1500000
5    T105   Arpit      2500000
> r2<-read.csv("avg_members_per_team.csv")
> r2
  pid               pname team_id avg_members
1 P11             Airline    T105           1
2 P12                Bank    T104           1
3 P13      Stock Analysis    T101           1
4 P14 Drone Application    T102           1
5 P14 Drone Application    T103           1
> r3<-read.csv("projects_with_many_teams.csv")
> r3
[1] pid       pname     num_teams
<0 rows> (or 0-length row.names)
> r4<-read.csv("projects_with_many_members.csv")
> r4
[1] pid         pname       num_members
<0 rows> (or 0-length row.names)
> r5<-read.csv("role_distribution.csv")
> r5
  team_id   ename               role num_members
1    T101   Arjun    Project Manager           1
2    T102  Atharv    Project Manager           1
3    T103   Rehan   Business Analyst           1
4    T104 Shobhit Senior Programmer           1
5    T105   Arpit          Team Lead           1

> # Close Database Connection
> dbDisconnect(mysqlconnection)
```

5.9 Summary

- R's file handling is a basic feature that makes it possible to manage and manipulate data that is stored in files.
- One of the fundamental tasks of file management in R is creating a file. Understanding how to construct files is crucial for a variety of data-related tasks, as they function similarly to containers for data.
- R has a wide range of functions for importing and exporting many types of data, including databases (like MySQL) and CSV, Excel, and JSON. These functions provide customization options for data types, encoding, headers, and delimiters. Examples of these functions include read.csv(), read_excel(), fromJSON(), and dbReadTable().
- R can import data from CSV files for manipulation and analysis. Make use of the read.csv() function, which produces a data frame by reading a file in table format.
- read.csv() method is popular for importing tabular data because of its ease of use and cross-platform compatibility.
- Make use of the write.csv() function, which creates a CSV file from a R data frame. R data frames can be exported to CSV files for storing and sharing data.
- Options include controlling quoting, handling decimal points, adding/removing row names, and specifying delimiters with the write.csv() method.
- Import data from Excel spreadsheets by using the read_excel() function from the readxl package. Customization options include specifying data types, ranges, and sheet names.
- Data frames can be exported to Excel format using the write_xlsx() function from the writexl package. It is possible to write data with formatting to several sheets.
- Employ the fromJSON() function from the jsonlite package to import hierarchical data from JSON files. Additionally, manage arrays and hierarchical structures.
- Use the toJSON() function from the jsonlite package to export lists or data frames to JSON format.
- Options for handling special characters and attractive printing are available toJSON() function from the jsonlite package.
- R offers the ability to connect to structured data formats. R provides the ability to connect to numerous databases, including PostgreS, Oracle, and MySQL.
- Creating a connection to a MySQL database is the initial step in loading data from MySQL. To establish a connection, use the RMySQL package's dbConnect() function.

- The dbGetQuery() function is used to manage joins and execute intricate queries.
- The dbWriteTable() method is used to write data to MySQL and is useful for inserting or updating data from R into MySQL tables.
- R has the capability to maintain data integrity, control mistakes that arise during file operations, send out educational messages, and back up plans.
- R's data manipulation features can be used to clean and convert a variety of file types.
- Large data set read/write operations can be optimized with R and employ effective data handling strategies to control memory consumption.
- R's extensive package library, which includes tools like dplyr, tidyr, and stringr, allows for thorough data transformation and cleaning. With the help of these packages, data frames can be filtered, aggregated, reshaped, and altered, making it possible to deal with missing values, outliers, and inconsistencies in an efficient manner.
- R programming versatility can help data scientists to quickly access and edit data from a wide range of sources, which makes it easier to integrate and preprocess data in numerous forms.

EXERCISES

I. Multiple-Choice Questions

1. Which of the subsequent functions requires more than one column?
   a. gather()
   b. separate()
   c. spread()
   d. All of the above
2. Dplyr's operational capabilities are comprised of which of the following?
   a. a set of tools for effectively altering data sets
   b. quicker, more reliable API, and easier to use
   c. direct database access and analysis enabling more straightforward big data management
   d. All of the above

3. Which R function allows the import of a CSV file?
   a. read.csv.data()
   b. read.table()
   c. read.csv()
   d. read_csv()
4. Which R function reads a single line from a CSV file?
   a. read.csv.data()
   b. read.table()
   c. read.csv()
   d. read_csv()
5. How can the file structure be checked in R?
   a. str()
   b. data.structure()
   c. structure()
   d. struct()
6. What kind of output does the summary() function produce?
   a. a vector with numbers
   b. a list with each column's summary statistics
   c. a thorough report within the console
   d. All of the above
7. Which R package is most frequently used to manipulate data?
   a. tidyr
   b. openxlsx
   c. vector
   d. dplyr
8. Which R package is most frequently used to manipulate data?
   a. tidyr
   b. openxlsx
   c. vector
   d. dplyr
9. Which parameter in the read.csv() function indicates whether column names appear on the CSV file's first line?
   a. header
   b. sep
   c. quote
   d. file

10. What is the method for connecting to a MySQL database using the RMySQL package?
    a. connectMySQL()
    b. openConnection()
    c. dbConnect()
    d. mySQLConnect()
11. What function is used to convert R objects to JSON format within the jsonlite package?
    a. connect_json()
    b. toJSON()
    c. to_JSON()
    d. write_JSON()
12. What function is used to convert R objects to JSON format within the jsonlite package?
    a. connect_json()
    b. toJSON()
    c. to_JSON()
    d. write_JSON()
13. Which argument determines whether row names are included in the output file when using the write.csv() function?
    a. col.names
    b. file
    c. x
    d. row.names
14. Which R package is most frequently used to query and connect to MySQL databases?
    a. MySQLR
    b. R_MySQL
    c. RMySQL
    d. Sqlite

II. Brief Questions

1. Describe the significance of file management in workflows related to data science. Give instances of the file formats that are frequently seen in data science projects.

2. Explain the differences between the 'openxlsx' and 'readxl' package to import Excel data.
3. Describe the essential R functions for reading and writing CSV files. Discuss their parameters and how they make data manipulation easier.
4. Compare and contrast the read.table() and read.csv() functions in R. Which functions would you utilize and when, and what are the main distinctions between them?
5. Explain the use of R packages such as readxl, writexl, and openxlsx for working with Excel files. What benefits do these packages offer over the standard R functions?
6. Describe procedure for establishing a connection from R to a MySQL database. Which R packages are most frequently used for this, and what are the main purposes of each package?
7. Describe the factors to keep in mind while using R to work with big data sets. How may efficient file handling methods and packages such as data.table help with the management of massive data sets?
8. Describe the steps involved in importing JSON data into R. Explain the jsonlite package's purpose and the main methods used to parse JSON data.
9. What features of R enable the smooth integration of various data sets into R for in-depth analysis without the need for additional tools?
10. List the relevant packages in R that are needed to expedite processes like cleaning, converting, and combining data sets from various sources, guaranteeing data consistency and quality prior to analysis.

III. Coding Exercise

1. Develop a user-defined function to remove data from a specific range of cells in an Excel file. This function will accept parameters such as the workbook, sheet number or name, starting row, ending row, starting column, and ending column. For instance, the function clearRange (wb, sheet, startRow, endRow, startCol, endCol) will modify the Excel sheet by excluding the specified range of cells. As an example, calling clearRange (wb, sheet="Sheet1", 3, 5, 6, 7) would remove the data from rows 3 to 5 and columns 6 to 7 in 'Sheet1' (Figures 5.33 and 5.34).

```
   S.No    Name Age Department Salary      City Experience
1     1    Jaya  42    Finance  50000     Noida         20
2     2   Tanya  26         IT  65000     Delhi          2
3     3   Tanya  26         IT  65000 Ghaziabad          2
4     4  Atharv  27        R&D 300000  Haridwar          2
5     5  arpita  34    Finance  50000  Dehradun          7
6     6  akshit  56         IT  65000 Bangalore         25
7     7    neha  22         IT  65000 Hyderabad          0
8     8   sneha  20        R&D 300000   Chennai          0
9     9  snehil  40    Finance  50000     Delhi         10
10   10  gunjan  50        R&D  34556     Noida         20
11   11 girdhar  33         IT  37899 Bangalore          8
```

FIGURE 5.33
Input

```
   S.No    Name Age Department Salary      City Experience
1     1    Jaya  42    Finance  50000     Noida         20
2     2   Tanya  26         IT  65000
3     3   Tanya  26         IT  65000
4     4  Atharv  27        R&D 300000
5     5  arpita  34    Finance  50000  Dehradun          7
6     6  akshit  56         IT  65000 Bangalore         25
7     7    neha  22         IT  65000 Hyderabad          0
8     8   sneha  20        R&D 300000   Chennai          0
9     9  snehil  40    Finance  50000     Delhi         10
10   10  gunjan  50        R&D  34556     Noida         20
11   11 girdhar  33         IT  37899 Bangalore          8
```

FIGURE 5.34
Output

2. Create a custom function to transform a 'Data Frame' object into 'XML' format using the 'xml2' package.
3. Write a user-defined function to convert 'List' type to 'XML' type using package 'xml2'.
4. A CSV file with the name 'heart.csv' that includes information on a number of patient characteristics and heart health has been provided to a data engineer for analysis. The data set has columns pertaining to target (existence of heart disease), sex, chol (cholesterol level), thalach (highest heart rate reached), age, and cp (kind of chest pain). Load this data set into R, carry out the following data operations and analysis, and produce insights.
   a. Load 'heart.csv' file into R and save it as the heart_data data frame. Make sure that each column has the proper data type.
   b. Find statistics like total count of records and summarize data.
   c. Determine the percentage of patients in the data set with heart disease (class_variable = 1). Output the percentage of the proportion.

   d. Examine the connection between average cholesterol (chol) and the type of chest pain (cp). Create R code to compute the average cholesterol level (chol) for every type of chest pain, group the data by cp, and provide a summary of your results.

5. You have been provided with an XML file named 'patient_data.xml' that contains detailed information about patients' medical records. Nodes representing the patient's age, gender, medical history (diagnosis, treatment), and other relevant information are included in the XML structure. Perform the mentioned analytical tasks, extract pertinent data, carry out particular data changes, and produce insights.
   a. Extract pertinent data from the XML file 'patient_records.xml' into R, including patient demographics (age, gender), medical history (diagnosis, treatment), and any other relevant information. Store the information in a format that will allow for analysis.
   b. Determine and print the total number of patients in the data set.
   c. Determine which medical ailment is most frequently diagnosed in patients. Print the ailment and the total number of individuals who have been diagnosed with it.
   d. Analyse the data set's age distribution by gender to get a better understanding of any potential age-related trends or distinctions between patients who are male and female.

6. A 'products.json' file is provided to an analytical team, containing information about products from an online retailer. The structure of the JSON data is as follows (Figure 5.35).
   a. Extract the product data into R by using the appropriate package and methods.
   b. Determine and print the total number of patients in the data set.
   c. Determine the stock value in total for every category. Price * stock is how stock value is determined.
   d. Determine which product has the most stock value.
   e. Generate a summary table that lists all of the products and their average prices broken down by category.

7. Imagine a data set named Student.csv containing attributes such as Student_ID, Name, Age, and Grade. Fill in the blanks in the provided R code below to perform an analysis on this CSV data.
   a. students <- ....("students.csv", header = TRUE) # Read data
   b. average_age <- ....(students$Age) # Finding average age of students
   c. write.csv(.........(Average_Age = average_age), ".........", row.names = FALSE) # Write average age to new csv file named average_new.csv

```
[
  {
    "id": 1,
    "name": "Laptop",
    "category": "Electronics",
    "price": 75000,
    "stock": 55
  },
  {
    "id": 2,
    "name": "Smartphone",
    "category": "Electronics",
    "price": 40000,
    "stock": 50
  },
  {
    "id": 3,
    "name": "Laptop Table",
    "category": "Furniture",
    "price": 4000,
    "stock": 25
  },
  {
    "id": 4,
    "name": "TV",
    "category": "Electronics",
    "price": 380000,
    "stock": 35
  }
]
```


FIGURE 5.35
Product .json sample data.

d. summary_grade <- summary(..............) # Find out statistical summary of grade
e. average_age_by_grade <- ....(Age ~ Grade, data = students, FUN = ...) # Compute average age by grade level

References

Boggaram Gopinath, C. M. (2009). Efficient in-database analytics through embedding MySQL into R. NC State University Libraries.

Brahmia, Z., Hamrouni, H., & Bouaziz, R. (2020). XML data manipulation in conventional and temporal XML databases: A survey. Computer science review, 36, 100231, Elsevier.

Chan, B. K., & Chan, B. K. (2018). Data analysis using R programming. In Biostatistics for Human Genetic Epidemiology.Springer (pp. 47–122).
Daróczi, G. (2015). Mastering data analysis with R. Packt Publishing Ltd.
Gardener, M. (2017). Statistics for ecologists using R and Excel: Data collection, exploration, analysis and presentation. Pelagic Publishing Ltd.
Golding, N. (2019). greta: Simple and scalable statistical modelling in R. Journal of open source software, 4(40), 1601.
Jones, E., Harden, S., & Crawley, M. J. (2022). The R book. John Wiley & Sons.
Lanzetta, V. B., Dasgupta, N., & Farias, R. A. (2018). Hands-on data science with R: Techniques to perform data manipulation and mining to build smart analytical models using R. Packt Publishing Ltd.
Liu, S., Tian, C., Sun, J., Wang, R., Lu, W., Zhao, Y., ... & Du, X. (2024). Conformance Testing of Relational DBMS Against SQL Specifications. arXiv preprint arXiv:2406.09469.
Liu, Z. H. (2019). JSON data management in RDBMS. In Emerging Technologies and Applications in Data Processing and Management (pp. 20–44). IGI Global.
Maindonald, J., & Braun, W. J. (2010). Data analysis and graphics using R: An example-based approach (Vol. 10). Cambridge University Press.
Marrs, T. (2017). JSON at work: practical data integration for the web. O'Reilly Media, Inc.
Nolan, D., & Lang, D. T. (2014). XML and web technologies for data sciences with R (Vol. 1). Springer.
Ooms, J. (2014). The jsonlite package: A practical and consistent mapping between json data and r objects. arXiv preprint arXiv:1403.2805.
Ooms, J., James, D., DebRoy, S., Wickham, H., Horner, J., & Ooms, M. J. (2024). Package 'RMySQL'. CRAN.
Pathak, M. A. (2014). Beginning data science with R. Springer.
Rockoff, L. (2021). The language of SQL. Addison-Wesley Professional.
Sievert, C. (2020). Interactive web-based data visualization with R, plotly, and shiny. Chapman and Hall/CRC.
Spector, P. (2008). Data manipulation with R. Springer Science & Business Media.
Team, R. C. (2000). https://cran.r-project.org/doc/manuals/r-release/R-data.html.
Van der Loo, M., & De Jonge, E. (2018). Statistical data cleaning with applications in R. John Wiley & Sons.
Villalobos Alva, J. (2024). Import and export. In Beginning Mathematica and Wolfram for Data Science: Applications in Data Analysis, Machine Learning, and Neural Networks (pp. 147–186). Apress.
Visne, I., Yildiz, A., Dilaveroglu, E., Vierlinger, K., Nöhammer, C., Leisch, F., & Kriegner, A. (2012). speedR: An R package for interactive data import, filtering and ready-to-use code generation. Journal of statistical software, 51, 1–12.
Walesiak, M., Dehnel, G., & Dudek, A. (2024). A dynamic approach to relative taxonomy and robust measures of central tendency. Communications in statistics-simulation and computation, 53(6), 2645–2661.

6

Data Visualization

Learning Objectives:

Following their study of this chapter, scholars and students will be able to:

- Learn the essential ideas behind data visualization
- Understand how important it is to communicate information through data visualization effectively
- Get acquainted with the most important R packages for data visualization, including lattice, ggplot2, and plotly
- Understand how to use R to construct simple plots, such as histograms, bar charts, line graphs, and scatter plots
- Discover how to add titles, labels, and colours to simple plots
- Develop knowledge of ggplot2's sophisticated customization features, like themes, facets, and scales
- Investigate the interactive visualization tools in R, especially plotly
- Develop the ability to make interactive graphs so that consumers may dynamically examine data
- Learn how to use R's map function to visualize geospatial data
- Utilize packages such as ggmap and leaflet to generate both static and dynamic maps
- Apply visualization techniques to real-world data sets

6.1 Introduction

Information and data are represented graphically with the help of data visualization techniques. Data visualization tools and techniques offer an easy-to-use means of seeing and comprehending trends, outliers, and patterns in data through the use of visual elements such as charts, graphs, and maps

DOI: 10.1201/9781003489382-6

(Mauludina et al., 2024). From the standpoint of data analytics, data visualization is a crucial procedure that converts complicated data sets into understandable and informative graphical representations. It does several things, such as making data interpretation easier, revealing hidden patterns and trends, and making it easier to communicate discoveries. Data scientists can display data in a way that is easily understood by stakeholders by utilizing visual elements like charts, graphs, and maps (Healy, 2018). This allows stakeholders to make decisions based on the visualized information. Accuracy, clarity, simplicity, and aesthetic appeal are the cornerstones of good data visualization, guaranteeing that the information is accurately depicted and easily comprehensible (Marriott, 2024). With R packages for data visualization, users can discover new elements and improve their comprehension by engaging interactively with the data through interactive visualizations (Kabacoff, 2024). In essence, proficient data visualization involves narrating a story using data and guiding the audience through it with well-crafted images that emphasize significant discoveries and facilitate data-driven choices. The principal ideas and concepts of data visualization are listed below:

1. **Data Visualization's Objective:**
    - Complex Data Simplification: Complex data sets are made easier to comprehend and analyse through the use of visualizations.
    - Discern Trends and Patterns: Trends, correlations, and patterns that may not be visible in raw data can be found with the aid of graphical representations.
    - Efficient Transmission of Information: Effective and clear visuals can communicate information to a large audience.
2. **Forms of Information Visualization:**
    - Charts: Bar charts, line charts, pie charts, area charts, and scatter plots.
    - Graphs: Network graphs, tree maps, and hierarchical diagrams.
    - Maps: Geographical maps, heat maps, and choropleth maps.
    - Infographics: Combine multiple visualization types with textual information to tell a comprehensive story.
3. **Components of an Effective Visualization:**
    - Accuracy: Properly and distortion-free portray data.
    - Clarity: Ensure that any labels, legends, and annotations in the visualization are easily comprehensible.
    - Simplicity: Stay away from superfluous detail and concentrate on the most important details.
    - Aesthetics: Carefully choose your colours, forms, and sizes to improve readability and attractiveness.

4. **Principles of Design:**
    - Choosing the Correct Chart Style: Take into account the facts and the message you wish to portray when choosing the right style of chart.
    - Colour Theory: Make sure that colours are readable and that they emphasize important information.
    - Layout and Composition: Organize visual components in a logical and harmonious way to direct the viewer's attention through the content.
5. **Visualization Techniques:**
    - Choosing the Correct Chart Style: Take into account the facts and the message you wish to portray when choosing the right style of chart.
    - Colour Theory: Make sure that colours are readable and that they emphasize important information.
    - Layout and Composition: Organize visual components in a logical and harmonious way to direct the viewer's attention through the content.

6.2 Basic Plot Creation in R

Creating basic plots in R entails utilizing a variety of packages and functions to visualize data in an easy-to-understand manner (Zamora Saiz et al., 2020). R offers robust data visualization features with its built-in graphing system and a number of add-on packages, such as ggplot2. Any data analysis work requires an understanding of fundamental charting functions and principles.

6.2.1 Histogram

A graphical tool used to arrange a collection of data points into user-specified ranges is called a histogram. It is employed to illustrate a data set's distribution as well as the frequency of data points falling inside particular ranges (Medley-Rath et al., 2024). Each bar's height indicates how many data points are in each of the bins into which the data has been divided. Histograms are helpful in determining any skewness or kurtosis, locating outliers, and comprehending the underlying distribution of the data. It has a unique characteristic of displaying no gaps between the bars, resembling a vertical bar graph. The R hist() function from the base R package can be used to create a histogram. The syntax for generating a histogram is shown below.

```
hist(x, breaks, main, xlab, ylab, col, border, xlim, ylim, ...)
```

where 'x' specifies the data values for which the histogram is needed, 'breaks' indicates the total number of bins and is optional, 'main' specifies the title of the graph and is optional, 'xlab' provides labelling for the x-axis and is optional, 'ylab' provides labelling for the y-axis and is optional, 'col' is used to specify the bar colours and is optional, 'border' indicates the colour of the bar borders and is optional, 'xlim' sets the x-axis limits and is optional, and 'ylim' sets the y-axis limits and is optional. The complete example R code for histogram creation is illustrated below and the output is visualized in Figure 6.1:

```
> # Histogram Creation in R
> # Sample data
> data <- c(10, 12, 12, 13, 13, 13, 14, 14, 14, 14, 15, 15,
15, 15, 15)
>hist(data, breaks = 5, main = "Histogram of Sample Data",
xlab = "Value", ylab = "Frequency", col = "orange",border = "black")

> # Save the histogram as a PNG file
>png("histogram.png")
>hist(data, breaks = 5, main = "Histogram of Sample Data",
xlab = "Value", ylab = "Frequency", col = "orange",border =
"black")
>dev.off()
```

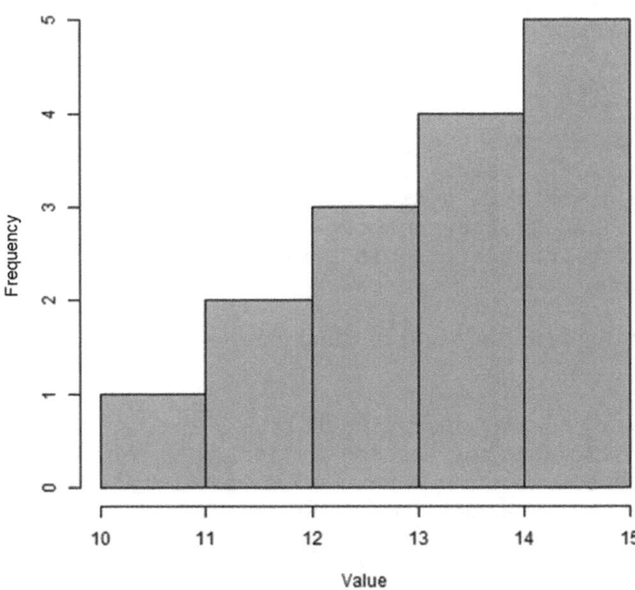

FIGURE 6.1
Histogram of sample data.

A numeric vector called data with a range of values is the sample data. The data vector is passed to the hist() function. 'breaks = 5' indicates how many bins to include in the histogram. The function png("histogram.png") creates a new PNG device in order to store the plot. The PNG device is closed by dev. off(), which saves the file to disk.

6.2.2 Strip Chart

A strip chart, sometimes referred to as a strip plot, is a straightforward scatter plot in which distinct categories are represented by one axis and data values are represented by the other (Grant, 2018). When displaying the distribution of data points on a single axis, it is helpful. Strip charts work especially well for comparing distributions across various categories, showcasing individual observations, and displaying all the data points in a small data set. The strip chart construction syntax is described below:

```
stripchart(x, method="overplot", vertical=TRUE, jitter=TRUE,
xlim=NULL, ylim=NULL, main=NULL, xlab=NULL, ylab=NULL, ...)
```

A numerical vector of the data points to be plotted is denoted by 'x'; 'method' describes the points to be plotted ('overplot,' 'stack', 'side,' 'centre', 'jitter'); 'vertical' specifies that the strip chart is horizontal if FALSE and vertical if TRUE; 'jitter' if TRUE, then each point has a tiny bit of random noise added to it to prevent overlap; '"xlim", "ylim"'the x- and y-axis limits; the plot's title is given by 'main'; and '"xlab", "ylab"' are the x and y axes' labels. The following sample R code shows how to create a strip chart in its entirety:

```
> # Strip Chart Creation Example
>set.seed(123)
> data <- data.frame(group = rep(letters[1:3], each = 20),
value = rnorm(60, mean = 0, sd = 1))
> # Create a vertical strip chart
>stripchart(data$value, method = "jitter", vertical = TRUE,
main = "Strip Chart of Sample Data", xlab = "Value",
ylab = "Group", col = "green")
```

Figure 6.2 displays the result of the code used to create the vertical strip chart.

```
> # Create a horizontal strip chart
>stripchart(data$value, method = "jitter", vertical = FALSE,
main = "Horizontal Strip Chart of Sample Data", xlab =
"Group", ylab = "Value", col = "green")
```

Figure 6.3 displays the result of the code used to create the horizontal strip chart.

Using rnorm(), create a sample data set with three groups (a, b, and c) and random values. To make a strip chart, use the stripchart() function. data$value is the numeric vector of values to be plotted, method = 'jitter'

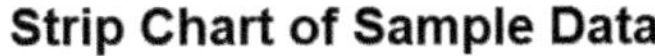

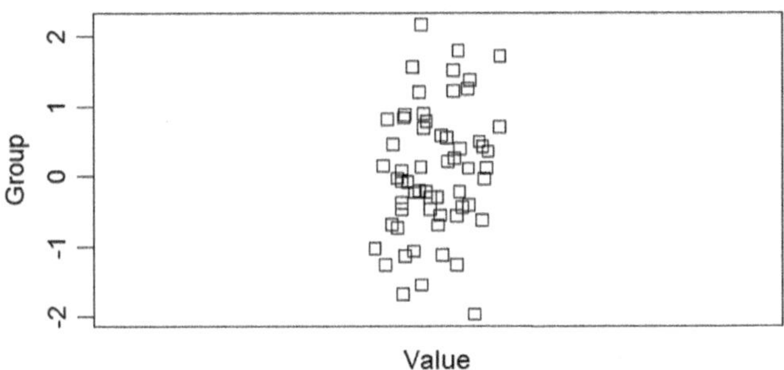

FIGURE 6.2
Vertical strip chart.

jitters up the points so they are more visible, vertical = FALSE produces a horizontal strip chart, while TRUE produces a vertical one, and col indicates the point colours.

6.2.3 Pie Chart

A pie chart is a circular image that has been cut into slices to show numerical quantities (Hehman & Xie, 2021). A category is represented by each slice, and the size of each slice reflects the category's importance in relation to the total. Pie charts work well for displaying data with few categories and are frequently used to illustrate the relative sizes of parts to a whole. But when there are a lot many categories or when there aren't many variations between them, they could lose their effectiveness. The syntax for making a pie chart in R is shown below:

```
pie(x, labels=names, main=NULL, col=NULL, clockwise=FALSE,
init.angle=0, ...)
```

where 'x' represents the data that has to be turned into a pie chart, 'labels' specify the labels of each slice and is optional, the title of the chart is supplied using 'main,' and 'col' is used to provide colours to the graph, The slice direction can be specified using the term 'clockwise.' If set to TRUE, the slices are drawn clockwise; if set to FALSE, they are drawn counterclockwise;

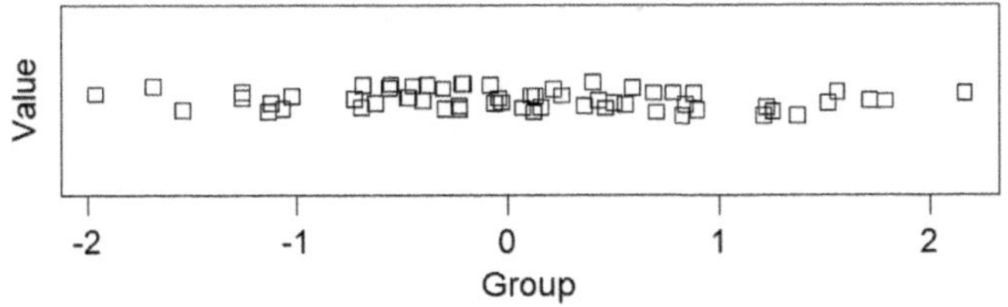

FIGURE 6.3
Horizontal strip chart.

'initangle' provides the initial angle for the initial slice. The full code for generating a pie chart is demonstrated below:

```
> #Pie Chart Creation Example
> data <- c(25, 15, 20, 15,25)
> names <- c("Category X1", "Category X2", "Category X3",
"Category X4","Category X5" )
> # Create a pie chart
>pie(data, labels = names, main = "Pie Chart of Sample Data",
col = rainbow(length(data)))
```

The generated pie chart can be seen in Figure 6.4.

Sample data is produced by establishing a vector data set with numbers that represent the values for every category. Establish character vector names for the labels that belong to each category. A pie chart can be made with the pie() method. The numerical values that will be shown as pie chart slices are contained in the data. labels = names gives each slice a label. The pie chart's title is specified in the main, and Col = rainbow(length(data)) uses a rainbow colour palette to establish the colours for each slice.

Here is an intricate illustration of creating a pie chart in R that involves further customization and data manipulation. This example shows how to use R to make an aesthetically pleasing and educational pie chart with adjustments like colours, exploding slices, and adding a legend. This is visualized in Figure 6.5.

```
> # Another Example of Pie Chart
> # Sample data: Sales data for different electronic product
categories
> categories <- c("TV", "Washing Machine", "Mixer Grinder",
"Dish Washer", "Air Conditioner")
> sales <- c(500, 700, 1000, 100, 2000)

> # Colors for each category
>colors<- c("#1f77b4", "#ff7f0e", "#2ca02c", "#d62728", "#9467bd")
> # Explode slices (pull out one slice)
> explode <- c(0, 0, 0, 0.1, 0)
```

Pie Chart of Sample Data

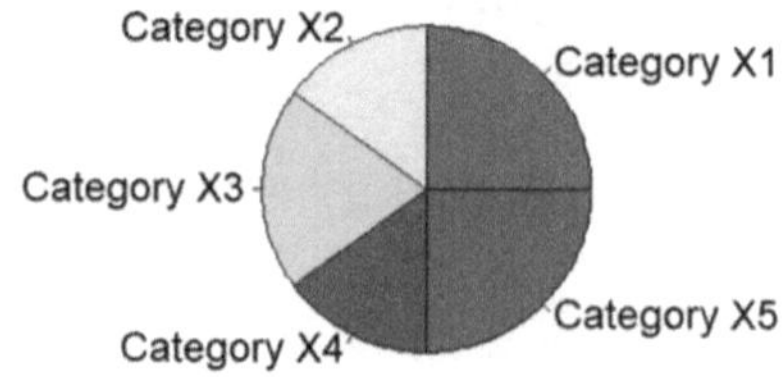

FIGURE 6.4
Pie chart.

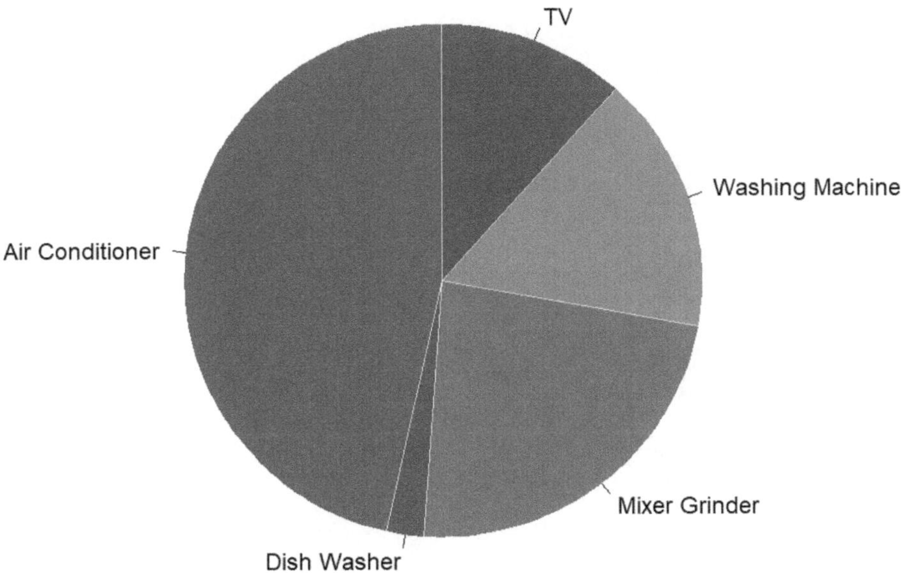

FIGURE 6.5
Pie chart showing sales of electronics product category.

```
> # Pie Chart Creation
>pie(sales, labels = categories, col = colors, main = "Sales
Distribution by different Electronics Product Category",
clockwise = TRUE,init.angle = 90,explode = explode,
border = "white",cex.main = 1.2,cex.lab = 0.8,cex.axis = 0.8)
```

Sample data is created using two variables: the first one is 'categories,' which holds the titles of several product groups and the second one is 'sales,' which is storing actual sales of category-wise electronic products. For easier distinction, each category is given a hexadecimal colour code using 'colours,' 'explode' is used for a vector indicating how much of the pie should be cut into portions, and clockwise = ACTUAL determines the clockwise orientation of the slice drawing. The pie chart is rotated by init.angle = 90 so that the top (12 o'clock) position is where the first slice begins. The legend() function adds a legend to the pie chart to explain the colours used for each category, as reflected in Figure 6.6.

```
> # Add a legend
>legend("topright", categories, fill = colors, cex = 0.8,
title = "Categories")
```

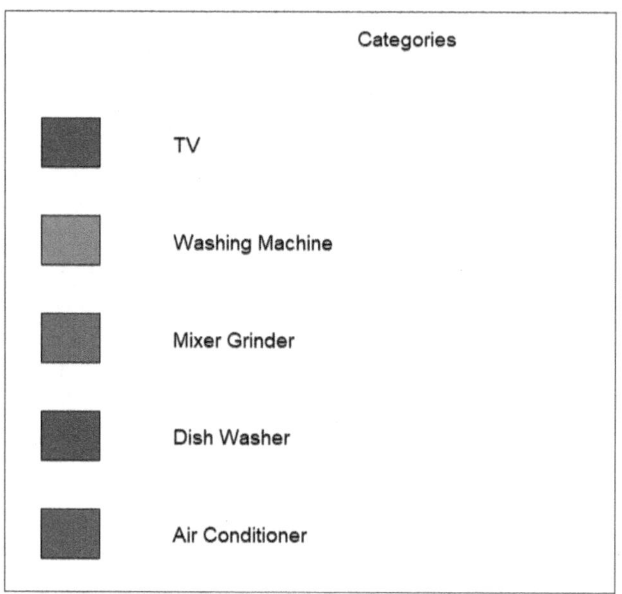

FIGURE 6.6
Electronic product data colour description using legend().

6.2.4 Scatter Plot

A scatter plot is a kind of graphic or mathematical diagram that shows the values of two variables, given a collection of data using Cartesian coordinates. Patterns, trends, and outliers can be found; correlations can be found; and linkages between variables are seen with it (Healy, 2018). Every dot on the scatter graphic denotes a dataset observation. When determining the type and intensity of relationships—such as linear or nonlinear relationships—between variables, scatter plots are very helpful. The syntax for the generation of a scatter plot in R is given below:

```
plot(x,y,main,xlab,ylab,pch,col)
```

where 'x' and 'y' supply the data values, 'main' provides the scatter plot title, 'xlab' and 'ylab' offer the labels for the x- and y-axes, 'pch' plots the points in a particular style, and 'col' determines the colour of the points. Using the code listed below, a scatter plot is produced after values x1 and x2 for the axes x and y are supplied:

```
> x1<- c(1, 2, 3, 4, 5)
>x2<- c(2, 3, 5, 7, 11)
>plot(
+      x1, x2,
+      main="Customized Scatter Plot,
+      xlab="X Axis",
```

```
+       ylab="Y Axis",
+       pch=19,
+       col="blue",
+       cex=1.5,
+       xlim=c(0, 6),
+       ylim=c(0, 12)
+ )
>
> # Add grid lines for better readability
>grid(nx = NULL, ny = NULL, col = "red", lty = "dotted")
```

The result of the above code is displayed in Figure 6.7.

6.2.5 Line Chart

Information that is shown as a sequence of data points, referred to as 'markers,' connected by straight line segments in a line chart (Dougherty & Ilyankou, 2021). Data points are frequently visualized in chronological sequence using this technique. When comparing several data sets, illustrating patterns across time, and emphasizing the rate of change, line charts work well. When the focus is on illustrating the evolution of values across

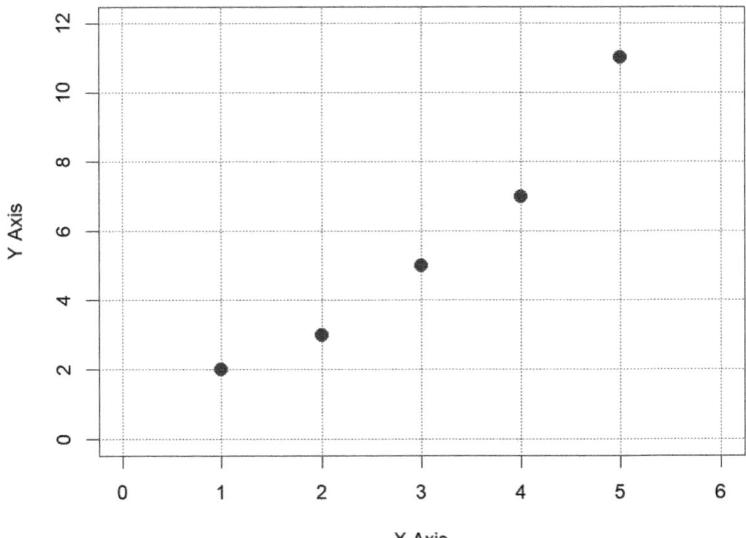

FIGURE 6.7
Scatter plot. (Data for the x- and y-axes are provided by x1 and x2; pch=19 specifies the plotting point style, which is a solid circle for value 19; cex=1.5 modifies the point size; ylim = c(0, 12) and xlim = c(0, 6) set an x- and y-axis limit; and grid()incorporates a grid.)

intervals, they are especially helpful for time-series data. To generate a line chart in R, the basic syntax is:

```
plot(x,type,col,xlab,ylab)
```

The description of the parameters used is as follows: The numerical values are contained in the vector 'x'; 'type' accepts the values 'p' to draw only points, 'l' to draw only lines, and 'o' to draw both lines and points; the labels for the x- and y-axes are 'xlab' and 'ylab'; 'col' is used to give the lines and points their colours; 'main' is used to provide the title of the graph; and 'lwd' is used to specify the line width. Figure 6.8 presents the outcome.

```
> # Line Chart Example
># Sample data
> x1<- c(1, 2, 3, 4, 5)
>x2<- c(2, 3, 5, 7, 11)
> # Create line chart
>plot(x1, x2,type="l",main="Simple Line Chart",xlab="X Axis",
ylab="Y Axis", col="blue",
lwd=2)
```

In another example, to make a more intricate line graph in R, we will add comments, a legend, and several lines plotted in various ways.

```
# Sample data
> x <- 1:10
> y1 <- c(2, 3, 5, 7, 11, 13, 17, 19, 23, 29)
> y2 <- c(1, 4, 6, 8, 10, 12, 14, 16, 18, 20)
> y3 <- c(30, 27, 24, 21, 18, 15, 12, 9, 6, 3)

> # Create an empty plot
>plot(
+     x, y1, type="n",                    # Data Values, "n" type
                                          for no plot
+     main="Complex Line Chart Example",  # Plot Title
+     xlab="X Axis",                      # X-axis Label
```

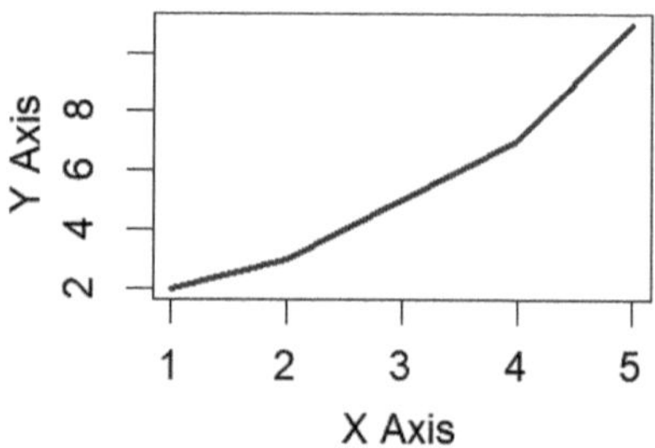

FIGURE 6.8
Line chart.

```
+     ylab="Y Axis",                     # Y-axis Label
+     xlim=c(1, 10),                     # X-axis Limits
+     ylim=c(0, 35)                      # Y-axis Limits
+ )

> # Add several lines in various styles.
>lines(x, y1, type="o", col="blue", lwd=2, pch=19)
# Line number one
>lines(x, y2, type="b", col="red", lwd=2, lty=2, pch=17)
# Line Two
>lines(x, y3, type="o", col="green", lwd=2, lty=3, pch=15)
# Line Three
> # Legend Addition
>legend(
+     "topright",                        # Legend Position
+     inset=c(-0.1, 0),                  # To place the legend inside
the plot area, adjust the inset
+     legend=c("Series 1", "Series 2", "Series 3"),  # Labels
+     col=c("blue", "red", "green"),# Line Colors
+     lty=c(1, 2, 3),                    # Line types
+     pch=c(19, 17, 15),                 # Plotting points style
+     lwd=2,                             # Width of the Line
+     cex=0.8,                           # Shrink the legend text.
+     bty="n"                            # Take down the box
enclosing the legend.
+ )

> # Add grid lines for better readability
>grid(nx = NULL, ny = NULL, col = "lightgray", lty = "dotted")

> # Add annotations
>text(5, 30, "Annotation for Series 1", col="blue", pos=4)
>text(5, 15, "Annotation for Series 2", col="red", pos=4)
>text(5, 10, "Annotation for Series 3", col="green", pos=4)

> # Highlight a specific point with a different color
>points(5, 11, col="purple", pch=19, cex=2)
>text(5, 11, "Key Point", col="purple", pos=3)
```

Above, R will produce a more complex line chart with several lines, a legend, and annotations; the result is shown in Figure 6.9.

A description of supplementary components used in the creation of a complex line chart is as follows (Kabacoff, 2024):

- plot(x, y1, type="n"): Produces a blank plot with the given boundaries in x and y.
- type="n": Makes sure that at first, no lines or points are drawn.
- lines(): Expands the plot by adding new lines.

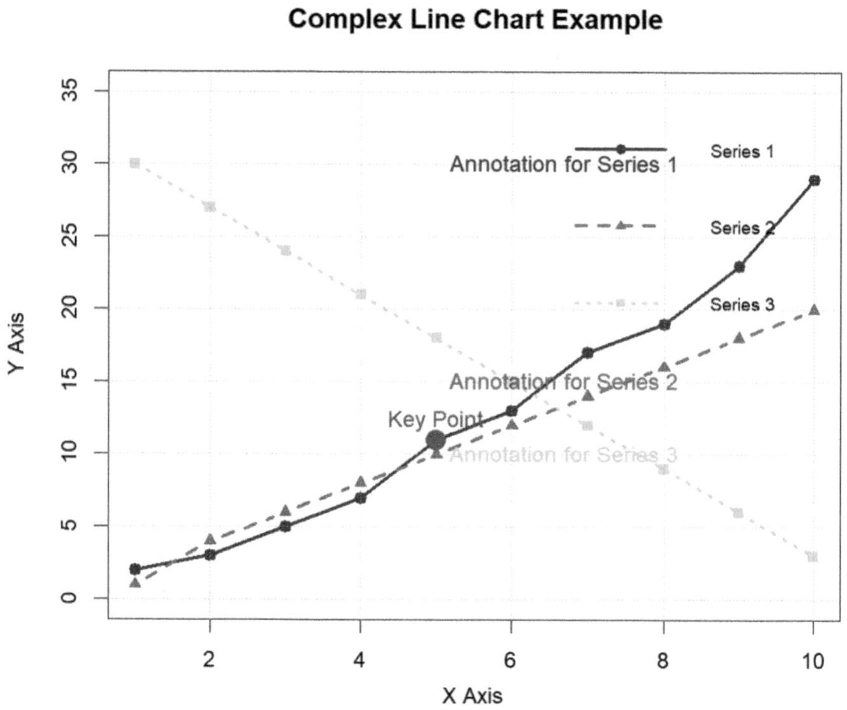

FIGURE 6.9
Complex line chart with annotations.

- The parameters type, col, lwd, lty, and pch allow you to adjust each line's appearance.
- legend(): Enhances the story with a legend. The labels are specified by the legend parameter, the colours by col, the line types by lty, the plotting characters by pch, and the line width by lwd.
- bty="n": For a neater appearance, remove the box surrounding the legend.
- The legend's location is adjusted to the left by inset=c(-0.1, 0) to make sure it fits inside the plot area. These settings can be changed to best suit your particular plot.
- cex=0.8: Shrinks the text in the legend.
- text(): Annotates the plot with text. The position in relation to the coordinates is specified by the pos argument.
- points(): Inserts a point to draw attention to a particular spot on the plot. The point's appearance can be customized using the col, pch, and cex properties.

6.3 Advanced Plot Creation in R

Using robust visualization tools to create advanced plots in R offers a more flexible and easy way to create intricate graphs. Creating complex graphical displays with R by altering the base plotting system or functions from base R is a common way to create advanced plots. Plots can be made in many different ways with base R, ranging from straightforward scatter plots to intricate multipanel arrangements (Nordmann et al., 2022). You can create complex visuals by varying settings and combining various plotting functions. The numerous sophisticated plotting graphs available in R are listed here.

6.3.1 Box Plot

A box plot, sometimes referred to as a box and whisker plot, offers a visual representation of a set of data. The quartiles, median, and possible outliers are shown. When comparing distributions across many groups or data sets, box charts are helpful (Hehman & Xie, 2021). The interquartile range (IQR), which includes the middle 50% of the data, is represented by the box. In the box, the median is shown by the line. The first and third quartile's 1.5 * IQR values are where the whiskers terminate, and the values outside of this range are regarded as outliers.

The 'mtcars' data set, a built-in data set in R that includes details on a variety of 32 distinct car models, is what we used to create the box plot. An overview of the attributes and a few values of 'mtcars' are shown in Table 6.1.

The following provides information on box plot syntax and parameter descriptions:

```
boxplot(data, main, ylab, xlab, col, border, notch,
horizontal, notchwidth, outline, names)
```

TABLE 6.1

Data Description of 'mtcars' Data Set

	mpg	cyl	disp	hp	drat	wt	qsec	vs	am	gear	carb
Mazda RX4	21.0	6	160	110	3.90	2.620	16.46	0	1	4	4
Mazda RX4 Wag	21.0	6	160	110	3.90	2.875	17.02	0	1	4	4
Datsun 710	22.8	4	108	93	3.85	2.320	18.61	1	1	4	1
Hornet 4 Drive	21.4	6	258	110	3.08	3.215	19.44	1	0	3	1
Hornet Sportabout	18.7	8	360	175	3.15	3.440	17.02	0	0	3	2
Valiant	18.1	6	225	105	2.76	3.460	20.22	1	0	3	1

- boxplot(data,...): Using the supplied data, this function produces a box plot.
- data: The data to be utilized in the box plot creation process is specified by this option.
- main: A string containing the plot's primary title.
- ylab: A string that indicates the y-axis label.
- xlab: A string that indicates the x-axis label.
- col: The colour or colours to be used in the box filling. Either a single colour or a vector of colours can be supplied. For the boxes, 'white' is the default colour.
- border: The colour or colours used for the box borders; you can supply a single colour or a colour vector.
- notch: A logical value that indicates if the median should have a notch drawn around it. When comparing medians, notches are utilized; if there is no overlap between the notches on two boxes, the medians differ considerably at the 5% significance level.
- horizontal: A logical value that indicates if the box plots should be plotted horizontally.
- notchwidth: The notch's width in relation to the box's width; usually expressed as a scalar value.
- outline: A logical value that indicates whether or not the box outlines should be drawn.
- names: Plot names beneath each box plot. These names are used as labels for each box in the event that the data is a matrix or data frame.

The following code creates a box plot based on the weight, horsepower, and fuel economy of 'mtcars' data set, and the output is visualized in Figure 6.10.

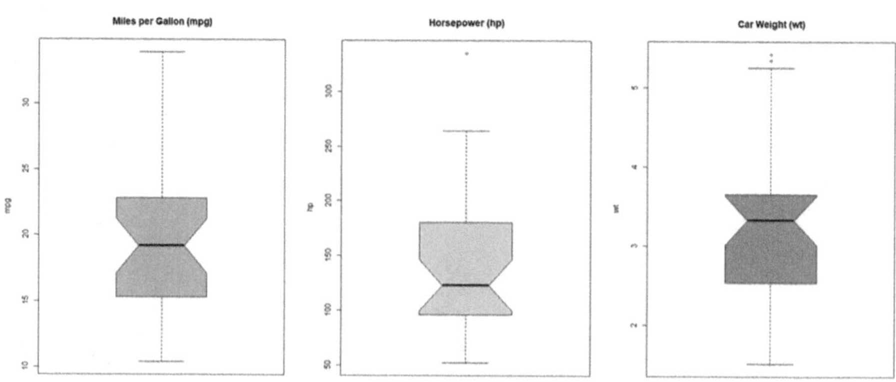

FIGURE 6.10
Box plot for the miles per Gallin, horsepower, and weight features on the 'mtcars' data set.

```
># Boxplot Example on "mtcars" dataset
># Load the mtcars dataset (if not already loaded)
> data(mtcars)

> # Create a boxplot for mpg, hp, and wt
> par(mfrow=c(1,3))  # Setting up a 1x3 grid for plots

> # Boxplot for mpg
>boxplot(mtcars$mpg, main = "Miles per Gallon (mpg)",
ylab = "mpg",col = "skyblue",border = "black",notch = TRUE)

> # Boxplot for hp
>boxplot(mtcars$hp, main = "Horsepower (hp)", ylab = "hp",
col = "lightgreen",border = "black",notch = TRUE)

> # Boxplot for wt
>boxplot(mtcars$wt, main = "Car Weight (wt)", ylab = "wt",
col = "lightcoral",border = "black",notch = TRUE)
```

As an additional illustration to make the plots more useful, we will develop box plots for many variables and provide extra customization.

```
> # Load the mtcars dataset
> data(mtcars)

> # Set margins
>par(mar = c(3, 3, 2, 1))  # Adjust bottom, left, top,
right margins

> # Create a multi-panel plot for mpg, hp, wt, and qsec
>par(mfrow=c(2, 2))  # Setting up a 2x2 grid for plots

> # Boxplot for mpg
>boxplot(mpg ~ cyl, data = mtcars,
+  main = "Miles per Gallon (mpg) by Number of Cylinders",
+  xlab = "Number of Cylinders",
+  ylab = "mpg",
+  col = c("skyblue", "lightgreen", "lightcoral"),
+  border = "black",
+  notch = TRUE)

> # Boxplot for hp
>boxplot(hp ~ cyl, data = mtcars,
+  main = "Horsepower (hp) by Number of Cylinders",
+  xlab = "Number of Cylinders",
+  ylab = "hp",
+  col = c("red", "lightgreen", "orange"),
+  border = "black",
+  notch = TRUE)
```

```
> # Boxplot for wt
>boxplot(wt ~ cyl, data = mtcars,
+  main = "Car Weight (wt) by Number of Cylinders",
+  xlab = "Number of Cylinders",
+  ylab = "wt",
+  col = c("red", "lightgreen", "orange"),
+  border = "black",
+  notch = TRUE)

> # Boxplot for qsec
>boxplot(qsec ~ cyl, data = mtcars,
+  main = "Quarter Mile Time (qsec) by Number of Cylinders",
+  xlab = "Number of Cylinders",
+  ylab = "qsec",
+  col = c("red", "lightgreen", "orange"),
+  border = "black",
+  notch = TRUE)

> # Reset graphics parameters
>par(mfrow=c(1, 1))  # Reset to default single-panel plot

> # Increase plot device size
>pdf("boxplots.pdf", width = 10, height = 8)  # Example for PDF
output
> # Reset plot device
>dev.off()  # Close the PDF device
```

Figure 6.11 displays more a customizable box plot where in the above code par(mar=c(3, 3, 2, 1)) arranges the margins bottom, left, top, and right.

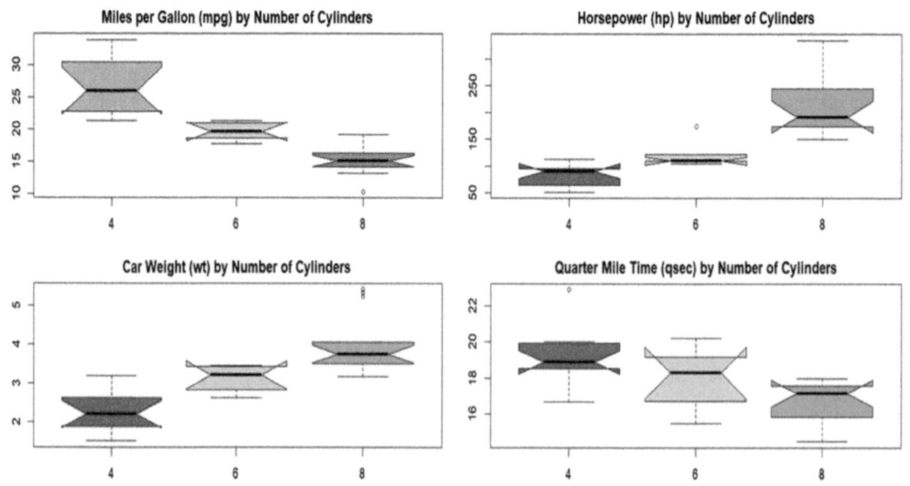

FIGURE 6.11
Box plot with multiple variables and more customization.

To ensure that the multipanel setting has no effect on subsequent plots, par(mfrow=c(1, 1)) returns the plotting parameters to default, and pdf() is used to save the graph as a PDF file. You can increase the charting area by giving pdf() the width and height parameters. Adapt these values to meet your own requirements.

6.3.2 QQ Plot

In statistics and data analysis, a QQ plot, also known as a quantile-quartile plot, is a graphical tool used to determine if a set of data follows a particular distribution or not. The quantiles of the data set are compared to those of a theoretical distribution, usually a normal distribution, using this technique. QQ charts are a useful tool for evaluating whether a specific theoretical, normal, exponential, distribution, etc. is followed by the data. Specifically, QQ plots are frequently employed to determine whether the data conforms to a normal distribution. The steps below can be used to make a QQ plot.

Step I: Data Sorting: The data points are first arranged in ascending order.

Step II: Quantile Computation: It involves calculating the quantiles for both the data set and the theoretical distribution.

Step III: Plotting: The data set's quantiles (vertical axis) are graphed against the theoretical distribution's quantiles (horizontal axis).

The qqnorm() function in R can be used to make a normal QQ plot, while the qqplot() function can be used to create QQ plots against different distributions. Here is a simple example using the 'mtcars' data set that comes with R:

```
> # Load the mtcars dataset (if not already loaded)
> data(mtcars)
> # Q-Q plot for mpg against normal distribution
>qqnorm(mtcars$mpg, main = "Q-Q Plot of mpg (Miles per Gallon)")
>qqline(mtcars$mpg)   # Add a line to the Q-Q plot
```

A QQ plot of the data versus a theoretical normal distribution is produced using the qqnorm()function. 'qqline()' is used to visually evaluate the deviation from normalcy, as shown in Figure 6.12; it adds a line to the QQ plot. If the data is properly distributed, a straight line should pass through each point. The points for analysing QQ plots are as follows:

1. Straight Line: The data is considered regularly distributed if it is located near a straight diagonal line.
2. Curved/Deviating Line: The terms 'curved' or 'deviating line' refer to a graph that shows departures from the expected distribution or non-normality.

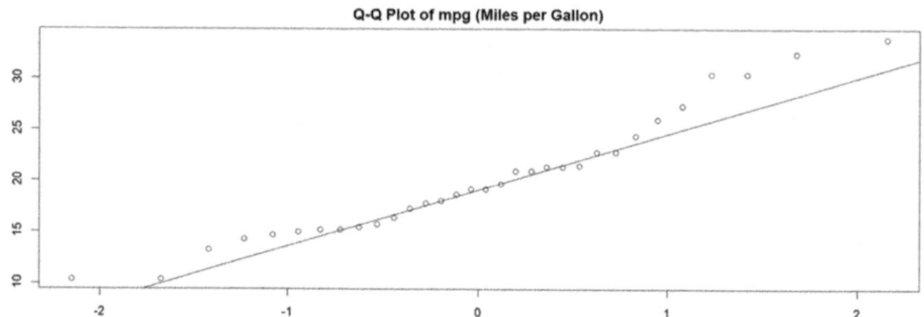

FIGURE 6.12
QQ plot.

QQ plots are powerful tools for assessing distributional assumptions in statistical analysis. To display and evaluate your data's distribution versus theoretical distributions, such as the normal distribution, you can quickly and easily construct and modify QQ charts with R (Paquot & Larsson, 2021). Plot scale adjustments, reference line additions, and data set comparisons are standard procedures used to obtain insights into the distribution features of the data.

To demonstrate QQ plots for the mpg variable in the mtcars data set against a theoretical normal distribution and compare it with another data set (disp), let us first show how to modify plot scales, add reference lines, and use QQ plots in R to compare different data sets. Additionally, we will alter the narratives to improve comprehension, as reflected in the code below. The graph is highlighted in Figure 6.13:

```
> # Load the mtcars dataset
> data(mtcars)

> # Create Q-Q plot for mpg against a theoretical normal
distribution
>qqnorm(mtcars$mpg, main = "Q-Q Plot of mpg (Miles per Gallon)
vs Normal Distribution")
>qqline(mtcars$mpg, col = "red")  # Add a line to the Q-Q plot
(red color)
```

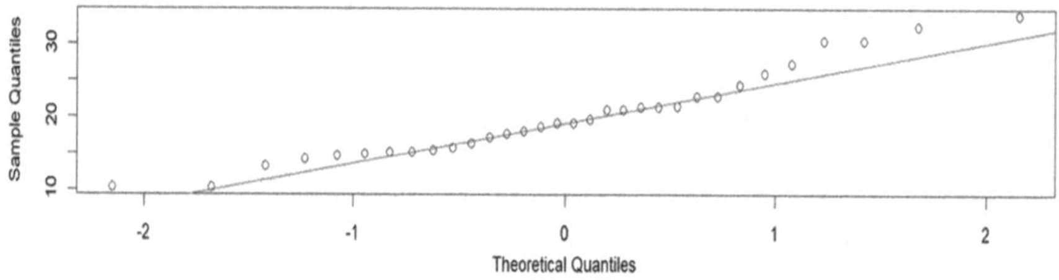

FIGURE 6.13
QQ plot with reference line for comparison.

```
> # Adjusting plot scales for better visualization
>par(mfrow = c(1, 2))  # Setting up a 1x2 grid for plots

> # Q-Q plot for mpg vs disp (Engine Displacement)
>qqplot(mtcars$mpg, mtcars$disp, main = "Q-Q Plot: mpg vs
disp", xlab = "mpg (Miles per Gallon)", ylab = "disp (Engine
Displacement)",col = "blue", pch = 20)  # Customizing point
color and shape

> # Adding reference line for comparison
>abline(a = 0, b = 1, col = "red")  # Adding a 45-degree
reference line

> # Reset graphics parameters
>par(mfrow = c(1, 1))  # Reset to default single-panel plot
```

Plot Interpretation:

1. The first plot (qqnorm() and qqline()) aids in determining whether or not mpg has a normal distribution. Points that are in close proximity to the red line indicate normalcy.
2. Mpg and disp are contrasted in the second figure (qqplot() with abline()). Similar distributional properties are shared by both variables, as seen by points aligning along the red line.

6.3.3 Heat Map

In R, a heat map is a graphical data representation where each colour corresponds to a particular value within a matrix. When displaying intricate data sets and patterns, including correlations, clustering, or distributions across several variables, heat maps can be quite helpful (Bessler, 2023). A colour gradient is commonly used in heat maps to depict the matrix's values. The colour of each cell in the matrix is determined by its value, making it simple to spot patterns. Generally, lighter hues indicate lower values, whereas darker hues imply higher values. Key factors for using heat maps to visualize data are displaying how variables are correlated, that is correlation analysis; examining data on gene expression; spatial data mapping and analysis; and putting heat maps and dendrograms side by side.

The 'heatmap()' function is indeed part of the default installation of base R (Rahlf, 2017). It is a straightforward yet powerful tool for creating heat maps to visualize relationships and patterns in data. With the help of the R heat map function, we can plot a heat map of the data set using the syntax:

```
heatmap(data,main, xlab, ylab, ...)
```

where 'data' is the data matrix that we wished to plot a heat map on is specified by data, 'main' details the plot's title, and '"xlab", "ylab"' give the labels

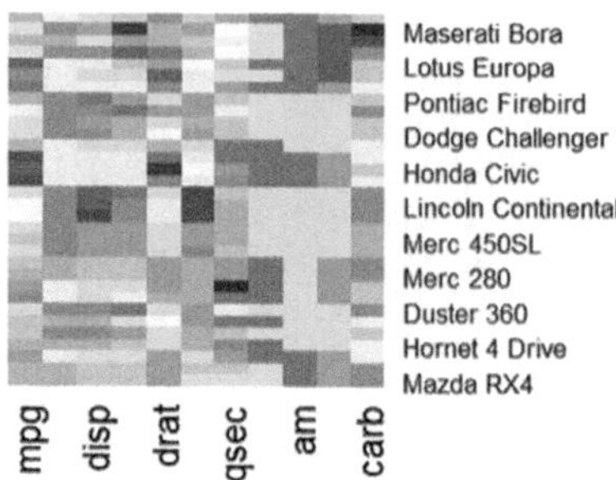

FIGURE 6.14
Heat map.

for the x- and y-axes. The code to generate a heat map on 'mtcars' is shown below. The following code generates a heat map using the 'mtcars' data set.

```
> # HeatMap Creation
> d <- as.matrix(mtcars)
>heatmap(d, Colv = NA, Rowv = NA, scale="column")
```

Figure 6.14 illustrates the creation of a heat map in the absence of any column or row dendrogram creation; also, colour specifications are not provided. To normalize, use 'scale'.

Use the command below to generate a dendogram alongside a heat map with custom colours; the code's output is displayed in Figure 6.15.

```
> # Heat Map with Dendogram and custom color
>data<- as.matrix(mtcars)

> # custom colors
```

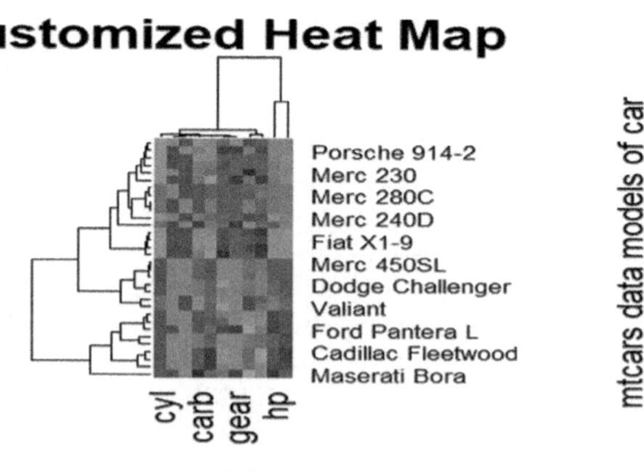

FIGURE 6.15
Customized heat map.

```
>nc<- colorRampPalette(c("orange", "darkred"))
>
> # plotting the heatmap
>plt<- heatmap(data,col = nc(100),main = "Customized Heat
Map",margins = c(5,10),xlab = "variables",ylab = "mtcars data
models of car",scale = "column")
```

There are several methods to personalize the colour palette. Use one of the R native palettes, terrain.color(), rainbow(), heat.colors(), topo.colors(), and cm.colors(), as the following code example illustrates:

```
>heatmap(data, scale="column", col = cm.colors(256))
>heatmap(data, scale="column", col = terrain.colors(256))
>heatmap(data, scale="column", col = topo.colors(256))
>heatmap(data, scale="column", col = heat.colors(256))
```

6.3.4 Tree Map

A data visualization method called a tree map uses a series of nested rectangles to show hierarchical data. The dimensions of each rectangle in the data set correspond to a particular measure, such as a proportion or numerical value. Tree maps are helpful for comparing proportions between categories at various levels of the hierarchy and for visualizing complex hierarchical data structures. Key tree map features are as follows:

1. A node in the hierarchy is represented by each rectangle in a tree map, which shows hierarchical data structures. Every rectangle's size represents a metric (population, revenue, etc.).
2. Rectangles have a hierarchical structure whereby they are layered within one another. Usually parent nodes are represented by bigger rectangles that are split into smaller rectangles that represent child nodes.
3. Colours can be used to encode other factors or categories in addition to size. For instance, within the same hierarchy, distinct colours may stand for various values or categories.
4. Every rectangle's area corresponds to the metric that is being displayed. This makes it possible to compare the relative sizes and relevance of the various categories within the hierarchy quickly.
5. To make exploring big data sets easier, several contemporary tree map implementations allow interactive features like tooltips, zooming, and filtering.

You can use a variety of packages in R to produce tree maps. A well-liked package that offers adaptable functionalities for creating tree maps is 'treemap.' The treemap() method is used to create a tree map and the full syntax is written below (Kabacoff, 2024):

```
treemap(index, vsize, palette, title)
```

where hierarchical structure is defined by the parameter 'index,' 'vsize' can help to resize rectanges, the colour palette can be defined by 'palette,' and 'title' can print the plot title. An example of creating a tree map in R is provided here and the output is graphically presented in Figure 6.16.

```
>install.packages("treemap")
>library(treemap)
> # Sample Data
> data <- data.frame(group = c(rep("A", 3), rep("B", 4),
rep("C", 3), rep("D", 5)), subgroup = c("X", "Y", "Z", "P",
"Q", "R", "S", "T", "U", "V", "W", "X", "Y", "Z", "P"),
value = c(10, 20, 15, 5, 12, 18, 25, 30, 8, 6, 12, 20, 15, 10, 25))
> # Create treemap
>treemap(data,
+   index=c("group", "subgroup"), # Define hierarchical structure
+   vSize="value",    # Metric to size rectangles
+   palette="Set3",   # Color palette
+   title="Example Treemap"
+ )
```

In this example, the tree map package in R is used to generate a simple tree map that displays hierarchical data as layered rectangles with sizes determined by a given measure. You may adjust the tree map representation to your unique data and analytical needs by modifying the parameters and discovering new features.

6.3.4.1 Interpretation of the Tree Map

Hierarchical data from the data set is visualized in the tree map above. Various groups and their corresponding subgroups, proportionally scaled according to the {value} measure, are displayed. These are some important

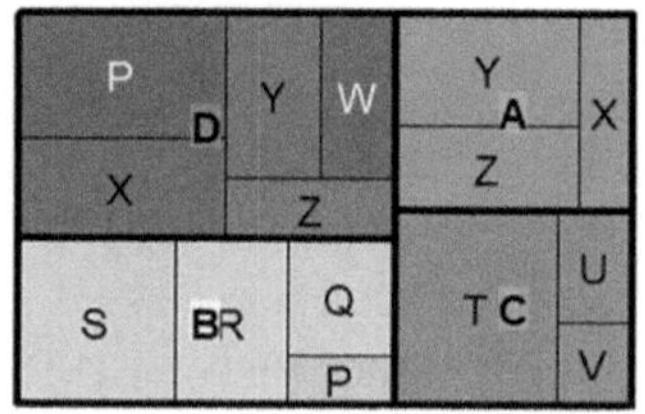

FIGURE 6.16
Tree map.

realizations. With subgroups X, Y, and Z displaying a range of values, group A has the highest representation. Despite having fewer members, group D makes a substantial contribution to the total value because its subgroups P and Q have higher values. The colour scheme (Set 3) aids in visually distinguishing between various groups.

In another implementation of tree map, the data set under consideration is a simulation of a situation in which sales data is gathered for multiple product subcategories across multiple regions. It is organized hierarchically, with Region as the top-level category, followed by Category, and Subcategory as the most detailed level. Each combination of Region, Category, and Subcategory is linked to a sales value.

Step I: Create Sample Data Set

```
>set.seed(123)
> data <- data.frame(Region = rep(c("North America",
"Europe", "Asia"), each = 9),Category = rep(c("Electronics",
"Clothing", "Furniture"), times = 9),Subcategory =
rep(c("Phones", "Laptops", "Tablets", "Shirts", "Pants",
"Shoes", "Chairs", "Tables", "Sofas"), times = 3),Sales =
runif(27, 1000, 10000))
> head(data)

         Region    Category Subcategory    Sales
1 North America Electronics      Phones 3588.198
2 North America    Clothing     Laptops 8094.746
3 North America   Furniture     Tablets 4680.792
4 North America Electronics      Shirts 8947.157
5 North America    Clothing       Pants 9464.206
6 North America   Furniture       Shoes 1410.008
```

Step II: Load Libraries

```
> library(treemap)
> library(dplyr)
```

Step III: Generate Tree Map

```
> # Define a custom color palette
>custom_palette<- colorRampPalette(c("lightblue", "blue"))(100)

> # Creating the tree map
>treemap(data,index = c("Region", "Category",
"Subcategory"),vSize = "Sales",vColor = "Sales",type =
"value",palette = custom_palette,title = "Sales Distribution
by Region, Category, and Subcategory")
```

Step IV: Generated Tree Map: Presented in Figure 6.17 is the tree map of Sales Distribution.

Distribution by Region, Category, and Subcat

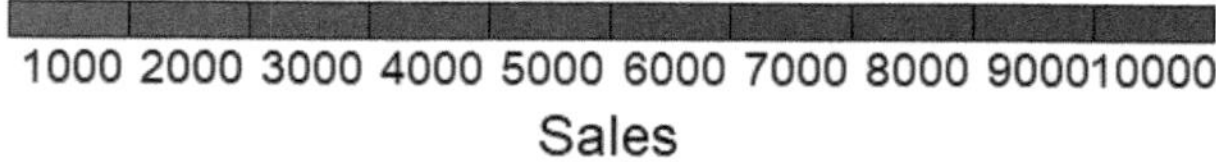

FIGURE 6.17
Tree map of sales distribution.

Step V: Tree Map Interpretation

At the top level of the tree map are regions, each nested within categories and subcategories. The size and colour intensity of the rectangles represent sales values. You can quickly determine which categories, subcategories, and sales regions are high and low with a tree map. Additionally, we can find research insights such as the following:

Question: What is the overall value of sales in each region?

Insight: Identifying the region with the highest sales can be done by summarizing sales by region. The R code for determining overall sales value is written here:

```
>total_sales_region<- data %>% group_by(Region) %>%
summarize(Total_Sales = sum(Sales))
> print(total_sales_region)
# A tibble: 3 × 2
  Region          Total_Sales
<chr>             <dbl>
1 Asia                66671.
2 Europe              48671.
3 North America       56933.
```

6.3.5 Contour Plot

A contour plot is a three-dimensional surface that is shown graphically using contours, or lines, to show the levels of a third variable as a function of two independent variables. By showing constant Z slices, or contours, in a 2-D format, a contour plot is a graphical way to visualize a 3-D surface (Kassambara, 2017). An alternative to a 3-D surface plot is a contour plot. Contour plots are made in R using the 'contour' function.

```
Syntax: contour(x, y, z, nlevels = 10, levels =
pretty(range(z, na.rm = TRUE), nlevels), ...)
```

where the vectors x and y stand for the grid line positions where the values in z are measured; the values to be contoured are contained in a matrix 'z'; the quantity of contour layers is given by 'nlevels'; and the number vector indicates the levels at which the contours should be drawn, indicated by 'levels.'

To use some sample data to create a contour plot, here is how we will move forward:

Step I: Generate an exemplar data set.

Step II: For each of the two independent variables, make a grid of values.

Step III: The third variable should be manually interpolated across the grid.

Step IV: Create a contour plot using the base R functions.

```
> # Generate sample data
>set.seed(123)
> x <- seq(-pi, pi, length.out = 50)
> y <- seq(-pi, pi, length.out = 50)
> z <- outer(x, y, function(x, y) sin(x) * cos(y))

 > # Plot the contour
>contour(x, y, z,
+  nlevels = 10,
+  col = "red",
+  main = "Contour Plot of sin(x) * cos(y)",
+  xlab = "X",
+  ylab = "Y")
```

The produced contour plot is shown in Figure 6.18.

```
> # Adding filled contour for better visualization
>filled.contour(x, y, z, color.palette = terrain.
colors,plot. Title = title(main = "Filled Contour Plot
of sin(x) * cos(y)",xlab = "X", ylab = "Y"),key.
Title = title(main = "Z"))
```

To generate a filled contour plot with a given colour scheme (Zamora Saiz et al., 2020), use the filled.contour() function, as shown in Figure 6.19.

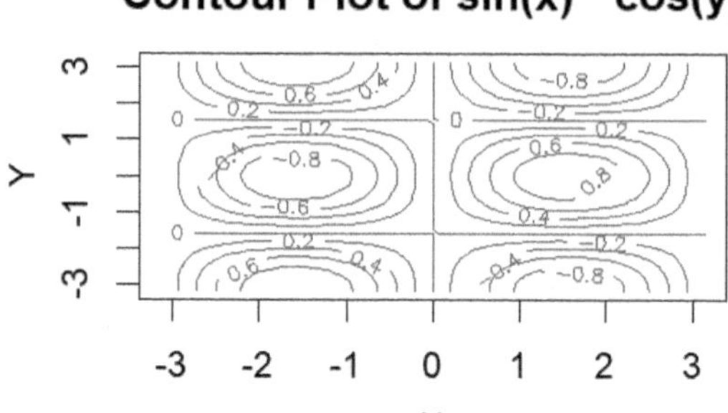

FIGURE 6.18
Contour plot.

6.3.6 Violin Plot

Plotting numerical data can be done with a violin plot, which is essentially a box plot combined with a kernel density plot. It displays the data distribution for a categorical variable at various levels (Takahashi & Takahashi, 2023). The violin plot's tighter areas show a lower density of data points, while the larger regions show a higher density. The distribution of data amongst several groups can be compared with the use of violin plots. The violin plot's important features are the following:

1. The data at various values of the categorical variable are represented by a kernel density estimate on either side of the 'violin.'
2. The centre line of the plot is symmetrical.
3. To display summary data, the plot may additionally incorporate box plot components (such as the median and quartiles).

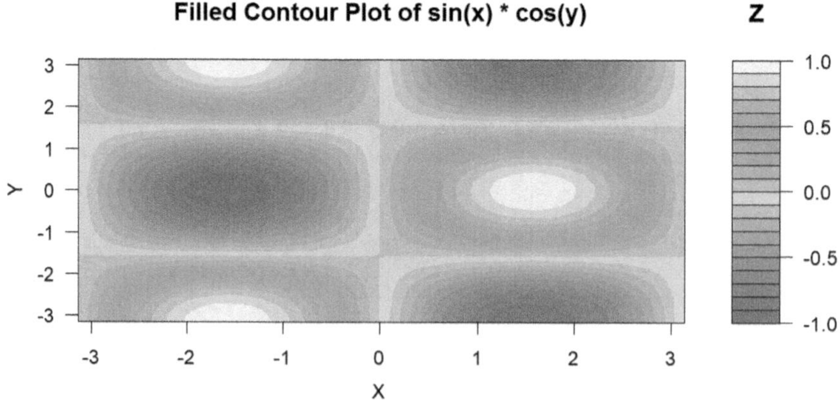

FIGURE 6.19
Filled contour plot.

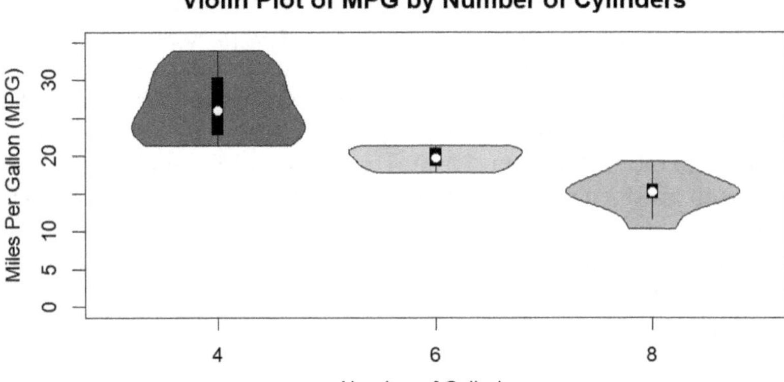

FIGURE 6.20
Violin plot.

4. The frequency of data points is represented by the width of the 'violin' at various values.

Plotting violins is not a feature of base R. But, for this, we may make use of the 'vioplot' package. The 'vioplot' package must first be installed and loaded (Adler & Adler, 2024).

```
Syntax: vioplot(data, col, main, xlab, ylab)
```

The following R code generates a violin plot, which is seen in Figure 6.20.

```
> # Violin Plot Example
> data(mtcars)

> # Define colors for violins
>colors<- c("red", "lightgreen", "lightpink")

> # Create the violin plot using vioplot
>vioplot(mtcars$mpg ~ as.factor(mtcars$cyl),col =
colors,horizontal = FALSE,  main = "Violin Plot of MPG by
Number of Cylinders",xlab = "Number of Cylinders",ylab =
"Miles Per Gallon (MPG)",ylim = c(0, 35))
```

6.4 Customized Plotting

Plotting in R can be customized to create customized representations that satisfy particular aesthetic and analytical requirements. Plots can be improved through customization to emphasize important findings and improve clarity.

Graphical components including colours, labels, annotations, axes, and general layouts are usually adjusted throughout this process. Choosing colour schemes that are appropriate for data interpretation, modifying axis labels to make data ranges more understandable, and adding annotations or legends to clarify intricate relationships or groupings within the data are some examples of customization (Hilfiger, 2015). This adaptability enables analysts and researchers to produce aesthetically pleasing and educational plots that facilitate data-driven decision-making processes and successfully convey findings.

Plotting with customization in R gives you fine control over many aspects of your visualizations, making them visually beautiful and functionally informative areas is provided below.

6.4.1 Chart Labelling and Documentation in Base R

To improve the readability and interpretability of your visualizations, labelling and documenting charts in base R require utilizing a variety of functions and arguments. It is crucial to provide the main title, axis labels, and annotations for data analysis. For each component, the following provides a thorough explanation and code samples:

- Primary Title (main): It gives the story a main title that summarizes its goal or content. In functions that plot data, such as plot(), the main parameter is set.
- Axis Labels: Indicates the corresponding data dimensions by labelling the x- and y-axis.

  Plotting functions' xlab and ylab parameters can be set.
- Annotations: Adds extra words or labels to particular plot points. Utilize the text() function and supply labels; pos, x, and y are parameters.

```
># Histogram with Main Title, Axis Labels, and Annotations
> data <- rnorm(100)

> # Plotting the histogram
>hist(data, main = "Histogram of Random Data",  # Main title
+       xlab = "Values",       # X-axis label
+       ylab = "Frequency",    # Y-axis label
+       col = "pink")     # Fill color of bars

> # Adding annotations (optional)
>text(0, 10, labels = "Distribution of Values", col =
"darkred", pos = 3)
```

The main title (main) is used to describe the contents of the histogram; set it to 'Histogram of Random Data.' The x-axis labelling data values is set to 'Values' in xlab, while the y-axis label is the count or frequency of data

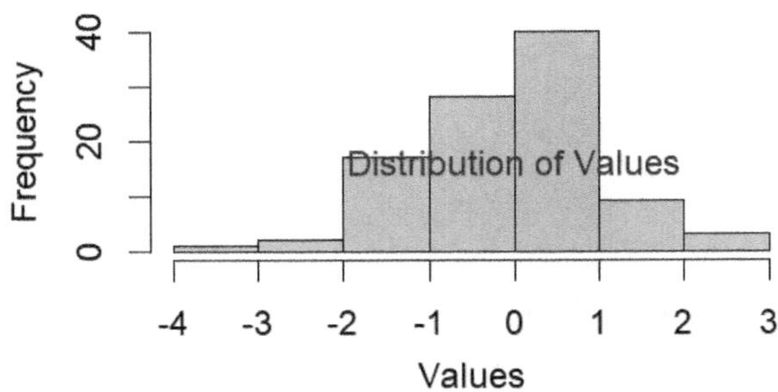

FIGURE 6.21
Title, axis, and annotations in histogram.

values and is set to 'Frequency' in ylab, pink is used to fill the bars in the histogram (col = 'pink'), and at coordinates (0, 10) with a dark red color (col = 'darkred'), an optional annotation labelled 'Distribution of Values' is inserted and placed to the right (pos = 3). The output is displayed in Figure 6.21.

6.4.2 Colour Specification in Base R

Customizing the colour of points, lines, text, and other graphical elements in your plots is possible with base R's simple colour specification feature. The following techniques and examples show you how to specify colours.

1. Employing Named Colours: You can directly specify a set of predefined named colours in base R by just calling them by name. Typical named colours include 'red,' 'blue,' and so on.

```
> # Example: Scatter plot with named colors
> x <- 1:10
> y <- x^2

> # Plot with red points
>plot(x, y, col = "red", main = "Scatter Plot with Named
Color")

> # Adding lines with different colors
>lines(x, y - 10, col = "blue")  # Blue line
>lines(x, y + 10, col = "yellow") # Yellow line
```

2. Using RGB Colour Specification: RGB (Red-Green-Blue) values can also be used to determine colours. RGB values are from 0 to 1, with red being (0, 0, 0), green being (0, 1, 0), blue being (0, 0, 1).

```
> # Example: Scatter plot with RGB color specification
> x <- 1:10
> y <- x^2

> # RGB colors: Red, Green, Blue
>plot(x, y, col = rgb(1, 0, 0), main = "Scatter Plot with
RGB Color")
>points(x, y - 10, col = rgb(0, 1, 0))  # Green points
>points(x, y + 10, col = rgb(0, 0, 1))  # Blue points
```

3. Using Hexadecimal Colour Codes: More precise control over colour hues is possible with hexadecimal colour codes. Any two-digit hexadecimal number between 00 and FF represents any of the three colour channels (red, green, and blue).

```
> # Example: Scatter plot with hexadecimal color codes
> x <- 1:10
> y <- x^2

> # Hexadecimal color codes: Red, Green, Blue
>plot(x, y, col = "#FF0000", main = "Scatter Plot with Hex
Color")
>points(x, y - 10, col = "#00FF00")  # Green points
>points(x, y + 10, col = "#0000FF")  # Blue points
```

4. Employing Grey Levels: Grey levels can be defined with a value between 0 (black) and 1 (white). For instance, setting the colour to a mid-grey hue uses col = "gray50".

```
> # Example: Scatter plot with gray levels
> x <- 1:10
> y <- x^2

> # Gray levels: Black, Gray50, White
>plot(x, y, col = "black", main = "Scatter Plot with Gray
Levels")
>points(x, y - 10, col = "gray50")  # Gray50 points
>points(x, y + 10, col = "white")   # White points
```

Incorporating different colour levels into a single scatter plot with base R is demonstrated in this example. Various data points can be efficiently shown and certain elements of interest can be highlighted by mixing named colours, RGB values, hexadecimal colour codes, and grey levels. The result is displayed in Figure 6.22.

```
> # Generate data
> x <- 1:10
> y <- x^2
```

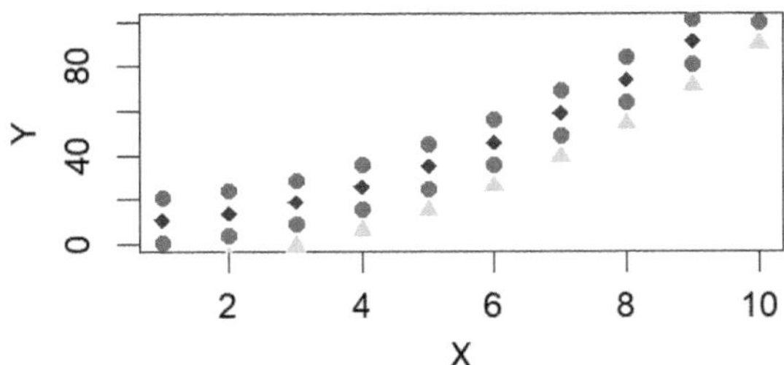

FIGURE 6.22
Usage of different colour labels in one plot.

```
> # Plot with different color levels
>plot(x, y, col = "red", pch = 16, main = "Scatter Plot with
Different Color Levels", xlab = "X", ylab = "Y")
>
> # Add points with different colors
>points(x, y - 10, col = rgb(0, 1, 0), pch = 17)# Green points
with RGB color
>points(x, y + 10, col = "#0000FF", pch = 18)# Blue points
with hexadecimal color code
>points(x, y + 20, col = "gray50", pch = 19) # Gray50 points
```

6.4.3 Font Specification in Base R

You can alter the appearance of text elements in your plots, including titles, axis labels, annotations, and legend texts, using base R's font specification feature. Here are some techniques and code samples to help you specify typefaces in different ways:

1. Tuning the Font Size: The size of text elements in relation to the default size is controlled by the 'cex' parameter. It adjusts the text size for annotations, axis labels, and titles.

```
> # Example: Scatter plot with adjusted font size
> x <- 1:10
> y <- x^2

>plot(x, y,
+       main = "Scatter Plot with Adjusted Font Size",#
Main title
+       xlab = "X Axis",   # X-axis label
+       ylab = "Y Axis",   # Y-axis label
+       cex.main = 1.5,    # Main title font size
+       cex.lab = 1.2,     # Axis labels font size
+       cex.axis = 0.8,    # Axis numbers fontsize
+       cex.sub = 0.8)     # Subtitle font size
```

2. Configuring Font Styles: The font parameter in base R allows you to specify font styles for text elements. Common values include 1 (plain), 2 (bold), 3 (italic), and 4 (bold italic).

```
> # Example: Scatter plot with different font styles
> x <- 1:10
> y <- x^2

>plot(x, y,
+       main = "Scatter Plot with Different Font Styles",#
Main title
+       xlab = "X Axis",    # X-axis label
+       ylab = "Y Axis",    # Y-axis label
+       font.main = 2,      # Bold main title
+       font.lab = 3)       # Italic axis labels
```

3. Modifying the Font Family: You can choose different font families for text components by using the family option. Prominent choices consist of 'serif,' 'sans,' 'mono,' and particular font names, in case they are accessible.

```
> # Example: Scatter plot with different font families
> x <- 1:10
> y <- x^2

>plot(x, y,
+       main = "Scatter Plot with Different Font Families",
# Main title
+       xlab = "X Axis",    # X-axis label
+       ylab = "Y Axis",    # Y-axis label
+       family = "mono")    # Monospaced fontfamily
```

6.4.4 Plot Size Specification in Base R

In R, 'plot size specification' means modifying the general layout, text and symbol sizes, plot dimensions, and other graphical element proportions. Here are some different ways in which you can declare sizes in R:

1. Calibrating Plot Dimensions: Functions like png(), jpeg(), and pdf() allow you to specify the width and height of your plot when saving it to a PNG, JPEG, or PDF file format.

```
> # Example: Saving a plot with specific dimensions
> x <- 1:10
> y <- x^2
```

```
> # Save as PNG with width = 800 pixels and height = 600 pixels
>png("plot.png", width = 800, height = 600)
>plot(x, y, main = "Scatter Plot", xlab = "X", ylab = "Y")
>dev.off()
```

2. Managing Plot Layout: Plot margins, axis orientation, and other graphical features can be adjusted using base R's par() function, which affects the plot's overall size and arrangement.

```
> # Example: Adjusting plot margins and layout
> x <- 1:10
> y <- x^2

>par(mar = c(5, 5, 4, 2))  # Set plot margins (bottom,
left, top, right)
>plot(x, y, main = "Scatter Plot with Adjusted Margins",
xlab = "X", ylab = "Y")
```

3. Aspect Ratio: Plot aspect ratio is controlled via the 'asp' option in base R, which preserves proportionate scaling between x- and y-axis units.

```
> # Example: Scatter plot with specific aspect ratio
> x <- 1:10
> y <- x^2

>plot(x, y, main = "Scatter Plot with Specific Aspect
Ratio", xlab = "X", ylab = "Y", asp = 1)
```

6.4.5 Symbol Plotting

The capacity to alter and present data points using different symbols or forms is known as symbol charting. When separating out groups or categories inside a data set, these symbols can be helpful. In plots, you can designate several charting symbols for data points by using the pch parameter in base R. Table 6.2 list the various symbols and corresponding 'pch' value.

```
> # Example: Scatter plot with symbols and colors
> x <- 1:10
> y <- x^2

>plot(x, y, pch = 16, col = "blue", main = "Scatter Plot with
Symbols and Colors", xlab = "X", ylab = "Y")
>points(x, y - 10, pch = 17, col = "green") # Triangle points
in green
>points(x, y + 10, pch = 18, col = "red")   # Cross points in
red
```

TABLE 6.2
Plot Symbols

pch value	Shape
1	Circle
2	Point Up Triangle
3	Cross
4	Plus
5	Diamond
6	Point Down Triangle
7	Square
8	Cross

6.4.6 Plot Settings

Plot margins, axis limitations, grid lines, background colour, and other visualization features can all be customized in R by adjusting the plot parameters. Several graphical parameters can be set using R. To manage the distance between the plot area and the plotting device's edge, you can modify the plot margins (mar). To focus on particular ranges of your data, you can adjust the x- and y-axis limits (xlim and ylim, respectively). Using the grid() method and parameters like col for colour and lty for line type, you can add grid lines to the plot. By utilizing the 'bg' parameter in par(), you can change the plot's background colour. Additionally, use the par() method to change the location of the title and axis labels. Below is a consolidated example that uses functions to display several plot settings. Figure 6.23 displays the output.

```
> # Example: Scatter plot with various plot settings
> x <- 1:10
> y <- x^2
```

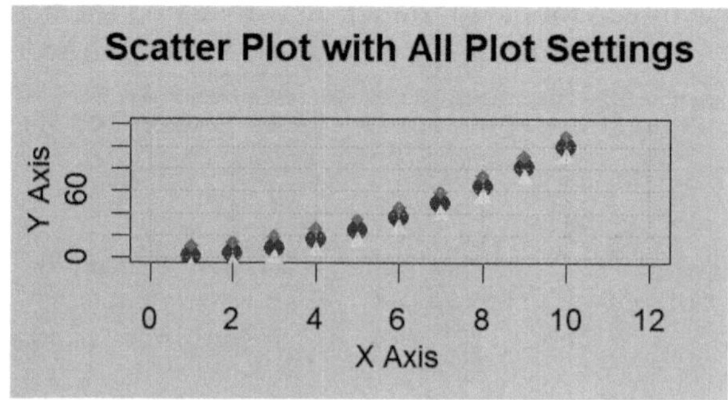

FIGURE 6.23
Plot settings in a scatter plot.

```
> # Set plot margins, adjust axis limits, add grid lines,
change background color, adjust axis labels and title position
> par(mar = c(5, 5, 4, 2), mgp = c(2, 1, 0), oma = c(0, 0, 2, 0),
bg = "lightblue")
>plot(x, y, pch = 16, cex = 1.5, col = "blue", xlim = c(0, 12),
ylim = c(0, 120),main = "Scatter Plot with All Plot Settings",
xlab = "X Axis", ylab = "Y Axis")
>points(x, y - 10, pch = 17, col = "green")
>points(x, y + 10, pch = 18, col = "red")
>grid(col = "gray", lty = 3)
```

6.5 Packages for Data Visualization

Data visualization is a vital component of data analysis, offering a means to understand and convey data insights effectively. Effective comprehension and communication of data insights are made possible by data visualization, which is an essential component of data analysis (Kabacoff, 2022). Data visualization allows for immediate comprehension of information, which would otherwise take significantly longer to extract from spreadsheets or text reports. The R programming language is intended for use in scientific research, graphical data analysis, and statistical computation. It is typically chosen for data visualization since, thanks to its packages, it provides flexibility and little coding. With a plethora of packages built to provide a diverse range of visualizations, R is widely recognized for its broad skills in data visualization. Listed here are the most popular and potent R data visualization packages:

- ggplot2
- lattice
- plotly
- leaflet
- dygraphs
- esquisse

The ensuing sections will contain a thorough explanation of each visualization technique as well as a list of pros and cons for each.

6.5.1 ggplot2

The Grammar of Graphics, a data visualization framework created by Leland Wilkinson, provides the foundation for the ggplot2 R package (Wickham, 2011). The Grammar of Graphics offers a methodical approach to defining and creating visual aids. It disassembles graphics into individually manipulable semantic components. Using this method, users can combine basic graphical elements to build advanced visualizations. With a logical and consistent approach to plot construction, it is intended to produce intricate, multilayered images. Using high-level API, ggplot2 can produce data visualizations including scatter plots, bar charts, error charts, and more. Because of its versatility, strength, and ease of use, ggplot2 is highly acclaimed (Wilkinson, 2011). The following are some of its main benefits:

1. This package offers a method to data visualization that is well organized and cohesive.
2. It enables simple customization and progressive plot creation.
3. It makes it easier to create intricate charts, which lowers the learning curve and boosts output.
4. A workflow for data analysis is made easier by seamless connection with existing Tidyverse packages like tidyr and dplyr (Kabacoff, 2022).
5. Interactive plots that are made by users can improve their experience with exploratory data analysis.

A ggplot2 consists of the following fundamental parts:

- data: Plotting data
- geom: Geometric patterns that are utilized to visualize
- aes: Aesthetics of geometric and statistical items, such as transparency, size, form, colour, and position
- sacle: Contains a collection of values for every plot's aesthetic mapping
- coord: Used to map data coordinates and arrange geometric objects
- stat: Determines the various data values that are plotted
- theme: The general graphic components of a plot, such as the backdrop, typefaces, colours, and grids and axes
- facet: For each collection of data, a grid of charts is shown

Let us utilize the 'mtcars' data set to show how to create different kinds of charts with ggplot2. One famous data set that comes with R is called mtcars. It has information on several automobile models, such as weight, horsepower, number of cylinders, miles per gallon (mpg), and more. Below is sample code that uses ggplot2 to create different kinds of complex graphs.

1. Scatter Plot: A scatter plot (Figure 6.24) can illustrate the correlation between horsepower (hp) and miles per gallon (mpg).

```
> # Load necessary library
> library(ggplot2)
> # Load the mtcars dataset
```

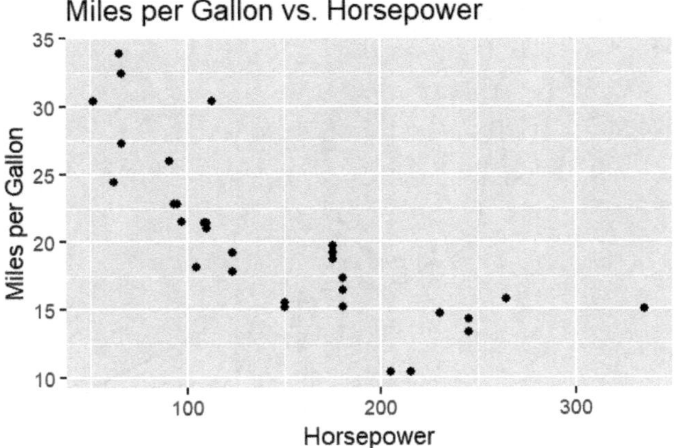

FIGURE 6.24
Scatter plot using ggplot2.

```
> data(mtcars)
> # Preview the data
> head(mtcars)
                   mpg cyl disp  hp drat    wt  qsec vs am gear carb
Mazda RX4         21.0   6 160 110 3.90 2.620 16.46  0  1    4    4
Mazda RX4 Wag     21.0   6 160 110 3.90 2.875 17.02  0  1    4    4
Datsun 710        22.8   4 108  93 3.85 2.320 18.61  1  1    4    1
Hornet 4 Drive    21.4   6 258 110 3.08 3.215 19.44  1  0    3    1
Hornet Sportabout 18.7   8 360 175 3.15 3.440 17.02  0  0    3    2
Valiant           18.1   6 225 105 2.76 3.460 20.22  1  0    3    1

>ggplot(mtcars, aes(x = hp, y = mpg)) +geom_point() + labs(title
= "Miles per Gallon vs. Horsepower",x = "Horsepower",y = "Miles
per Gallon")
```

2. Line Chart: A line graph (Figure 6.25) can illustrate the relationship between car weight (wt) and miles per gallon (mpg).

```
ggplot(mtcars, aes(x = wt, y = mpg)) + geom_line() +
labs(title = "Miles per Gallon by Car Weight",x = "Weight
(1000 lbs)",y = "Miles per Gallon")
```

3. Bar Graph: A bar graph (Figure 6.26) can display the number of vehicles in each category of cylinders.

```
>ggplot(mtcars, aes(x = factor(cyl))) + geom_bar(fill =
"orange") + labs(title = "Count of Cars by Number of
Cylinders",x = "Number of Cylinders",y = "Count of Cars")
```

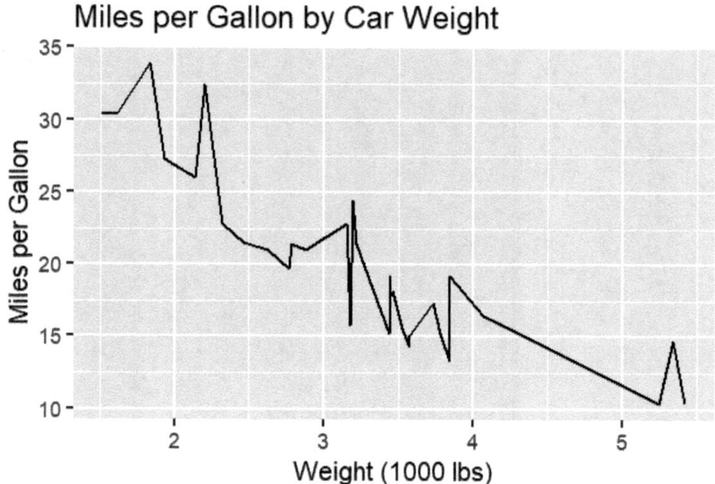

FIGURE 6.25
Line graph using ggplot2.

4. Histogram: A histogram (Figure 6.27) shows the distribution of miles per gallon (mpg).

```
>ggplot(mtcars, aes(x = mpg)) + geom_histogram
(binwidth = 2, fill = "orange", color = "black"),labs
(title = "Distribution of Miles per Gallon",x = "Miles
per Gallon",y = "Frequency")
```

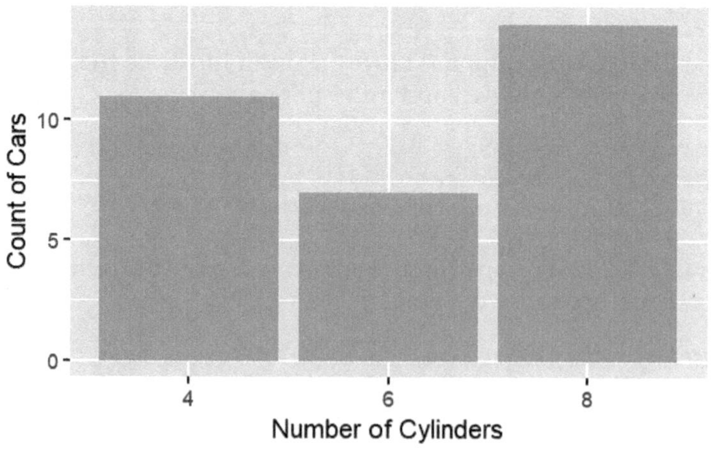

FIGURE 6.26
Bar graph using ggplot2.

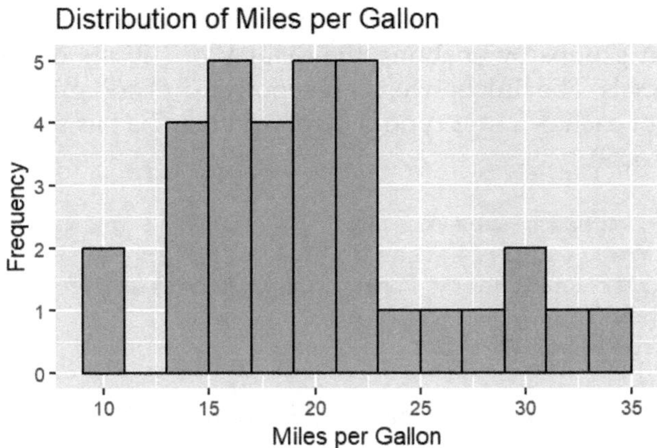

FIGURE 6.27
Histogram using ggplot2.

5. Box Plot: A box plot (Figure 6.28) shows how miles per gallon (mpg) are distributed among various cylinder classifications.

```
>ggplot(mtcars, aes(x = factor(cyl), y = mpg,
fill = factor(cyl))) + geom_boxplot() + labs
(title = "Miles per Gallon by Number of Cylinders",
x = "Number of Cylinders",y = "Miles per Gallon")
```

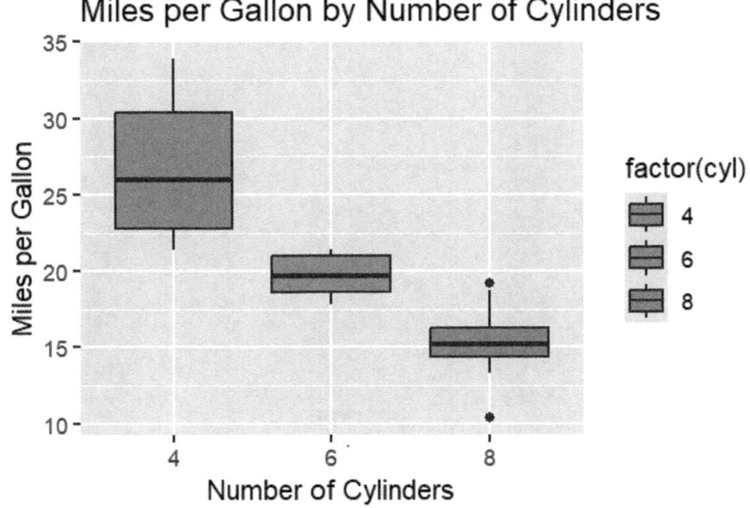

FIGURE 6.28
Box plot using ggplot2.

6. Faceted Plot: Faceting makes it simpler to compare distributions between groups by enabling the display of subsets of data in distinct panels. The link between horsepower (hp) and miles per gallon (mpg) for each cylinder type is displayed using a faceted scatter plot (Figure 6.29).

```
>ggplot(mtcars, aes(x = hp, y = mpg)) + geom_point() +
facet_wrap(~ cyl) + labs(title = "Miles per Gallon vs.
Horsepower Faceted by Number of Cylinders", x =
"Horsepower",
y = "Miles per Gallon")
```

7. Density Plot: A density diagram (Figure 6.30) illustrates the weight distribution of cars (wt).

```
>ggplot(mtcars, aes(x = wt)) + geom_density(fill =
"orange") + labs(title = "Density Plot of Car Weights",
x = "Weight (1000 lbs)",y = "Density")
```

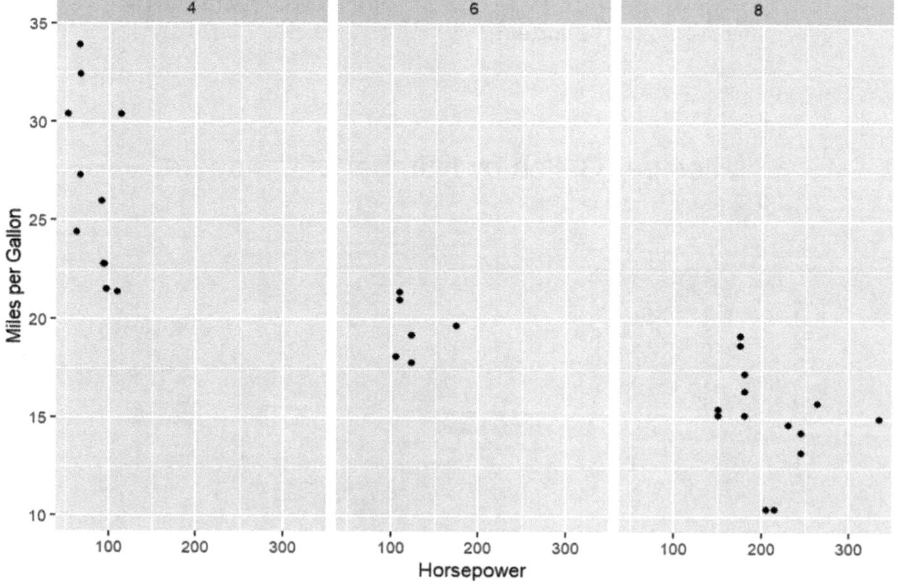

FIGURE 6.29
Faceted plot using ggplot2.

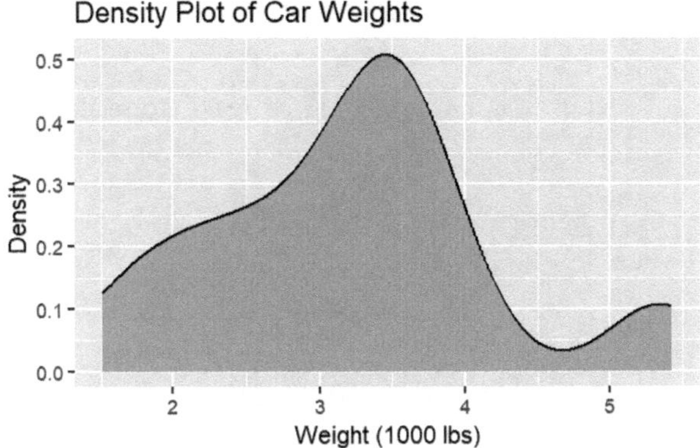

FIGURE 6.30
Density plot using ggplot2.

8. Violin Plot: A violin plot that displays the horsepower (hp) distribution among various cylinder classes is visualized in Figure 6.31.

```
>ggplot(mtcars, aes(x = factor(cyl), y = hp,
fill = factor(cyl))) + geom_violin() + labs(title =
"Distribution of Horsepower by Number of Cylinders",
x = "Number of Cylinders",y = "Horsepower")
```

9. Heat Map: The correlations between the numerical variables in the 'mtcars' data set must be determined, and the correlation matrix must then be formatted so that ggplot2 can use it to produce a heat map. The pairwise correlations between each of the data set's numerical variables are determined by the 'cor' function. The wide-format

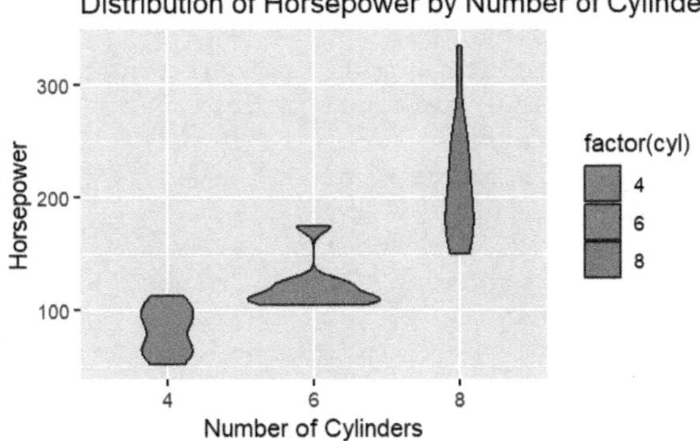

FIGURE 6.31
Violin plot using ggplot2.

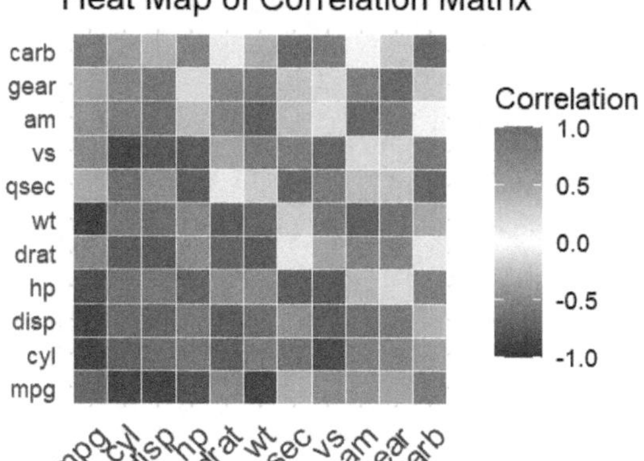

FIGURE 6.32
Heat map using ggplot2.

correlation matrix is transformed into a long-format data frame that is compatible with ggplot2 by the melt() function from the reshape2 package (Abedin, 2014). The heat map is produced by the geom_tile() function. A colour gradient for the correlations is defined using the scale_fill_gradient2() function; blue denotes negative correlations, red denotes positive correlations, and white denotes no correlation. An easy-to-understand method of comprehending the correlations between the many variables in the 'mtcars' data set is shown in Figure 6.32.

```
> # Load necessary libraries
> library(ggplot2)
> library(reshape2)
> # Load the mtcars dataset
> data(mtcars)
> # Calculate the correlation matrix
>cor_matrix<- cor(mtcars)
> # Melt the correlation matrix into long format
>cor_matrix_melted<- melt(cor_matrix)
> # Create the heatmap
>ggplot(cor_matrix_melted, aes(x = Var1, y = Var2,
fill = value)) +
+      geom_tile(color = "white") +
+      scale_fill_gradient2(low = "blue", high = "red",
       mid = "white",
+                            midpoint = 0, limit = c(-1, 1),
                             space = "Lab",
+                            name="Correlation") +
+      theme_minimal() +
+      theme(axis. Text.x = element_text(angle = 45,
       vjust = 1,
```

```
+       size = 12, hjust = 1)) +
+       coord_fixed() +
+       labs(title = "Heatmap of Correlation Matrix",
        x = "", y = "")
```

These illustrations demonstrate how ggplot2 can be used to create a wide range of charts for data analysis and visualization. By using different types of plots, you can gain multiple perspectives on the 'mtcars' data set, such as understanding relationships, distributions, and comparisons across different categories.

6.5.2 lattice

The R lattice package provides a flexible and strong foundation for creating trellis graphics, which are particularly useful for multivariate data visualization (Grothendieck, 2008). Its foundation lies in the ideas of 'trellis display,' a framework for visualizing data. The purpose of trellis graphics is to give conditional plots, which show data separated by the levels of one or more categorical variables, a logical framework. A panel is the name given to each plot in a trellis display. The plotting of each panel is controlled by the lattice package through the usage of panel functions. The lattice package specifies the variables to be plotted and the conditioning variables through a formula interface. Different panels, each showing a portion of the data, can be created by employing conditioning variables. Several benefits of using lattice packages include the following:

- A consistent presentation of multivariate data is guaranteed by trellis graphics.
- It is simple to show how correlations between variables alter when other variable levels vary.
- Incredibly configurable, with a number of pre-installed themes and the option to create unique panel features.

Table 6.3 gives a concise summary of the primary functions available in the lattice package.

Let us explore how to create a couple of the graph types listed in Table 6.3 using the lattice package. We will use the 'mtcars' data set for illustration.

1. Scatter Plot:A simple scatter plot is made with xyplot(), the main lattice function for making scatter plots. A scatter plot with conditioning on cylinder count is shown in Figure 6.33.

```
> # Scatter plot of mpg vs. hp
>xyplot(mpg ~ hp, data = mtcars, main = "Miles/Gallon vs.
Horsepower",xlab = "Horsepower",ylab = "Miles per Gallon")
```

TABLE 6.3

Primary Functions in the Lattice Package

Method	Elucidation
xyplot()	Scatter plot creation
cloud()	3-D scatter plot creation
stripplot()	Strip plot creation
histogram()	histogram creation
bwplot()	Box plot creation
barchart()	Bar graph creation
splom()	Matrix of scatter plot creation
dotplot()	Dot chart creation
contourplot()	3-D contour chart creation
qq()	Quantile plot creation
wireframe()	3-D wireframe graph creation

2. Histogram: A histogram in the lattice package is generated with the histogram() method and is depicted in Figure 6.34.

```
>histogram(~ mpg, data = mtcars, main = "Histogram of
Miles per Gallon",xlab = "Miles per Gallon")
```

3. Box Plot: The bwplot() function is employed to generate a box plot (Figure 6.35), and the complete code is provided below:

```
> # Box plot of mpg by number of cylinders
>bwplot(mpg ~ factor(cyl), data = mtcars, main = "Box
Plot of Miles/Gallon by Cylinders",xlab = "Number of
Cylinders",ylab = "Miles per Gallon")
```

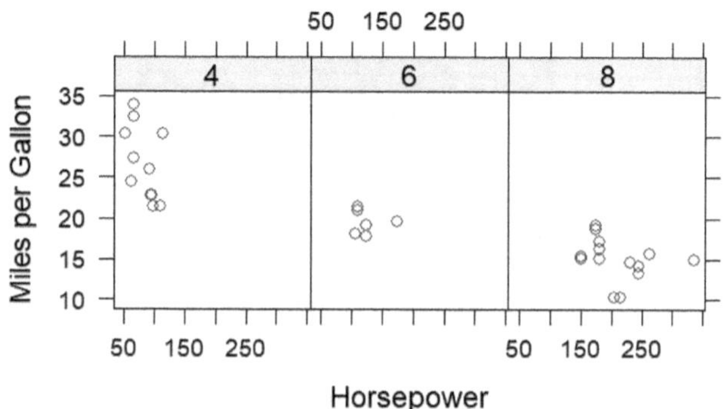

FIGURE 6.33
Scatter plot using lattice.

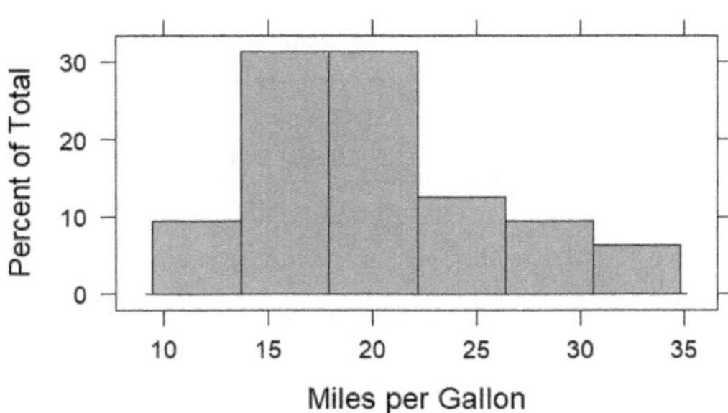

FIGURE 6.34
Histogram using lattice.

4. Density Plot: Density plots are created using the densityplot() method.

```
> # Density plot of mpg
>densityplot(~ mpg, data = mtcars, main = "Density Plot
of Miles per Gallon",xlab = "Miles per Gallon")
```

5. Multipanel Scatter Plot: Scatter plots for every pair of variables in a data set are shown in multipanel scatter plots, sometimes referred to as pairs plots or scatter plot matrices. A multipanel scatter plot's primary function is to show the pairwise correlations between the variables in a data set. It enables us to swiftly spot trends, connections, and possible outliers in the data across a variety of dimensions. Figure 6.36 illustrates a scatter plot with multiple panels.

```
> # Scatter plot conditioned on number of cylinders and gears
>xyplot(mpg ~ hp | factor(cyl) * factor(gear), data = mtcars,
main = "Miles per Gallon vs. Horsepower by Cylinders and
Gears",xlab = "Horsepower",ylab = "Miles per Gallon")
```

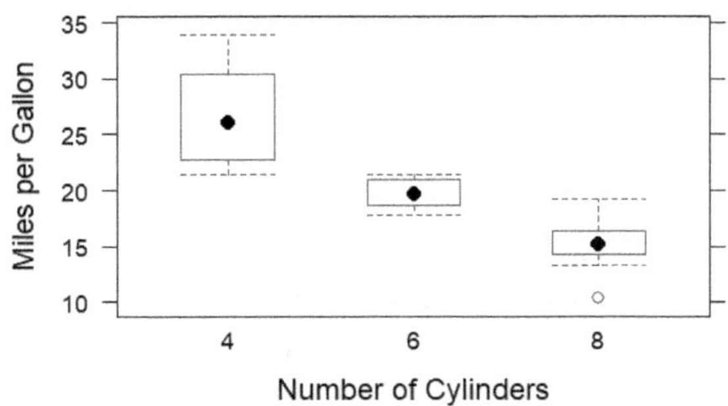

FIGURE 6.35
Box plot using lattice.

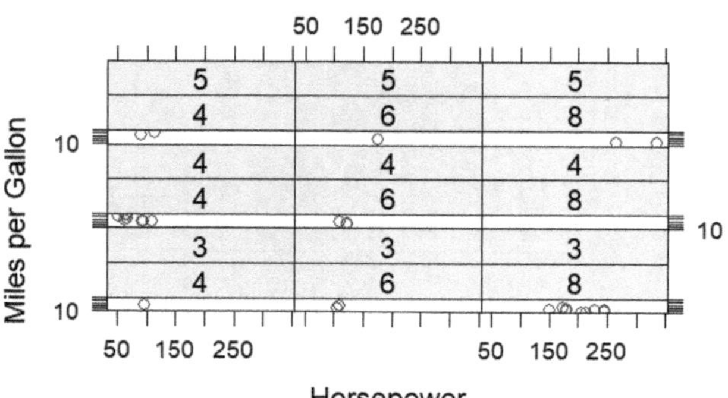

FIGURE 6.36
Multiple panel scatter plot using lattice.

6. Custom Panel Functions: The lattice package's custom panel functions offer an influential method for adding certain components to each panel in a plot, enabling extremely personalized visualizations. In lattice, a panel function is just a function that specifies what has to be drawn in each plot panel. Regression lines, custom points, text comments, and more can be added with these tools. The data points to be shown, x and y, along with additional parameters, are the arguments taken by each panel function. The common panel functions in this package are summarized in Table 6.4.

   Custom panel functions enable advanced customization, and the R code demonstrating a customized panel function is provided below.

```
> # Sample data creation
>set.seed(123)  # for reproducibility
>sample_data<- data.frame(var1 = rnorm(100),var2 = rnorm(100),
var3 = rnorm(100),var4 = rnorm(100))
>custom_panel_pairs<- function(x, y, ...) {
+     panel.xyplot(x, y, ...)
```

TABLE 6.4
Panel Functions Summary

Function_Name	Summary
panel.xyplot	Simple scatter plot of points
panel.abline	Includes a line in the panel
panel.lmline	Includes a regression line
panel.text	Integrates textual annotations
panel.smoothScatter	Incorporates a smooth density representation into the scatter plot

```
+      # Add text annotations (row numbers)
+      panel.text(x, y, labels = 1:length(x), pos = 1,
       cex = 0.6, col = "blue")
+ }
> # Create the scatter plot matrix
>splom(~sample_data,panel = custom_panel_pairs,main =
"Custom Panel Function Example",col.symbol = "red")
```

The custom_panel_pairs() function specifies how each panel in the scatter plot matrix (splom) should be customized. The default scatter plot points are plotted using the panel.xyplot() method. Meanwhile, labels = 1:length(x) fetches the row numbers from the data, and panel. text adds text annotations (more precisely, row numbers) to each panel. For the sample data, the above code will create a scatter plot matrix (pairs plot). It will feature custom annotations showing row numbers (1–100) and scatter plots illustrating relationships between various pairs of variables (var1, var2, var3, var4) in each panel, resembling Figure 6.37.

7. 3-D Plot: Three-dimensional plots are not natively supported by the lattice package. However, you can use the 'latticeExtra' package, which integrates with the 'rgl' package to provide 3-D plots and adds more functionality to lattice.

```
>install.packages("latticeExtra")
>install.packages("rgl")
> # Load necessary libraries
> library(lattice)
> library(latticeExtra)
> # Convert the mtcars dataset to a suitable format
```

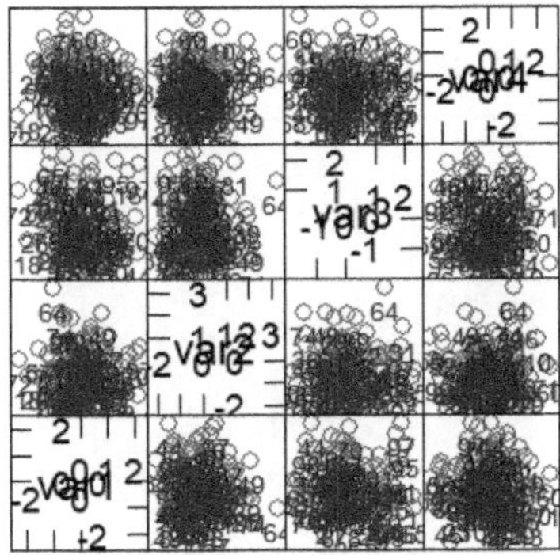

FIGURE 6.37
Customized plot.

```
>mtcars$car<- rownames(mtcars)
> # Create the 3D scatter plot
>cloud(mpg ~ hp * wt, data = mtcars, screen = list(z = 40,
x = -60),xlab = "Horsepower",ylab = "Weight",zlab = "Miles per
Gallon",main = "3D Scatter Plot of mtcarsData",col = "blue",
pch = 16)
```

With the help of the aforementioned code, you can investigate the correlations between mpg, horsepower, and wt in the mtcars data set by rotating and zooming in on the interactive 3-D scatter plot (Figure 6.38). Three-dimensional scatter plots are made using the cloud() function found in the latticeExtra package (Chambers, 2008). You can specify the angles for rotating the 3-D graphic using the screen argument.

6.5.3 plotly

Graphs of interactive quality can be created online with plotly, an interactive graphing library. Based on the D3.js and WebGL libraries, plotly is especially useful for making interactive graphs. Plotly is a library in R that can be used to build interactive plots by calling methods that produce various plot kinds (Sievert, 2020). In R, plotly is an effective tool for producing interactive visualizations that are suitable for publication. You can develop a variety of visualizations to meet your data analysis needs by being familiar with its fundamental elements and customization choices (Sunitha et al., 2023). Plotly's fundamental components include:

- Data: The information that you wish to display. It could be represented using matrices, data frames, or other types of data structures.
- Traces: The fundamental data units in plotly. A single visualization type, such as a scatter plot, bar graph, etc., is represented by each trace.
- Layout: The plot arrangement, including titles, axes, and notes.

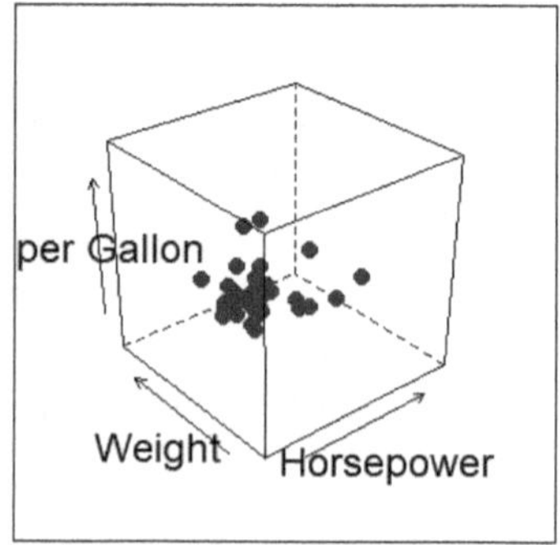

FIGURE 6.38
3-D scatter plot using lattice.

In R programming, a package needs to be installed before it can be used. To set up the complete plotly package, type this: >install.packages('plotly').

plot_ly() is a crucial function for creating graphs because it starts a plotly visualization. This method transfers R objects to the web-based interactive charting tool plotly.js (licenced under an MIT licence). Usually the first step in creating a plot is to define the data and the kind of layout you want. Here is a simple R example to create various plot types using this package:

1. Scatter Plot: A scatter plot to investigate the link between horsepower (hp) and miles per gallon (mpg) is presented in Figure 6.39.

```
> library(plotly)
> data(mtcars)
> # Scatter Plot using plotly
> fig <- plot_ly(mtcars, x = ~hp, y = ~mpg, type =
'scatter', mode = 'markers',text = ~paste("Car:",
rownames(mtcars)),marker = list(size = 10, color =
'blue'))

> fig <- fig %>% layout(title = "Horsepower vs. MPG",
xaxis = list(title = "Horsepower"),yaxis = list(title =
"Miles per Gallon"))

> fig
```

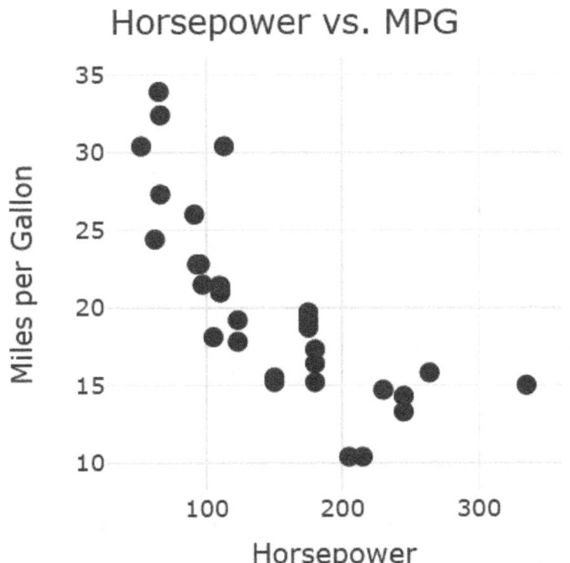

FIGURE 6.39
Scatter plot using plotly.

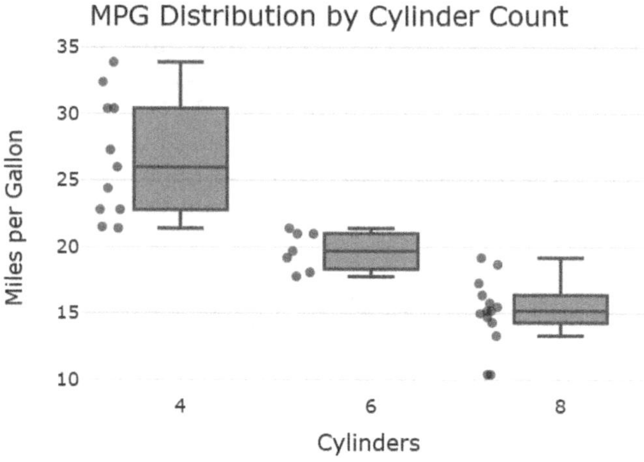

FIGURE 6.40
Box plot using plotly.

2. Box Plot: This R code can be used to create a box plot that compares the miles per gallon (mpg) distribution among various cylinder counts (cyl). The result is shown in Figure 6.40.

```
> fig <- plot_ly(mtcars, x = ~as.factor(cyl), y = ~mpg,
type = 'box',marker = list(color = 'rgba(7, 40, 89, 0.5)'),
boxpoints = 'all')

> fig <- fig %>% layout(title = "MPG Distribution by
Cylinder Count",xaxis = list(title = "Cylinders"),yaxis =
list(title = "Miles per Gallon"))

> fig
```

3. Histogram: To analyse the miles per gallon (mpg) distribution, this is shown in a histogram (Figure 6.41).

```
> fig <- plot_ly(mtcars, x = ~mpg, type =
'histogram',marker = list(color = 'rgba(100, 100, 200, 0.7)'))

> fig <- fig %>% layout(title = "MPG Distribution",xaxis =
list(title = "Miles per Gallon"),yaxis = list(title =
"Count"))

> fig
```

4. Bar Graph: Use R code to make a bar graph that displays the average miles per gallon (mpg) for each cylinder (cyl) that is listed below. The output is displayed in Figure 6.42.

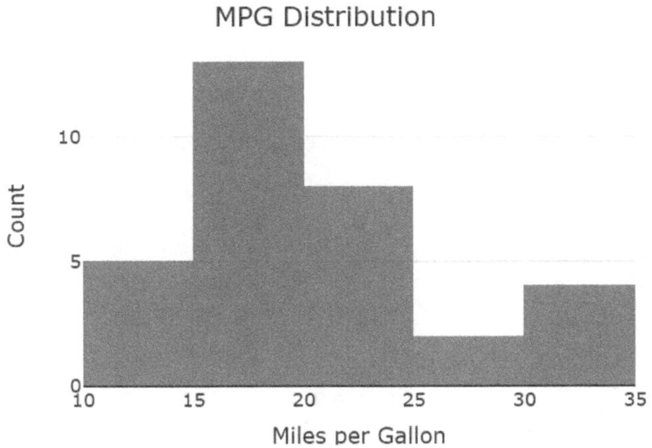

FIGURE 6.41
Histogram using plotly.

```
>agg_data<- aggregate(mpg ~ cyl, data = mtcars, FUN = mean)

> fig <- plot_ly(agg_data, x = ~as.factor(cyl), y = ~mpg,
type = 'bar',marker = list(color = 'rgba(204, 204, 204, 1)',
line = list(color = 'rgba(0, 0, 0, 1)', width = 1.5)))

> fig <- fig %>% layout(title = "Average MPG by Cylinder
Count",xaxis = list(title = "Cylinders"),yaxis =
list(title = "Average Miles per Gallon"))

> fig
```

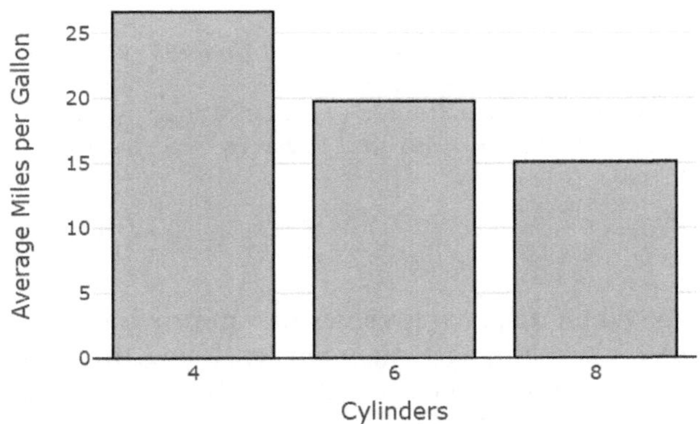

FIGURE 6.42
Bar graph using plotly.

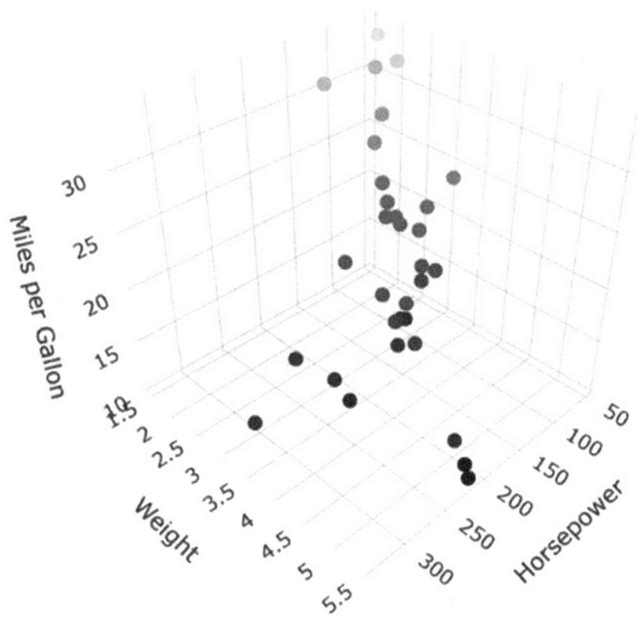

FIGURE 6.43
3D scatter plot using plotly.

5. 3-D Scatter Plot: A 3-D scatter plot to investigate the correlations amongst three variables, weight (wt), horsepower (hp), and miles per gallon (mpg), is illustrated in Figure 6.43.

```
> fig <- plot_ly(mtcars, x = ~hp, y = ~wt, z = ~mpg,
type = 'scatter3d', mode = 'markers',marker = list(size =
5, color = ~mpg, colorscale = 'Viridis'))

> fig <- fig %>% layout(title = "3D Scatter Plot: HP,
Weight,
and MPG",scene = list(xaxis = list(title = "Horsepower"),
yaxis = list(title = "Weight"),zaxis = list(title = "Miles
per Gallon")))

> fig
```

6. Heat Map: When displaying values in a matrix format, a heat map is helpful for visualizing their intensity. Using the code provided below, we will compute the correlation matrix of the mtcars data set; the resulting graph is displayed in Figure 6.44.

```
> # Calculate correlation matrix
>cor_matrix<- cor(mtcars)
```

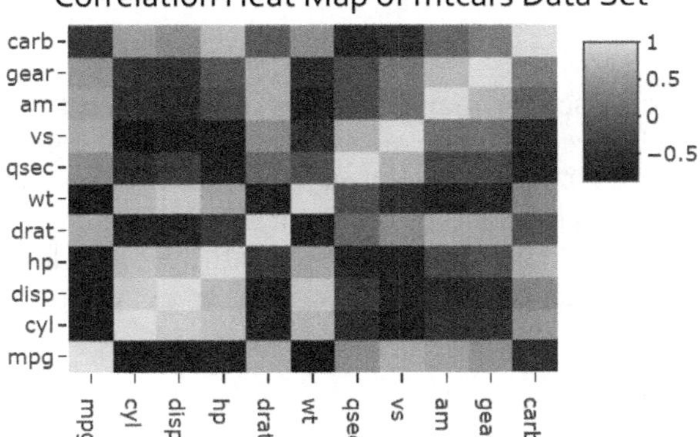

FIGURE 6.44
Heat map using plotly.

```
> # Convert correlation matrix to a format suitable for Plotly
> cor_data <- as.matrix(cor_matrix)

> # Create heatmap
> fig <- plot_ly(z = cor_data, x = colnames(cor_data), y =
rownames(cor_data), type = 'heatmap', colorscale = 'Viridis')

> fig <- fig %>% layout(title = "Correlation Heatmap of mtcars
Dataset",xaxis = list(title = ""),yaxis = list(title = ""))

> fig
```

7. Map: Robust support is offered by plotly for making maps, which can be aesthetically pleasing and quite interactive. The ability to visually display spatial data makes maps indispensable for data analysis. They make it possible to investigate regional patterns and trends that conventional charts and tables might not make clear. Maps have several benefits for data analysis, which are listed below:
   - The geographic distribution of data points can be seen with the aid of maps. This is especially helpful for locating outliers or other spatial patterns like clusters.
   - Thematic maps, such as choropleth maps, display the spatial variation of a specific variable or subject.
   - Maps can demonstrate how geographical patterns shift over time when paired with time data.
   - Using heat maps or density maps, one can identify trends and concentrations of data points

Here are some examples of how to use plotly in R to produce various kinds of maps:

a. Scatter Map: A scatter map is a form of map used to visualize the association between two or more variables with geographic coordinates (Dabbas, 2021). It is sometimes referred to as a scatter plot map or geographic scatter plot. Every point on the map denotes an observation with a particular longitude and latitude. This lets you visualize the distribution of data points over a certain region. Below is the code for how to use plotly in R to produce a scatter map:

```
> library(plotly)

> # Sample data: cities with latitudes and longitudes
> cities <- data.frame(city = c("New York", "Los Angeles",
"Chicago", "Houston", "Phoenix"),lat = c(40.7128, 34.0522,
41.8781, 29.7604, 33.4484),lon = c(-74.0060, -118.2437,
-87.6298, -95.3698, -112.0740),size = c(50, 100, 75, 40,
30),color = c('red', 'blue', 'green', 'purple', 'orange'))

> # Create the scatter map
> fig <- plot_ly(cities, lat = ~lat, lon = ~lon, text =
~city, type = 'scattergeo', mode = 'markers',marker =
list(size = ~size, color = ~color))

> fig <- fig %>% layout(title = 'City Locations with
Different Sizes and Colors',geo = list(scope = 'usa'))

> fig
```

A data frame called data_cities holds the names of the cities along with their latitudes and longitudes, as well as extra variables like size and colour. The plot_ly() function creates a scatter map with points plotted based on latitude and longitude. The points' appearance can be altered by adjusting the size and colour variables, and the layout() function sets the map's geographic scope and adds a title. The output of the above code is displayed in Figure 6.45.

b. Choropleth Map: Geographical regions are shaded or patterned on a choropleth map, a kind of theme map, according to the value of a given variable. Data that has been aggregated over geographical units, such as nations, states, counties, or districts, is visualized using it. Each region's colour or shading correlates to the value of the variable it represents, making cross-area comparisons simple. Choropleth maps are interactive, letting users click or hover

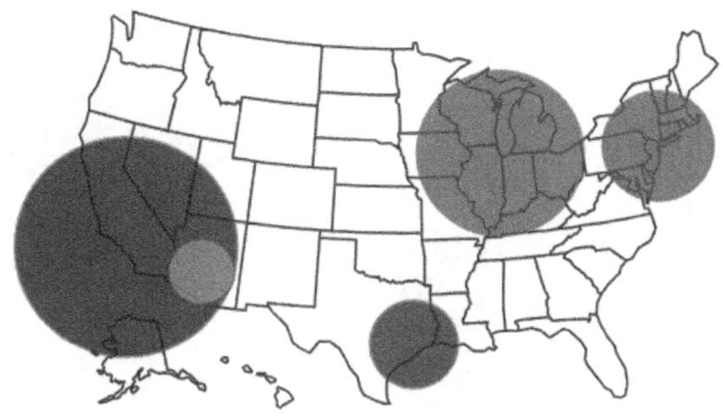

FIGURE 6.45
Scatter map using plotly.

over different regions to see more information or in-depth details (Dhande et al., 2023). The R code to generate a choropleth map is provided below:

```
> library(plotly)

> # Sample data: state populations
>state_populations<- data.frame(state = c("California",
"Texas", "Florida", "New York", "Pennsylvania"),population
= c(39538223, 29145505, 21538187, 20201249, 13002700),
state_code = c("CA", "TX", "FL", "NY", "PA"))

> # Create the choropleth map
> fig <- plot_ly(state_populations, type = 'choropleth',
locations = ~state_code, locationmode = 'USA-states',z =
~population, text = ~state, colorscale = 'Blues', colorbar
= list(title = 'Population'))

> fig <- fig %>% layout(title = 'state Populations in the
USA',geo = list(scope = 'usa'))

> fig
```

The sample data frame used for analysis is 'state_populations,' and it includes state names, population values, and state codes. The plot_ly() method creates a choropleth map, using different colours to indicate different populations; 'colorscale' describes the colour gradient that is used to show various demographic ranges; and 'colorbar' adds a colour bar to show the population's value range. Figure 6.46 displays a generated choropleth map.

FIGURE 6.46
Choropleth map using plotly.

c. Heat Map on Map: A heat map on a map is a type of visualization that represents the intensity or density of data points across a geographic area using colour gradients. In addition to highlighting data concentrations, it can highlight trends and hotspots in geographical data. By superimposing the heat map onto a geographic map, users can observe the distribution of data with respect to actual locations. This gives the data a spatial context. The following R code creates a heat map on a map.

```
> library(plotly)

> # Sample data: random points across the USA
>set.seed(123)
> points <- data.frame(lat = runif(100, 25, 50),
lon = runif(100, -125, -65))

> # Create the heatmap
> fig <- plot_ly(points, lat = ~lat, lon = ~lon,
type = 'densitymapbox',colorscale = 'YlOrRd',
radius = 10, z = ~lat + lon, showscale = FALSE)

> fig <- fig %>% layout(mapbox = list(style = 'open-
street-map'),title = 'Heatmap of Random Points in the
USA')

> fig
```

This methodology can be applied to the visualization of diverse forms of data, including heat maps pertaining to occurrences, sales figures, or population density, yielding insights into spatial patterns. Heat map visualization on a map is outlined in Figure 6.47.

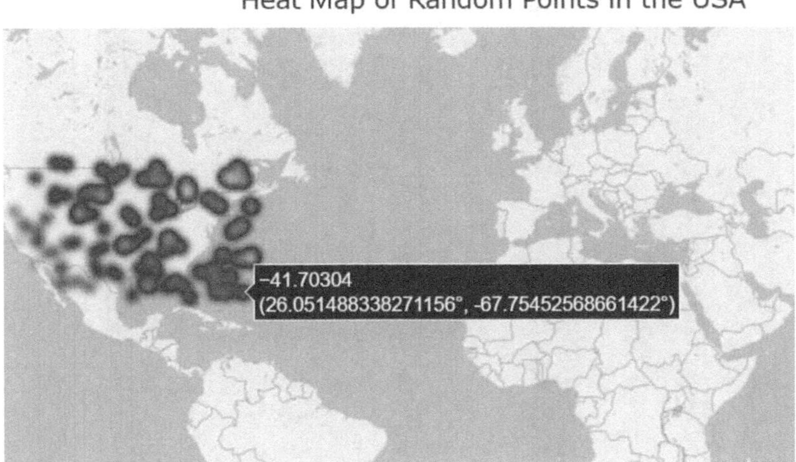

FIGURE 6.47
Heat map visualization on a map using plotly.

d. Line Map: A line map, often called a network map or a route map, is a kind of map that shows the routes or links between various geographic locations. These links can be networks of any form, including roads, railroads, and data transmission lines, or they can be routes, such as flight patterns, shipping lanes, migration routes, and so on. Line maps are useful for visualizing the flow of movement or the connectivity between various points on a map. The lines that link various spots are the primary characteristic of line maps. To symbolize various connection types or to draw attention to certain paths, these lines can have a range of thicknesses, colours, and styles.

This is how to use R's plotly to make a line map. The flight trajectories between the largest U.S. cities are shown in this example:

```
> library(plotly)

> # Sample data
> data <- data.frame(lat = c(37.7749, 34.0522, 40.7128),
lon = c(-122.4194, -118.2437, -74.0060))

> # Create the line map
> fig <- plot_ly(data = data,type = 'scattergeo',
mode = 'lines',line = list(width = 2),lat = ~lat,
lon = ~lon)

> fig <- fig %>% layout(geo = list(scope = 'usa'))
> fig
```

FIGURE 6.48
Line map using plotly.

Latitude (lat) and longitude (lon) values for three locations are contained in an example data frame called data, plot_ly() is the function used to build a line map, type = 'scattergeo' argument indicates that we are creating a geographic scatter plot, mode = 'lines' option indicates that lines linking the points are to be plotted, line = list(2 | width) parameter sets the lines' width to two units, and the lat = ~lat and lon = ~lon arguments map the latitude and longitude values from the data frame to the plot. This code snippet generates an interactive line map, illustrated in Figure 6.48.

6.6 Exploratory Data Analysis

Exploratory data analysis (EDA) is the process of identifying and summarizing a data set's salient features through the use of statistical methods and visualizations. It is an essential phase in the data analysis process that, before more formal statistical techniques or predictive modelling are applied, helps identify patterns, identify anomalies, test hypotheses, and verify assumptions (Monsen, 2024).

Discovering and preparing data for additional analysis or modelling requires the use of EDA, which has many benefits. The key advantages of EDA include the following:

- Comprehending the Data Structure: The structure of the data set, including the different variable types, the distribution of the data, and the existence of any anomalies or trends, may all be thoroughly understood with the use of EDA.

- Recognizing Anomalies and Outliers: EDA is useful for locating abnormalities and outliers in the data, which can have a big impact on the analysis's outcomes. Unusual data points, or outliers, have the ability to distort statistical metrics. Anomalies are irregularities in the data that could point to fraud, mistakes in data input, or other noteworthy occurrences.
- Managing Missing Data: EDA assists in locating and fixing missing data, preventing incomplete or biased data from influencing or invalidating later analysis.
- Finding Connections and Trends: EDA facilitates the identification of correlations and patterns between variables, which can yield insightful information for model construction and decision-making.
- Providing Guidance for Feature Engineering: Feature engineering, the act of adding new variables or altering current ones to enhance model performance, is guided by the insights gathered from EDA. It discovers new features that record crucial data using modifications (such polynomial and logarithmic) to more accurately depict the underlying data distribution.
- Enhancing the Quality of Data: By locating and resolving data quality problems like duplication, inconsistencies, and improper data types, EDA contributes to an improvement in data quality.
- Strengthening Communication: Establishing plots, graphs, and charts that clearly convey data insights. It involves stakeholders through the clear and accessible presentation of data findings. EDA improves stakeholder understanding of the data and its implications through the provision of simple and easy-to-understand visualizations.

EDA is a crucial phase in the data analysis process that has many benefits, including helping with model selection, enhancing data quality, and helping to understand data structure (Vigni et al., 2013). Analysts and data scientists can make more accurate and trustworthy analyses and decisions by utilizing EDA to get deeper insights into their data.

6.6.1 EDA Case Study 1—Diabetes Data

Gaining understanding of the data set through statistical techniques and visualizations is the aim of this EDA case study on diabetes data. Understanding the distribution of the variables, spotting trends, and getting the data ready for additional examination or modelling are the main goals. The attributes of diabetes data are presented in Table 6.5.

TABLE 6.5

Diabetes Data Description

Attribute_Name	Description
Pregnancies	Numerical attribute which describes the total count of pregnancies
Glucose	Numerical feature indicating the glucose value
BloodPressure (BP)	Numerical variable presenting the BP value
SkinThickness	Numerical attribute presenting the skin thickness value
Insulin	Numerical feature representing the insulin level
BMI	Indicates body mass index; numerical attribute
DiabetesPedigreeFunction (DPF)	Numerical feature that highlights the value for DPF
Age	Determines the age of the patient (numerical)
Outcome	Categorical variable (class variable) that indicates whether the person is diabetic or not

Objective: As a data scientist employed by a healthcare company, your job is to examine a data set that contains details about patients, such as different health indicators and whether or not they have been given a diabetes diagnosis. Your objective is to do EDA in order to obtain knowledge that may be useful in comprehending the variables influencing the prevalence of diabetes. Describe the initial steps you would take in performing EDA on this diabetes data set. What specific aspects of the data would you prioritize checking and why?

Solution:

Step I: Load the Required Libraries

```
> library(ggplot2)
> library(dplyr)
> library(plotly)
> library(corrplot)
> library(readr)
```

Step II: Load Data (Diabetes)

```
> #Load Data
> diabetes <- read_csv("diabetes.csv")
```

Step III: The Data Set's Overview

```
> # Display the first few rows of the dataset
> head(diabetes)
> # Summary statistics of the dataset
> summary(diabetes)
> # Structure of the dataset
> str(diabetes)
```

Step IV: Handling Missing Values

```
> # Check for missing values
>colSums(is.na(diabetes))

  Pregnancies      Glucose      BloodPressure
             0            0                 0
SkinThickness      Insulin               BMI
             0            0                 0
DiabetesPedigreeFunction      Age    Outcome
                       0        0          0

> # If there are missing values, you can handle them (e.g.,
by removing or imputing them)
> # Removing rows with missing values
> diabetes <- na.omit(diabetes)
```

Step V: Univariate Analysis: Analysis of Numerical and Categorical Features

```
> # Histograms for numerical variables
>ggplot(diabetes, aes(x = Glucose)) + geom_histogram
(binwidth = 10, fill = "red", color = "black") +
ggtitle("Distribution of Glucose Levels")
>ggplot(diabetes, aes(x = Age)) + geom_histogram(binwidth = 5,
fill = "red", color = "black") + ggtitle("Distribution of
Age")
```

Examine the distribution of each individual variable. Figures 6.49 and 6.50 analyse the distribution of the glucose and age features. Figure 6.51 examines the distribution of the target variable 'diabetes.'

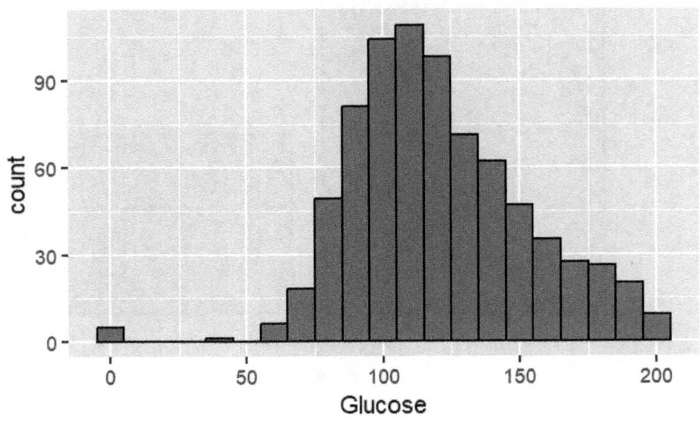

FIGURE 6.49
Glucose level distribution.

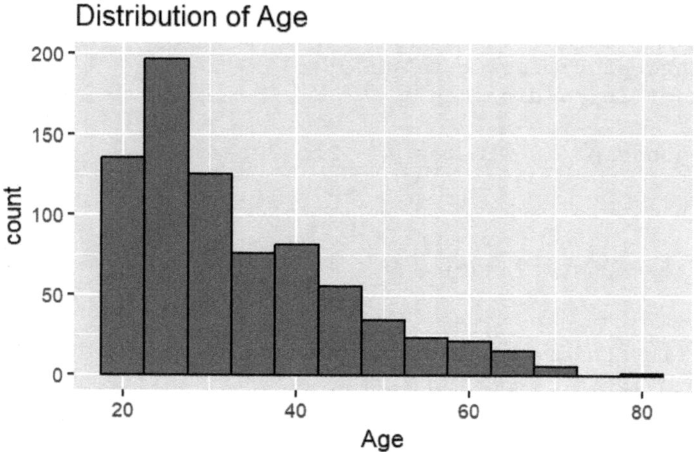

FIGURE 6.50
Age distribution.

```
> # Bar plots for categorical variables
>ggplot(diabetes, aes(x = Outcome)) + geom_bar(fill = "red",
color = "black") + ggtitle("Distribution of Outcome")
```

Step VI: Bivatiate Analysis

```
> # Scatter plot for numerical variables
>ggplot(diabetes, aes(x = Glucose, y = Insulin)) + geom_point
(color = "red") + ggtitle("Scatter Plot of Glucose vs. Insulin")
```

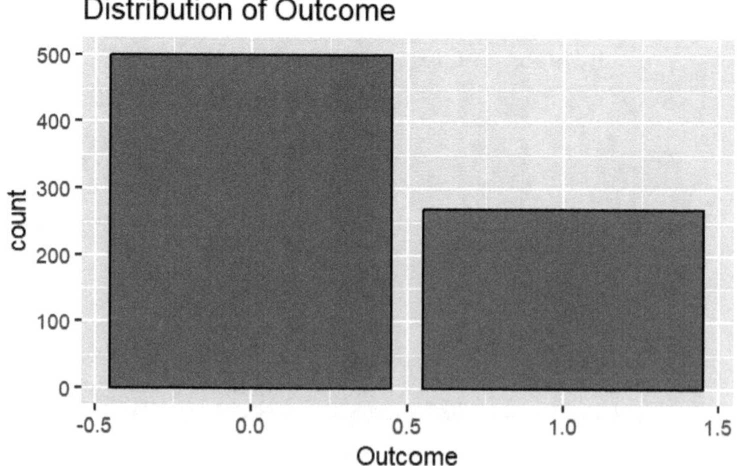

FIGURE 6.51
Target variable distribution.

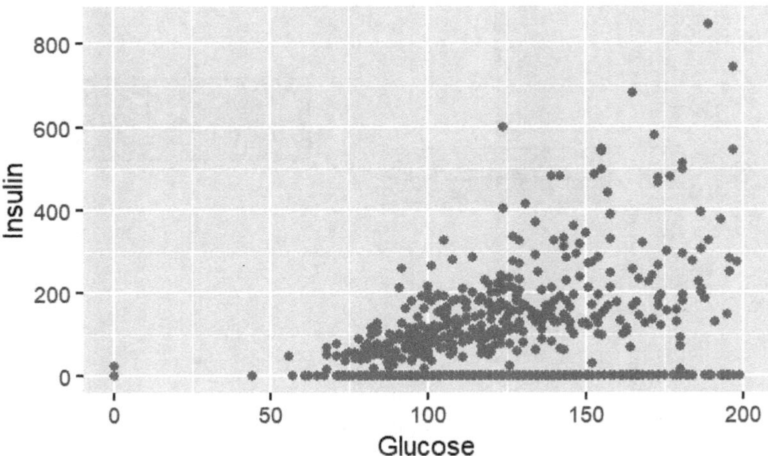

FIGURE 6.52
Scatter plot of glucose and insulin level.

In the above code, we analyse the relationship between two variables. Figure 6.52 outlines the distribution of glucose and insulin levels. In this example, we have plotted a numerical feature against another numerical feature.

```
> # Box plot for numerical vs. categorical variables
>ggplot(diabetes, aes(x = factor(Outcome), y = Glucose))
+ geom_boxplot(fill = "red", color = "black") + ggtitle
("Box Plot of Glucose Levels by Outcome")
```

In the above R code, we analyse the relationship between numerical and categorical features. Figure 6.53 presents a box plot of the glucose feature in relation to the outcome (a categorical target feature), highlighting any potential outliers.

Step VII: Multivariate Analysis

```
> # Pair plot
>pairs(diabetes[, 1:5], main = "Pair Plot of First 5
Variables")
> # Correlation matrix
>cor_matrix<- cor(diabetes[, sapply(diabetes, is.numeric)])
>corrplot(cor_matrix, method = "circle")
```

In this example, we are exploring the relationship between multiple features. A pair plot is generated for the initial five features of the diabetes data and is represented in Figure 6.54. Additionally, a correlation matrix is created to reflect the associations between features. Figure 6.55

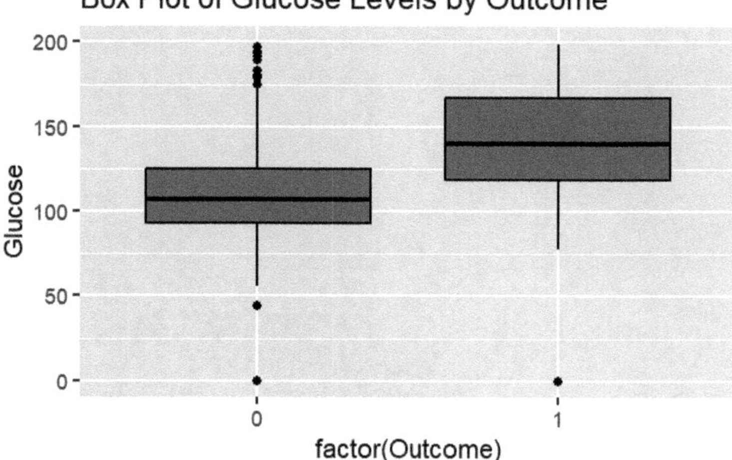

FIGURE 6.53
Box plot of glucose with respect to the target variable.

presents the correlation matrix, summarizing both positively and negatively correlated features in the graph.

Step VIII: Interactive Visualization

```
> # Interactive scatter plot
> fig <- plot_ly(data = diabetes, x = ~Glucose, y = ~Insulin,
color = ~Outcome, type = 'scatter', mode = 'markers')
> fig <- fig %>% layout(title = 'Interactive Scatter Plot of
Glucose vs. Insulin')
> fig
```

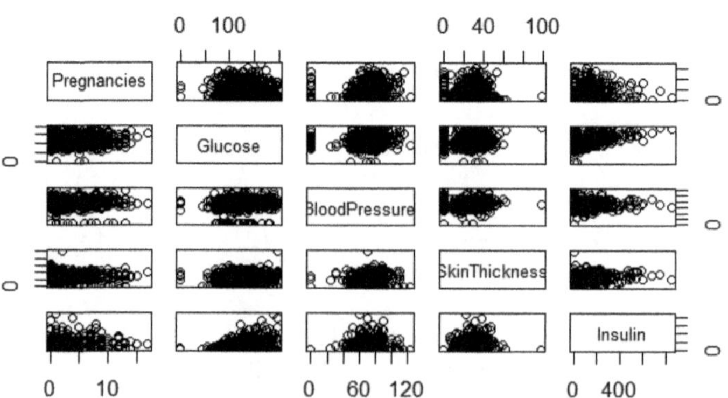

FIGURE 6.54
Pair plot.

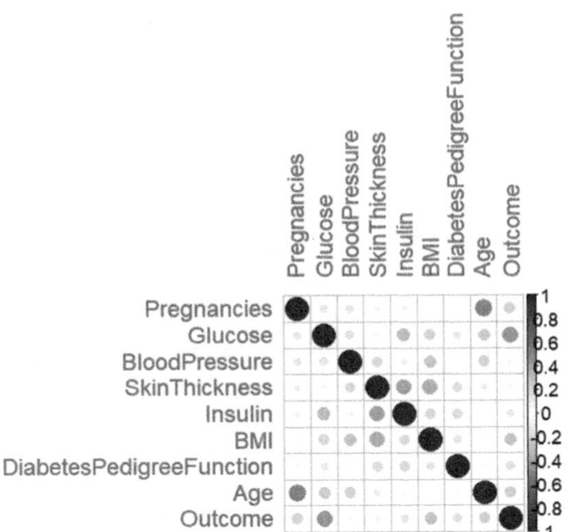

FIGURE 6.55
Correlation matrix of features.

Interactive plots are created using the plotly library. Figure 6.56 highlights a customized scatter plot generated for the variables Glucose and Insulin levels.

```
> # Interactive histogram
>fig_hist<- plot_ly(data = diabetes, x = ~Glucose,
type = 'histogram', nbinsx = 20)
>fig_hist<- fig_hist %>% layout(title = 'Interactive
Histogram of Glucose Levels')
>fig_hist
```

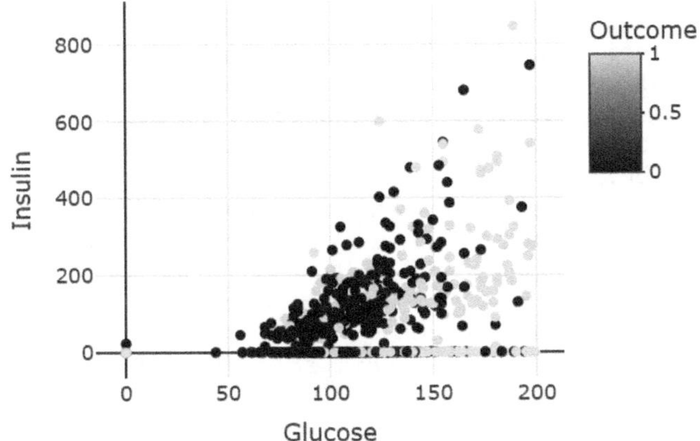

FIGURE 6.56
Customized scatter plot (glucose versus insulin).

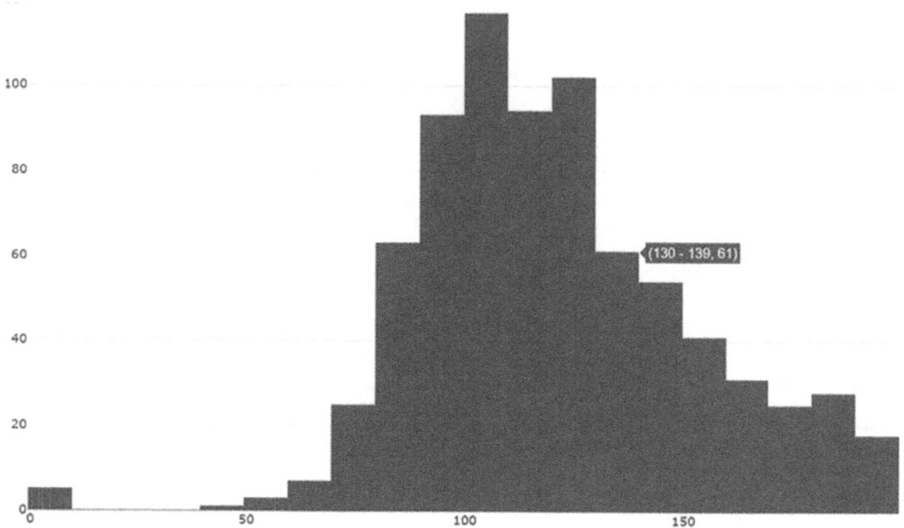

FIGURE 6.57
Customized histogram.

A customized histogram is created and shown in Figure 6.57. As you can see, clicking anywhere on the graph will display the coordinate points.

Step IX: Summary and Insights

It is also possible to summarize the insights obtained from the EDA. For instance, we have summarized the data by noting the total number of rows and columns present in the data set. Additionally, we can provide insights about diverse distributions and other key findings.

```
>cat("- The dataset contains", nrow(diabetes), "observations
and", ncol(diabetes), "variables.\n")
- The dataset contains 768 observations and 9 variables.
>cat("- There are no missing values after cleaning.\n")
- There are no missing values after cleaning.
>cat("- Glucose levels and age have a diverse
distribution.\n")
- Glucose levels and age have a diverse distribution.
>cat("- There is a noticeable difference in glucose levels
between the outcome groups.\n")
- There is a noticeable difference in glucose levels
between the outcome groups.
>cat("- The correlation matrix shows significant
correlations among some variables.\n")
- The correlation matrix shows significant correlations
among some variables.
```

6.6.2 EDA Case Study 2—Diabetes Data Analysis (MYSQL)

You are employed by a healthcare provider as a data analyst. To learn more about patients' health problems, the efficacy of their treatments, and possible areas for patient care improvement, you must evaluate a diabetic data set. R will be used for data analysis and visualization, and MySQL will be used for data storage and querying.

Objective: Conduct detailed analysis on the structured diabetic data with attributes (patient_id INT PRIMARY KEY, age INT, gender VARCHAR(10), bmi FLOAT, blood_pressure FLOAT, glucose_level FLOAT, insulin FLOAT, outcome INT) to address information regarding the health problems of patients, the efficacy of their treatments, and possible areas in which patient care could be improved.

Solution:

Step I: Data Preparation

Create a new database and table for the diabetic data by logging into MySQL. Figure 6.58 reflects the tabular structure of 'Diabetes_Data' for further analysis.

Step II: Linking R with MySQL

```
>install.packages(c("RMySQL", "dplyr", "ggplot2"))
>library(RMySQL)
> library(dplyr)
> library(ggplot2)

>mysqlconnection = dbConnect(RMySQL::MySQL(),dbname='db1',
host='localhost',port=3306,user='root',password='test123')
># List all tables
>dbListTables(mysqlconnection)
[1] "course"    "diabetes_data"    "exam"    "faculty"
[5] "stu"       "workman"
```

```
mysql> select * from diabetes_data;
+------------+------+--------+------+-------+---------------+---------+---------+
| patient_id | age  | gender | bmi  | bp    | glucose_level | insulin | outcome |
+------------+------+--------+------+-------+---------------+---------+---------+
|         11 |   56 | Male   | 18.5 |   110 |            90 |    25.5 |       0 |
|         12 |   66 | Male   | 21.5 |   170 |           110 |      30 |       1 |
|         13 |   47 | Female | 27.5 |   180 |           120 |      20 |       1 |
|         14 |   68 | Female | 19.5 | 175.6 |           110 |      19 |       0 |
|         15 |   75 | Male   | 21.5 | 195.5 |           120 |    22.5 |       1 |
+------------+------+--------+------+-------+---------------+---------+---------+
5 rows in set (0.00 sec)
```

FIGURE 6.58
Diabetes_Data relation.

Step III: Import Diabetes_Data from MySQL

```
> diabetes_data <- dbGetQuery(mysqlconection, "select * from
diabetes_data")
> diabetes_data
  patient_id age gender  bmi    bp glucose_level insulin outcome
1         11 56   Male 18.5 110.0            90    25.5       0
2         12 66   Male 21.5 170.0           110    30.0       1
3         13 47 Female 27.5 180.0           120    20.0       1
4         14 68 Female 19.5 175.6           110    19.0       0
5         15 75   Male 21.5 195.5           120    22.5       1
```

Step IV: Cleaning and Transforming Data

Data transformation and cleaning require 'dplyr' package. The Target column, (i.e., 'outcome') is converted to the categorical variable 'Yes/No,' and factors for attribute gender are added as indicated below.

```
> diabetes_data <- diabetes_data %>%mutate(gender =
factor(gender, levels = c ("Male", "Female")), outcome = factor
(outcome, levels = c(0,1), labels = c("No", "Yes")))
> diabetes_data
  patient_id age gender  bmi    bp glucose_level insulin outcome
1         11  56   Male 18.5 110.0            90    25.5      No
2         12  66   Male 21.5 170.0           110    30.0     Yes
3         13  47 Female 27.5 180.0           120    20.0     Yes
4         14  68 Female 19.5 175.6           110    19.0      No
5         15  75   Male 21.5 195.5           120    22.5     Yes
```

Step V: Data Visualization

Two graphs are generated. Figure 6.59 visualizes the age distribution of patients and Figure 6.60 visualizes BMI versus glucose level.

```
> #Plotting age distribution
>ggplot(diabetes_data, aes(x = age)) +geom_histogram
(binwidth = 5, fill = "blue", color = "black") +labs
(title = "Age Distribution of Patients", x = "Age",
y = "Count")

> # Plotting BMI vs Glucose Level
>ggplot(diabetes_data, aes(x = bmi, y = glucose_level, color
= outcome)) +geom_point() +labs(title = "BMI vs Glucose
Level", x = "BMI", y = "Glucose Level")
```

Step VI: Statistical Analysis

Finding an association between BMI and glucose level is presented in Figure 6.61, which includes code along with the output.

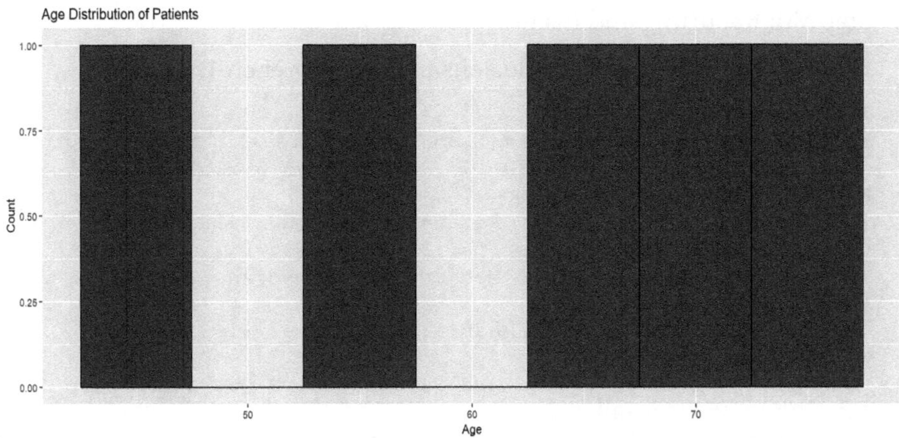

FIGURE 6.59
Data distribution of attribute 'age.'

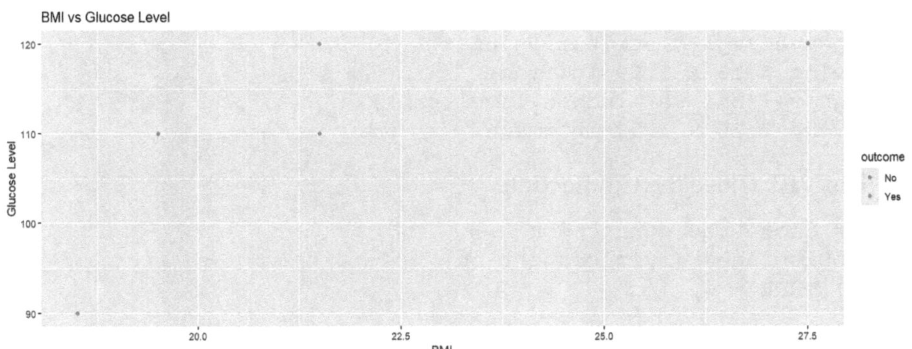

FIGURE 6.60
BMI versus glucose level.

```
> # Correlation between BMI and Glucose Level
> cor_test <- cor.test(diabetes_data$bmi, diabetes_data$glucose_level)
> cor_test

        Pearson's product-moment correlation

data:  diabetes_data$bmi and diabetes_data$glucose_level
t = 1.7039, df = 3, p-value = 0.187
alternative hypothesis: true correlation is not equal to 0
95 percent confidence interval:
 -0.4746654  0.9782739
sample estimates:
      cor
0.7012869
```

FIGURE 6.61
Association between BMI and glucose level.

Step VII: Reporting and Insights

Compile the results and conclusions into a comprehensive report.

```
> summary(diabetes_data)
 patient_id  age  gender  bmi  bp  glucose_level  insulin  outcome
 Min. :11  Min. :47.0  Male :3   Min. :18.5   Min. :110.0
Min.   : 90  Min. :19.0  No :2
 1st Qu.:12  1st Qu.:56.0  Female:2  1st Qu.:19.5  1st
Qu.:170.0  1st Qu.:110  1st Qu.:20.0  Yes:3
 Median :13  Median :66. 0 Median :21.5  Median :175.6
Median :110  Median :22.5
 Mean   :13  Mean   :62.4  Mean   :21.7  Mean   :166.2
Mean  :110  Mean  :23.4
 3rd Qu.:14  3rd Qu.:68.0  3rd Qu.:21.5  3rd Qu.:180.0  3rd
Qu.:120   3rd Qu.:25.5
 Max. :15  Max. :75.0 Max. :27.5  Max. :195.5  Max. :120  Max.
:30.0
```

Save the graphics, and then distribute them to medical practitioners.

```
> # Save plots as images
>ggsave("age_distribution.png")
Saving 2.15 x 1.99 in image
>ggsave("bmi_vs_glucose_level.png")
Saving 2.15 x 1.99 in image
```

Step VIII: Close the Connection

```
> # Close the connection
>dbDisconnect(mysqlconnection)
[1] TRUE
```

In this case study, you have created a MySQL database; linked R to the database; cleaned, visualized, and analysed the diabetes data; and produced insights from the data. This process can be modified for use in many different domains and is standard for data analysis jobs in the healthcare industry.

6.6.3 EDA Case Study 3—Earthquake Data Analysis

As a seismological research institution employee, you are responsible for examining a data set that includes data on earthquakes that have occurred globally between 1995 and 2023. This data set contains information about the magnitude, depth, date, and time of the earthquake as well as location coordinates. To learn more about the patterns, trends, and characteristics of earthquakes, you will use R to do EDA. The earthquake data set is described in Table 6.6.

TABLE 6.6
Earthquake Data Description

Attribute_Name	Description
title	The earthquake's event title
magnitude	The earthquake's magnitude
date_time	The timing and date of the earthquake
cdi	Map of internet intensity for communities
mmi	Mercalli intensity modified
alert	Level of alert
tsunami	Shows whether or not a tsunami was created
sig	Earthquake significance
net	Network that provided the information
nst	The number of seismic stations used to pinpoint the location of earthquakes
dmin	Horizontal distance between the closest station and the epicentre
gap	Azimuthal distance between stations that are adjacent to one another
magtype	Magnitude type
depth	The earthquake hypocentre's depth
latitude	The epicentre of the earthquake's latitude
longitude	The epicentre of the earthquake's longitude
location	Location details
continent	Continent on which the seismic event took place
country	Nation where the seismic event took place

Objective: Outline the process you used to utilize R for EDA on the seismic data set. To support your analysis, include detailed instructions, snippets of code, and visuals.

Solution:

Step I: Load Libraries

```
> library(readr)
> library(dplyr)
> library(ggplot2)
```

Step II: Data Loading and Inspection

```
> # Load earthquake dataset
>earthquake_data<- read_csv("earthquake_data.csv")
> # Check structure and summary of the dataset
> str(earthquake_data)
> summary(earthquake_data)
```

Step III: Temporal Data Analysis

Here, we are examining the historical distribution of earthquakes in terms of time.

```
> # Convert date/time column to proper datetime format
>earthquake_data$date_time<- as.POSIXct(earthquake_
data$date_time, format="%Y-%m-%dT%H:%M:%S")
> # Plot number of earthquakes over time
>ggplot(earthquake_data, aes(x = date_time)) +geom_
histogram(binwidth = "month", fill = "red", color = "black")
+labs(title = "Temporal Distribution of Earthquakes",x =
"Date", y = "Number of Earthquakes") +theme_minimal()
```

Step IV: Magnitude Analysis

```
> # Plot histogram of earthquake magnitudes
>ggplot(earthquake_data, aes(x = magnitude)) +geom_
histogram(binwidth = 0.5, fill = "red", color = "black")
+labs(title = "Distribution of Earthquake Magnitudes",
x = "Magnitude", y = "Frequency") +theme_minimal()
```

Figure 6.62 shows the result of the R code used to investigate the distribution of earthquake magnitudes.

Step V: Spatial Analysis

Plot the geographic distribution of earthquakes using maps and spatial visualization tools. The distribution of earthquake magnitude on a global map is highlighted in Figure 6.63.

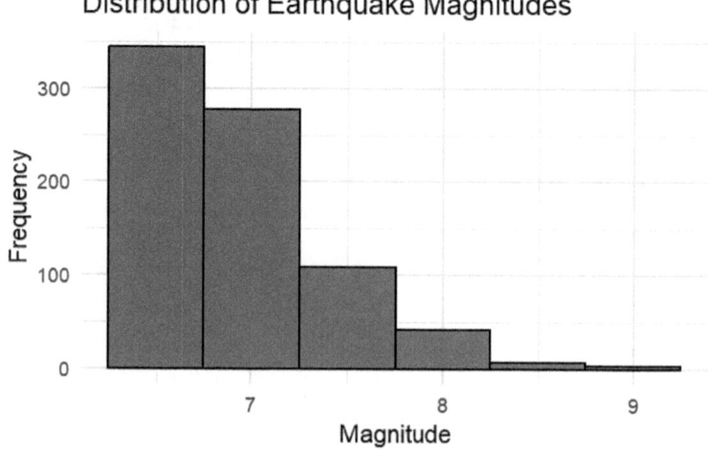

FIGURE 6.62
Earthquake magnitude distribution.

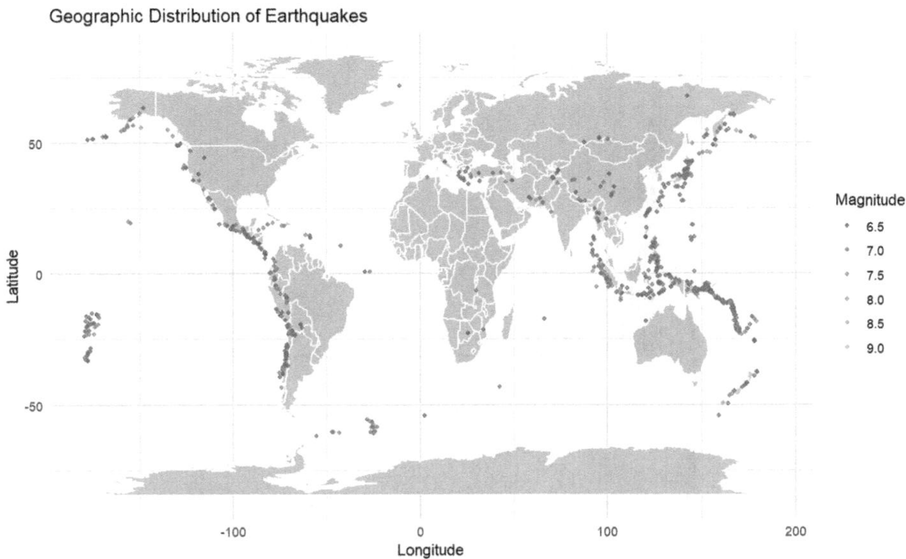

FIGURE 6.63
Distribution of earthquake magnitude on a world map.

```
> # Plotting earthquakes on a map using ggplot2 and maps
package
> library(maps)
> world <- map_data("world")

>ggplot() +
+     geom_polygon(data = world, aes(x = long, y = lat,
group = group), fill = "lightblue", color = "white") +geom_
point(data = earthquake_data, aes(x = longitude, y =
latitude, color = magnitude),size = 1, alpha = 0.6) +scale_
color_gradient(low = "red", high = "orange", guide =
"legend", name = "Magnitude") +labs(title = "Geographic
Distribution of Earthquakes",x = "Longitude", y =
"Latitude") +theme_minimal()

> # Plot histogram of earthquake depths
>ggplot(earthquake_data, aes(x = depth)) +geom_
histogram(binwidth = 10, fill = "red", color = "black")
+labs(title = "Distribution of Earthquake Depths",x =
"Depth (km)", y = "Frequency") +theme_minimal()
```

The depth distribution of earthquakes has been explored in Figure 6.64.

Step VI: Correlation Analysis

To investigate the possible links between earthquake features, a correlation matrix is created and the outcome is delineated in Figure 6.65.

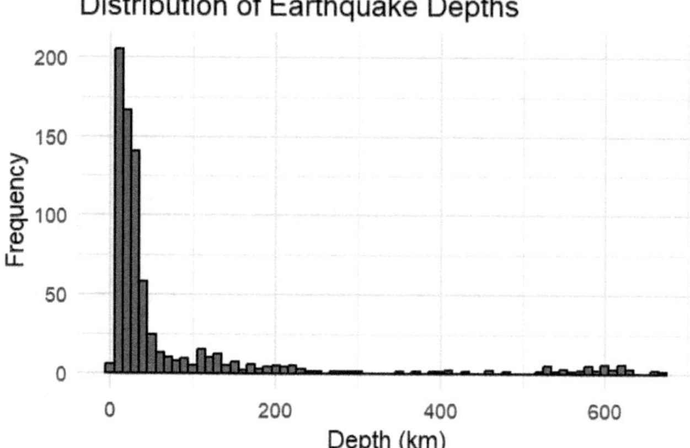

FIGURE 6.64
Distribution of earthquake depths.

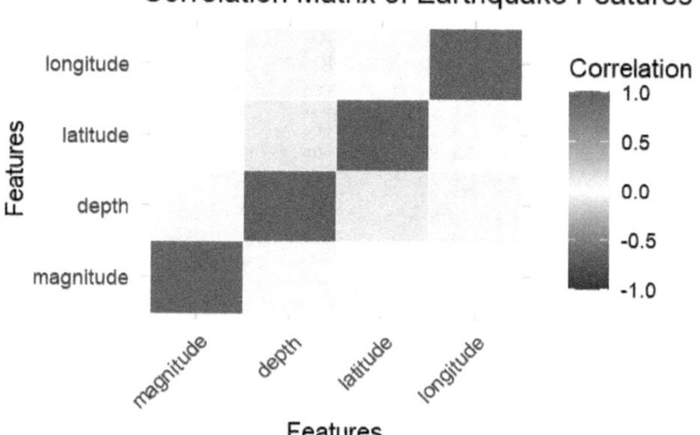

FIGURE 6.65
Correlation of earthquake features.

```
> # Plotting correlation matrix using ggplot2 and reshape2
> library(reshape2)
>library(ggplot2)
> # Calculate correlation matrix
>correlation_matrix<- cor(earthquake_data[c("magnitude",
"depth", "latitude", "longitude")])
>melt_corr<- melt(correlation_matrix)
>ggplot(data = melt_corr, aes(x = Var1, y = Var2,
fill = value)) +
+     geom_tile(color = "white") +scale_fill_gradient2
(low = "blue", mid = "white", high = "red", midpoint = 0,
limits = c(-1, 1), name = "Correlation") +
```

```
+      labs(title = "Correlation Matrix of Earthquake
Features",x = "Features", y = "Features") +
+      theme_minimal() +theme(axis. Text.x = element_text
(angle = 45, hjust = 1))
```

6.7 Summary

- Data visualization makes use of graphical representations to comprehend the linkages and distribution of data.
- Data visualization includes bar charts, scatter plots, box plots, and histograms. Additionally, it aids in locating outliers, trends, and patterns.
- Base R functions for graphics provides basic methods like plot(), hist(), and boxplot() for quick and easy visualizations.
- The Grammar of Graphics is implemented by the package ggplot2 to enable the creation of intricate, multilayered representations.
- The library lotly is utilized to enhance static plots with interactivity, enabling zooming, panning, and hover information.
- The 'lattice' package focuses on multivariate data visualization and conditioning plots using trellis graphs.
- 'Leaflet' constructs interactive maps with pop-ups, markers, supporting layers, and spatial data.
- The 'igraph' package is used for network analysis and visualization. It manages big graphs and offers methods for network structural study.
- Data cleaning; descriptive statistics; data visualization; correlation analysis; and univariate, bivariate, and multivariate analysis are the main steps, which are all included in exploratory data analysis (EDA).
- Descriptive statistics offers summary statistics such as quantiles, standard deviation, mean, and median. Functions like mean(), sd(), and summary() are frequently utilized.
- The direction and intensity of correlations between variables are measured by correlation analysis. Correlation coefficients are computed using the cor() function. These relationships are visualized using charts such as 'corrplot' and correlation matrices.
- The package 'plotly' facilitates the management of spatial data and geographic coordinates. It permits the layering of various map-based data items, such as polygons, lines, and markers.

- Choropleth maps use different colours or shades to represent data according to different geographic locations.
- The colour brewer palette library is called RColorBrewer. Map colour schemes are provided by it. It can be used to change the colours of graphs, maps, plots, and charts, among other things.

EXERCISES

I. Multiple-Choice Questions

1. Which of the following is the correct syntax to represent a 4 × 4 plot in a single page?
   a. par(row=(4,4))
   b. par(nrow=c(2,2,2,2))
   c. par(nrow=(4,4))
   d. par(mfrow=c(4,4))
2. ————————is a robust and advanced data visualization solution for R that draws inspiration from trellis graphics.
   a. Plotly
   b. Leaflet
   c. Lattice
   d. All of the above
3. In package ——- plotting the facets begins from the top left, and in package ——– the facets are plotted beginning from the bottom left.
   a. ggplot2,lattice
   b. ggplot2,plotly
   c. lattice,ggplot2
   d. plotly,ggplot2
4. Which of these is the proper way to save a graph as a PDF file?
   a. ggsave(filename = "Graph1.pdf", plot=Graph_plot)
   b. ggplotsave(filename = "Graph1.pdf", plot=Graph_plot)
   c. plotsave(filename = "Graph1.pdf", plot=Graph_plot)
   d. pdfsave(filename = "Graph1.pdf", plot=Graph_plot)

5. What kind of chart should be taken into account while attempting to demonstrate the link between three variables?
   a. scatter chart
   b. bubble chart
   c. bar chart
   d. All of the above
6. Write the missing method to generate a box plot using ggplot2 library.
   ```
   library(ggplot2)
   t1<-ggplot(data=x,aes(x=category,y=total_weight,fill=
   category))+———-?
   ```
   a. plot_boxplot()
   b. gg_boxplot()
   c. boxplot()
   d. geom_boxplot()
7. What is the procedure for cross-tabulating a table/value in R?
   a. table
   b. xtabs
   c. Both (a) and (b)
   d. None of the above
8. What R function is used to add points to a scatter plot from the ggplot2 package?
   a. geom_line()
   b. geom_point()
   c. geom_pt()
   d. geom_dot()
9. Which R package would you use to make a pop-up, marker, and layer-based interactive map?
   a. lattice
   b. base R methods
   c. leaflet
   d. ggplot
10. Which ggplot2 function is used to display a continuous variable's distribution as bars?
    a. geom_histogram()
    b. geom_hist()
    c. geom_bar()
    d. geom_line()

11. Using ggplot2 to create violin charts, fill the missing line of code to rotate a violin plot.

```
library(ggplot2)
# Basic violin plot
p <- ggplot(Data, aes(x=X1, y=X2)) + geom_violin()

# Rotate the violin plot
p + ....................??
ggplot(Data, aes(x=X1, y=X2)) + geom_violin(trim=FALSE)
```

   a. coord_violin()
   b. rotate_violin()
   c. coord_flip()
   d. coord_p()

II. Brief Questions

1. Explain the idea behind the Grammar of Graphics. How is this idea implemented in R using ggplot2?
2. Describe the process of using ggplot2 to construct a faceting (multi-panel plot). What are the advantages of data visualization faceting?
3. Elucidate how to improve ggplot2 visualizations using the plotly library. Which applications make interactive charts especially useful?
4. Explain how colour is used in data visualization. Which colour schemes are the best to use, and how might the RColorBrewer package help with that?
5. Describe the R heat map creation and customization process. Which methods and packages are frequently used to visualize heat maps?
6. Express in words the significance of data visualization for exploratory data analysis (EDA). How can patterns, trends, and anomalies in a data set be found using visualizations?
7. Expound how to make interactive maps in R using the leaflet package. What are some of the main characteristics that allow leaflet to be used with geospatial data visualization?
8. Describe how trellis graphics are used in lattice work. How may trellis graphics be used to compare different data subsets?
9. Explain the process of adding personalized tooltips and annotations to interactive plotly displays. Why do these characteristics matter?
10. Explore the benefits of using plotly to make interactive 3-D plots. Which data and analysis gain the most from 3-D visualizations?

III. Coding Exercise

1. You work at a public health system as a data analyst. A data set with statistics on the number of COVID-19 cases, recoveries, and fatalities over time in various regions has been sent to you. It is your responsibility to make the data's trends visually appealing in order to point out areas that might require more support. Consider the data with attributes (date, region, cases, number of COVID-19 total cases), recoveries (total count of disease recovery till date), and deaths. To explore the patterns of COVID-19 cases, recoveries, and fatalities over time for each location, use ggplot2 to create a line plot. For each category of data (cases, recoveries, fatalities), use a distinct colour and incorporate an aspect specific to each location.
2. As a data analyst, you are employed by an online store. A data set with attributes (date, category, and sales) comprising weekly sales information for several product categories over the last year has been given to you. It is your responsibility to visually represent the sales trends for every category in order to spot any general and seasonal tendencies. To see the weekly sales patterns for every product category, use the lattice package to create a plot. For every category, use a distinct panel.
3. A data set (Team [Team Name], Module [Software module], Score [Performance score]) comprising performance ratings for various software modules tested by several teams during the previous year has been given to you. It is your responsibility to visually represent the performance score distribution for every module in order to spot any irregularities or possible outliers. The distribution of performance scores for each module can be seen by making a violin plot. For each team, choose a different hue.
4. As a travel analyst, you have access to a data set that includes details on well-liked tourist destinations across different cities. The data set contains each attraction's geographic coordinates (latitude and longitude), the name of the city, and the total number of visitors each attraction has had in the previous 12 months. Making an interactive map to show the tourism destinations and visitor numbers is your duty. The structure of the data is as follows: City, Attraction (Tourist Spots), Latitude, Longitude, Visits (Number of visits in past years). With the leaflet package, make an interactive map to show the tourist attractions. To indicate how many people have visited each attraction, use varying marker sizes.
5. The addresses of several schools in an Indian city, as well as the total enrolment at each school, are included in the data set with

the features of school, latitude, longitude, and student. It is your responsibility to make an interactive map that shows these schools. To make it simpler to identify schools with higher enrolments, use marker sizes to denote the number of students.

6. A retail company is analysing data about the sales performance of various product categories and subcategories in distinct regions. Data set features are categories which specify the category of product, subcategory which provides details of subcategories of products, region is the location where the sales data is recorded, and sales is the total amount of sales. Generate a tree map to analyse the sales distribution, which will aid in determining the locations that are operating well and negatively. Each block's size in the tree map should correspond to the entire quantity of sales, and its colour should show the performance of sales.

References

Abedin, J. (2014). Data manipulation with R. Packt Publishing.

Adler, D., & Adler, M. D. (2024). Package 'vioplot'. CRAN.

Bessler, L. (2023). Heat maps. In Visual data insights using SAS ODS graphics: A guide to communication-effective data visualization (pp. 221–261). Apress.

Chambers, J. (2008). Data visualization and graphics. In Software for Data Analysis: Programming with R (pp. 237–288). Springer.

Dabbas, E. (2021). Interactive dashboards and data apps with plotly and dash: Harness the power of a fully fledged frontend web framework in Python–no JavaScript required. Packt Publishing Ltd.

Dhande, V. M., Kanale, P. R., Karlekar, M. M., Lodha, K. K., & Patil, A. (2023). Visualizing socioeconomic data in India: A comprehensive methodology for Choropleth map creation. International journal of software & hardware research in engineering (IJSHRE), ISSN-2347-4890. Vol 11 Issue 5, pp 59–65, May 2023.

Dougherty, J., & Ilyankou, I. (2021). Hands-on data visualization. O'Reilly Media, Inc.

Grant, R. (2018). Data visualization: Charts, maps, and interactive graphics. Chapman and Hall/CRC.

Grothendieck, G. (2008). Lattice: Multivariate data visualization with R. Journal of statistical software, 25, 1–3.

Healy, K. (2018). Data visualization: A practical introduction. Princeton University Press.

Hehman, E., & Xie, S. Y. (2021). Doing better data visualization. Advances in methods and practices in psychological science, 4(4). https://doi.org/10.1177/25152459211045334.

Hilfiger, J. J. (2015). Graphing data with R: An introduction. O'Reilly Media, Inc.

Kabacoff, R. (2022). R in action: Data analysis and graphics with R and Tidyverse. Simon and Schuster.

Kabacoff, R. (2024). Modern data visualization with R. CRC Press.

Kassambara, A. (2017). R graphics essentials for great data visualization: 200 practical examples you want to know for data science. STHDA, CreateSpace Independent Publishing Platform.
Marriott, K. (2024). The golden age of data visualization: How did we get here?. CRC Press.
Mauludina, M. A., Mulyani, S., Winarningsih, S., & Susanto, H. (2024). The role of data visualization in auditing: A systematic literature review. Cogent business & management, 11(1), 2358168.
Medley-Rath, S., Gillespie, M. D., Novosel, N., Combs, S., & Fearnow, D. (2024). Figures and charts and tables, Oh My!: A content analysis of textbook data visualizations. Teaching sociology, 52(3), 257–265.
Monsen, K. A. (2024). Exploratory data analysis. In Intervention Effectiveness Research: Quality Improvement and Program Evaluation in Healthcare: A Practical Guide to Real-World Implementation (pp. 105–115). Springer International Publishing.
Nordmann, E., McAleer, P., Toivo, W., Paterson, H., & DeBruine, L. M. (2022). Data visualization using R for researchers who do not use R. Advances in methods and practices in psychological science, 5(2), https://doi.org/10.1177/25152459221074654.
Paquot, M., & Larsson, T. (2021). Descriptive statistics and visualization with R. In A Practical Handbook of Corpus Linguistics (pp. 375–399). Springer International Publishing.
Rahlf, T. (2017). Data visualisation with R: 100 examples. Springer.
Sievert, C. (2020). Interactive web-based data visualization with R, plotly, and shiny. Chapman and Hall/CRC.
Sunitha, G., Sriharsha, A. V., Yalgashev, O., & Mamatov, I. (2023). Interactive visualization with plotly express. In Advanced Applications of Python Data Structures and Algorithms (pp. 182–206). IGI Global.
Takahashi, K., & Takahashi, L. (2023). Data visualization. In Materials informatics and catalysts informatics: An introduction (pp. 143–170). Springer Nature Singapore.
Vigni, M. L., Durante, C., & Cocchi, M. (2013). Exploratory data analysis. Data handling in science and technology, 28, 55–126.
Wickham, H. (2011). ggplot2. Wiley interdisciplinary reviews: Computational statistics, 3(2), 180–185.
Wilkinson, L. (2011). ggplot2: elegant graphics for data analysis by WICKHAM, H. Wiley. https://doi.org/10.1111/j.1541-0420.2011.01616.x.
Zamora Saiz, A., Quesada González, C., Hurtado Gil, L., Mondéjar Ruiz, D., Zamora Saiz, A., Quesada González, C., … & Mondéjar Ruiz, D. (2020). Visualization. An Introduction to Data Analysis in R: Hands-on Coding, Data Mining, Visualization and Statistics from Scratch (pp. 111–182). Springer.

7

Real-World Applications

Learning Objectives:

Following their study of this chapter, scholars and students will be able to:

- Explore how R can be used in corporate analytics to support the development of data-driven strategies and decision-making
- Recognize the wider ramifications of R's uses and the ways in which it advances other fields
- Discover instances from the real world where R has been essential in resolving challenging issues in business, science, and health
- Uncover how sharing and discussing analysis results is made easier using R's built-in collaboration facilities
- Study the best practices for embarking and carrying out R team-based analysis projects, including how to assign tasks and communicate with others
- Comprehend the importance of shared resources and documentation in team-based data analysis projects
- Investigate how R interacts with other programs and platforms (such as databases and web apps) to improve data analysis skills and expedite processes to address challenging real-world problems
- Understand how to work together on data analytic projects, manage changes, and monitor progress using version control systems (like Git) and R
- Understand the advantages of version control in guaranteeing reproducibility and preserving the integrity of data analytic operations

7.1 Introduction

In the ever-evolving landscape of data science and analysis, R has been a comprehensive tool for professionals and researchers in different fields. The usage

 DOI: 10.1201/9781003489382-7

of R in real life is presented in this chapter to display its impact on different industries like healthcare, finance, environmental science, and social sciences (Alexander, 2023). We will illustrate practical examples and case studies demonstrating how R's strong statistical and graphical features are used for decision-making, insight generation, and innovation fostering. Among others, predictive model development, dynamic visualization construction, and carrying out complex data analysis are just a few areas where R has helped address real problems and push forward understanding. R is a powerful tool that has made itself available to many professionals and researchers in various fields. This chapter will highlight the diverse applications of R by demonstrating its versatility, especially in areas such as healthcare, finance, and environmental studies (Sillero et al., 2023). By using practical examples and case studies, we can show how decision-making is made possible by using R's robust statistical and graphical capacities to extract insights or spur innovation. The use of R in developing predictive models generating visualizations dynamically or conducting comprehensive data analyses is invaluable when it comes to making sense of genuine world challenges and expanding knowledge (Prasad et al., 2023). Through practical illustrations as well as actual occurrences we will demonstrate how making choices is facilitated with R's sturdy empirical computations together with graphical representations revealing knowledge or encouraging originality. Concerning addressing real-life problems or enhancing comprehension, R has turned out to be invaluable, regardless of whether it entails creating predictive models, designing moving pictures, or sometimes executing complex analyses.

R has a lot of practical applications across various fields, each of which has its own problems with data and methods of analysis. For example, R is used in healthcare to analyse patient data model disease progression and optimize treatment, thus improving patient outcomes as well as making operations more efficient. Financial analysts use R to build complex risk model portfolio optimization and perform market trend analyses which are important when it comes to strategic planning and making investment decisions. Environmental scientists depend on R for spatial data analysis climate modelling and biodiversity studies contributing towards conservation efforts and informing policies (Yu et al., 2018). Survey research social network analysis (SNA) or behavioural studies on the other hand in social science benefit from R by unveiling causes of human behaviour.

7.2 R in Health Informatics

The field of health informatics is growing rapidly by integrating healthcare, information technology (IT), and data science to enhance patient outcomes, improve healthcare delivery, and assist in clinical decision-making. R has become a key player in this area by providing powerful tools and techniques for managing, analysing, and visualizing complex health data (Combi et al., 2023). In this section, we will examine how R is used in health informatics, including its applications to electronic health record (EHR) analysis, disease surveillance, predictive modelling, and personalized medicine.

Due to its ability to deal with mammoth data sets and perform complex statistical analyses, R is an indispensable tool in health informatics (Angelone & Vittorini, 2020). For example, EHRs are analysed by medical practitioners and scholars with the aim of identifying patterns and trends which can inform treatment strategies that enhance patient care. By incorporating machine learning algorithms into R, health professionals can predict outbreaks of diseases, determine risk factors for individual patients, as well as allocate resources better. Within the domain of personalized medicine, R allows genetic data analysis together with clinical parameters so that they can be used in customizing drugs for different individuals in order to make them more effective and safer. Additionally, R offers a plethora of packages like caret for machine learning, survival for survival analysis, and ggplot2 for data visualization which gives power to heath informatics' experts to hold extensive and complicated analyses (Ratnayake et al., 2024). With actual instances from cases occurring in practice, this chapter will demonstrate the profound impact R has had on health informatics through evidence-based decisions in support of innovation in general practice. R's applications in a range of health research problems are mentioned below:

- EHR Analysis
- Forecasting the Spread of Illnesses
- Analysis of Survival for Patient Outcomes
- Personalized Healthcare via Genetic Information Analysis
- Personalized Treatment
- Drug Discovery/Drug-Drug Interaction

Detailed analysis and explanation of the two case studies—the first analysing EHR health data and the second forecasting the development of the disease—are provided below.

7.2.1 Health Data Analysis (Case Study)

EHRs are essential parts of contemporary healthcare systems, including complete patient information in all its aspects like demographics, diseases, treatment histories, and payment details (Tang et al., 2024). EHR analysis is done by healthcare professionals with the aim of improving patient care and possibly leading to better operations as well as supporting clinical decision-making. This part explains how health data can be analysed for theoretical purposes by considering EHR attributes with some important research

TABLE 7.1
Health Data Description for EHR Analysis

Attribute_Name	Description
Name	Patient name
Age	Patient age
Gender	Patient gender
Blood Type	Blood group of patient
Medical Condition	Patient's diagnosis or state of health
Date of Admission	Hospital admission date of patient
Doctor	Medical specialist supervising the patient
Hospital	The hospital where the patient received care
Insurance Provider	The company that offered the patient health insurance
Billing Amount	Total amount invoiced to the patient or insurance for medical treatment
Room Number	The hospital room in which the patient was hospitalized
Admission Type	Type of admittance (e.g., elective, emergency)
Discharge Type	Patient discharge date
Medication	Prescribed drugs for the patient
Test Results	Results of the diagnostic tests performed

questions and possible findings and discovering patterns in patient readmissions in order to enhance medical results. R is used to examine EHR data, find trends, and display the findings (Abdelaziz et al., 2024). An overview and description of data used for analysis in 'Health Data' are provided in Table 7.1.

To carry out a detailed examination of EHRs concerning medical information by use of the stated factors, we will look into five research problems and their resultant visual aids created via R software. Below is a comprehensive methodology:

Research Problems:

1. How many individuals are of a certain age and of what gender?
2. What is the usual time spent in hospital for different illnesses?
3. In what way do charges differ depending on the insurance organization and sickness involved?
4. Which diseases do our clients frequently suffer from?
5. In any of the cases, does the blood group influence the medications given to patients?

Step-by-step health EHR analysis and solutions of the abovementioned research questions are outlined below.

Research Question 1: Showing distribution of individuals by age and gender

Examining patient demographics, hospitals can better understand their patient base and adjust their medical programmes accordingly.

```
# Load necessary libraries
library(tidyverse)
library(lubridate)
library(ggplot2)

# Load the dataset
ehr_data<- read.csv("health_data.csv")
# View the first few rows of the dataset
head(ehr_data)
```

Execution of this code will display the first few health data records. We shall then examine the patient demographics. The R code is presented below, and Figure 7.1 conveys the age and gender distribution of the patient data.

```
#Distribution of Patients by Age and Gender
# Age and Gender Distribution
age_gender_distribution<- ehr_data %>%group_by(Gender)
%>%summarise(Age = list(Age))

# Plot Age Distribution by Gender
ggplot(ehr_data, aes(x = Age, fill = Gender)) +
+     geom_histogram(bins = 20, alpha = 0.7, position =
'identity') +
+     labs(title = "Age Distribution by Gender",
+          x = "Age",y = "Count") +theme_minimal()
```

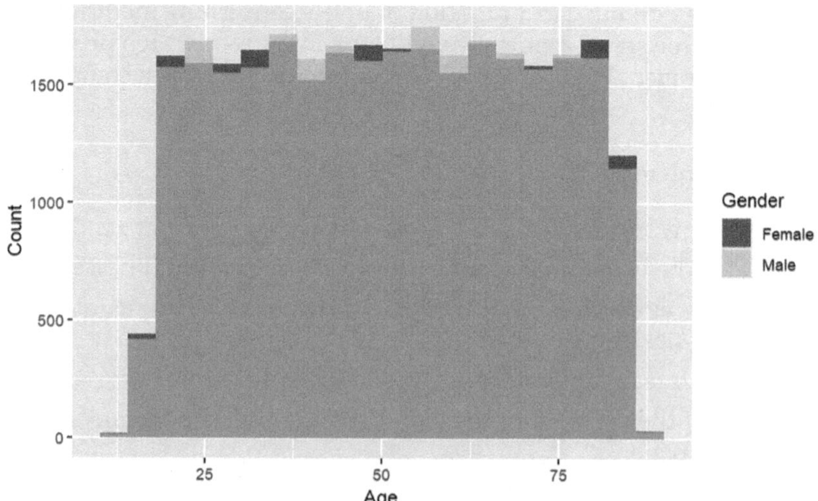

FIGURE 7.1
Age distribution gender wise.

Research Question 2: Mean length of stay based on medical illness

The average length of stay for various medical problems can be used to identify opportunities for process improvement and to help manage hospital resources.

```
# Average length of stay by medical condition
 ehr_data <- ehr_data %>%
  mutate(
    Date.of.Admission = as.Date(Date.of.Admission),
    Discharge.Date = as.Date(Discharge.Date),
    Length_of_Stay = as.numeric(Discharge.Date - Date.
of.Admission)
 )

avg_length_of_stay<- ehr_data %>%group_by(Medical.Condition)
%>%summarise(Average_Stay = mean(Length_of_Stay, na.rm =
TRUE))

# Plot average length of stay by medical condition
ggplot(avg_length_of_stay, aes(x = reorder(Medical.
Condition, Average_Stay), y = Average_Stay)) +geom_bar(stat
= "identity") +coord_flip() +labs(title = "Average Length of
Stay by Medical Condition",x = "Medical Condition",y =
"Average Length of Stay (days)") +theme_minimal()
```

Figure 7.2 demonstrates the output of the aforementioned code and shows the average length of hospital stay based on medical problems.

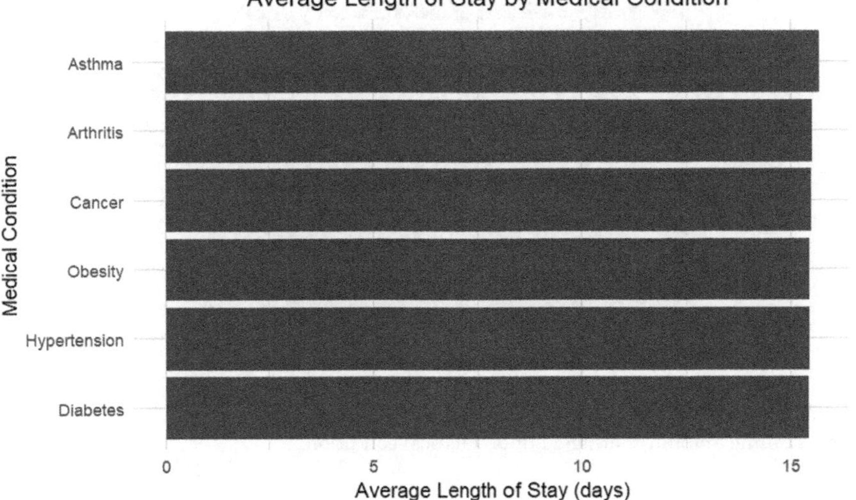

FIGURE 7.2
Average hospital stay based on illness.

Research Question 3: Showing billing amount based on insurance organization and sickness involved

Researching the billing trends of different insurance companies and medical problems provides information about potential discrepancies in billing methods as well as healthcare expenditures.

```
# Average billing amount by insurance provider and medical
condition
billing_data<- ehr_data %>%group_by(Insurance.Provider,
Medical.Condition) %>%summarise(Average_Billing =
mean(Billing.Amount, na.rm = TRUE))

# Plot billing amount by insurance provider and medical
condition

ggplot(billing_data, aes(x = reorder(Medical.Condition,
Average_Billing), y = Average_Billing, fill = Insurance.
Provider)) +geom_bar(stat = "identity", position = "dodge")
+coord_flip() +labs(title = "Billing Amount by Insurance
Provider and Medical Condition",x = "Medical Condition",
y = "Average Billing Amount") + theme_gray()+scale_fill_grey
(start = 0.2, end = 0.8)
```

The average amount of bills paid for all illnesses is illustrated in Figure 7.3.

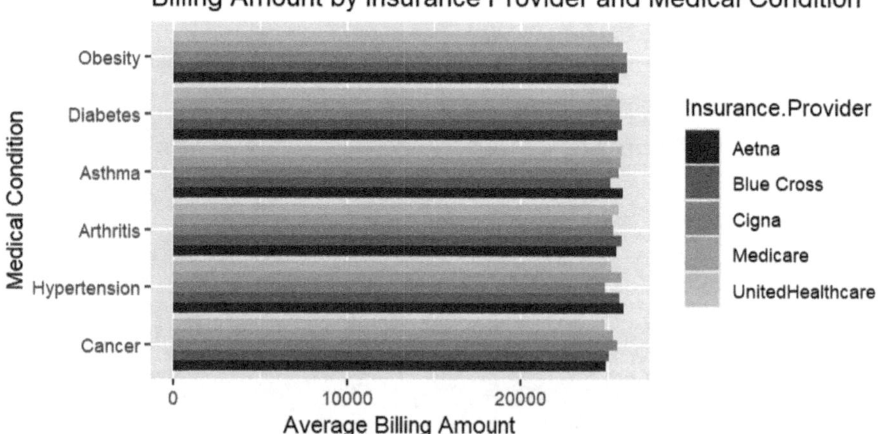

FIGURE 7.3
Average billing amount contingent upon medical condition.

Research Question 4: Medical conditions which are most prevalent among the patients

Identification of the most common medical diseases aids in the organization of clinical resources and focused medical interventions.

```
# Common Medical Conditions
# Frequency of medical conditions
medical_condition_freq<- ehr_data %>% group_by(Medical.
Condition) %> %summarise(Frequency = n()) %>% arrange
(desc(Frequency))
```

```
# Plot frequency of medical conditions
ggplot(medical_condition_freq, aes(x = reorder(Medical.
Condition, Frequency), y = Frequency)) +geom_bar(stat =
"identity") + coord_flip() +labs(title = "Common Medical
Conditions", x = "Medical Condition",y = "Frequency") +
theme_minimal()
```

Figure 7.4 highlights how identifying the most prevalent medical conditions helps with the coordination of clinical resources and targeted medical interventions.

Research Question 5: Relationship between prescribed medication and blood type

Exploring relationships between blood type and prescribed drugs can help to improve patient outcomes and individualized treatment plans.

```
#Correlation Between Blood Type and Medication
# Correlation between blood type and medication prescribed
blood_type_medication<- ehr_data %>%group_by(Blood.Type,
Medication) %>% summarise(Frequency = n()) %>% arrange(desc
(Frequency))
```

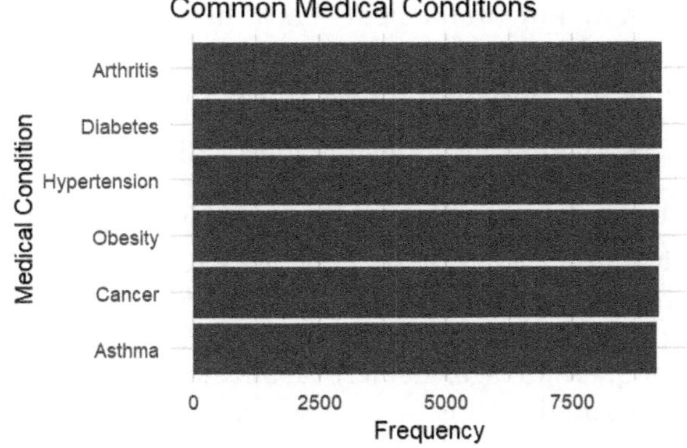

FIGURE 7.4
Frequency of dominating medical conditions.

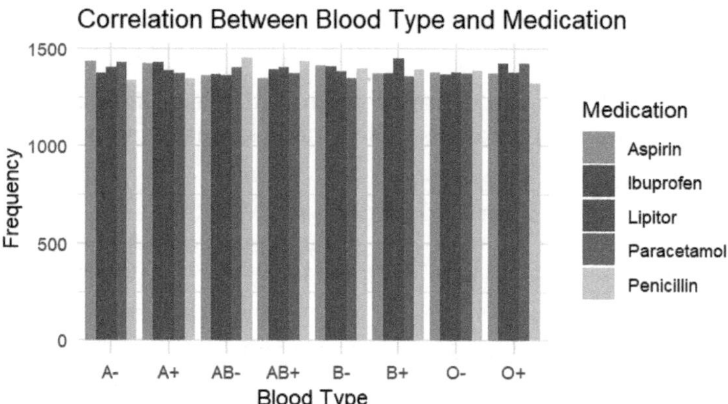

FIGURE 7.5
Correlation between prescribed drugs and blood type.

```
# Plot correlation between blood type and medication
ggplot(blood_type_medication, aes(x = Blood.Type, y = Frequency,
fill = Medication)) + geom_bar(stat = "identity", position =
"dodge") + labs(title = "Correlation Between Blood Type
and Medication",x = "Blood Type",y = "Frequency") +
theme_minimal()
```

Figure 7.5 discloses the correlation between blood type and prescription medication, which aids in the selection of appropriate medical drugs based on blood group type.

This exploration involves a plethora of factors regarding EHR information including understanding patient demographics, hospitalizations, health conditions in general, and billing. Additionally, it has visual tools which aid in knowing how the different attributes are related to each other and their data distributions. R is used to carry out significant deductions that guide informed decision-making in the field of healthcare.

7.2.2 Forecasting the Spread of Illness (Case Study)

The method of estimating the transmission of diseases via EHRs includes scrutinizing previous patient statistics in order to foretell forthcoming tendencies and calamities. Time series analysis, machine learning models such as Autoregressive Integrated Moving Average (ARIMA), as well as regression methods are employed to ascertain patterns and crucial elements affecting the dissemination of illnesses (Kramar & Alchakov, 2023). By looking at things like patient demographics (age, gender), health statuses, types of admissions, and treatment information, this assists medical practitioners in understanding how illnesses spread among people. Thus, it permits preventive healthcare administration from distributing resources effectively and creating targeted actions towards diminishing the consequences of future epidemics. Besides this, understanding and communicating these insights effectively could involve visualizing trends or using predictive models.

Based on the EHR data attributes (Table 7.1), we will predict how diseases will spread across various populations and tackle five distinct research problems related to that. This would entail time series analysis, machine learning tools, and visualization techniques to forecast what might happen in the future concerning the spread of diseases using past information. As healthcare professionals it is important for us to understand how diseases spread in order to provide better treatment options for our patients and ultimately improve their quality of life.

Research Problems

1. What are the historical patterns in the quantity of new cases?
2. Can we forecast how many new cases there will be in the upcoming month?
3. Which elements are most closely linked to the spread of disease?
4. How do certain medical problems affect the way that sickness spreads?
5. How do age and gender, two demographic characteristics, affect the transmission of illness?

Below is a step-by-step process for predicting the spread of sickness and the answers to the research questions raised above.

Research Question 1: Historical patterns in the number of new cases

Time series modelling is used for prediction in order to see the historical trend in new cases. The outcome is shown in Figure 7.6.

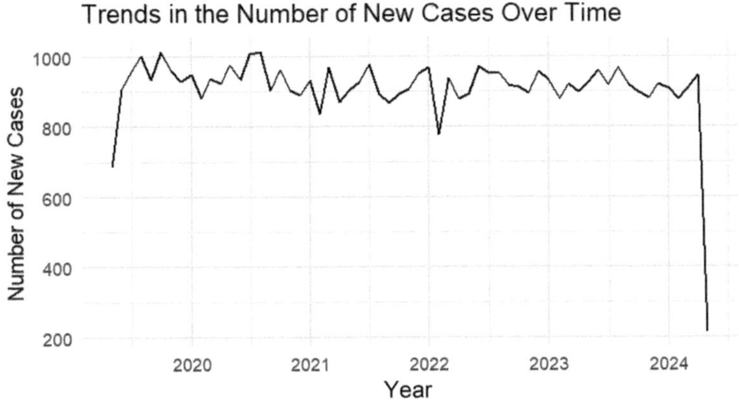

FIGURE 7.6
Trend analysis with time series modelling.

```
# Load necessary libraries
library(tidyverse)
library(lubridate)
library(forecast)

ehr_data<- read.csv("health_data.csv")

# Convert dates to Date type
ehr_data<- ehr_data %>% mutate(Date_of_Admission =
as.Date(Date.of.Admission, format = "%Y-%m-%d"),Discharge_
Date = as.Date(Discharge.Date, format = "%Y-%m-%d"))

#Trends in the Number of New Cases Over Time
# Count the number of new cases per month
case_trends<- ehr_data %>% mutate(Month = floor_date(Date_
of_Admission, "month")) %>% group_by(Month) %>% summarise
(New_Cases = n())

# Plot the number of new cases over time
ggplot(case_trends, aes(x = Month, y = New_Cases)) + geom_
line() + labs(title = "Trends in the Number of New Cases
Over Time", x = "Month", y = "Number of New Cases") +
theme_minimal()
```

Research Question 2: Next month new cases prediction

For time series forecasting, the ARIMA model is implemented (Kanagarathinam et al, 2024). The time series forecasting technique used widely is ARIMA (AutoRegressive Integrated Moving Average). It has three components, such as autoregression (AR), differencing (I), and moving average (MA). Therefore, it relies on past observations to predict future values by modelling their interdependencies and adjusting for non-stationarity via differencing. This makes ARIMA good at picking out linear trends and patterns found in time series data and hence a strong instrument for forecasting purposes. The R code to predict new cases using ARIMA is as follows:

```
# Create a time series object
case_ts<- ts(case_trends$New_Cases, start = c(year(min(case_
trends$Month)), month(min(case_trends$Month))), frequency = 12)

# Fit an ARIMA model
fit <- auto.arima(case_ts)

# Forecast the next 12 months
forecasted_cases<- forecast(fit, h = 12)

# Plot the forecasted cases
autoplot(forecasted_cases) +labs(title = "Forecasted Number
of New Cases for the Next Year",x = "Month",y = "Number of
New Cases") +theme_minimal()
```

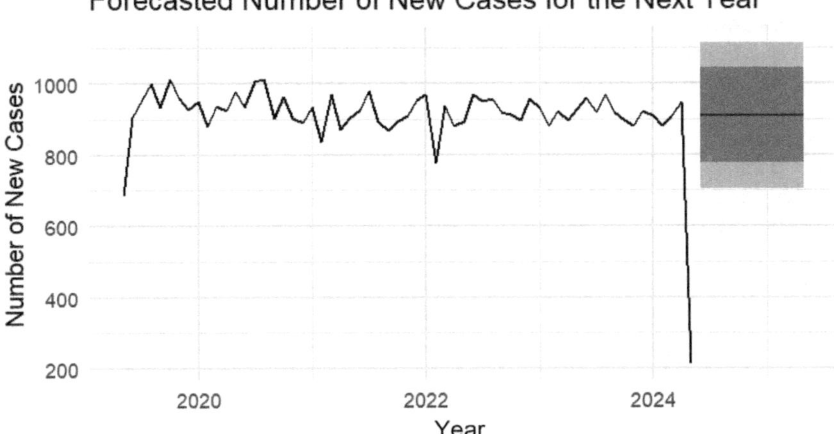

FIGURE 7.7
Prediction using ARIMA for the next 12 months.

As is apparent in Figure 7.7, ARIMA predicts future values by examining historical data points, taking dependencies between them into consideration, and using differencing to handle non-stationarity. IT is a useful tool for a variety of forecasting requirements since it is excellent at modelling linear patterns and trends.

Research Question 3: Relevant factors associated with the spread of the illness

Regression analysis is employed to locate significant variables (DeVore & Kaukis, 2003). A statistical technique for analysing the connections between a dependent variable and one or more independent variables is regression analysis. Regression analysis is useful in identifying important elements that contribute to the spread of sickness when applied to healthcare data. We can identify which characteristics (such as age, gender, blood type, medical condition, and admission type) are strongly correlated with the number of new cases by fitting a regression model to the data. The following describes how to use R code to do regression analysis on health data.

```
# Load necessary libraries

library(tidyverse)
library(lubridate)

# Load the dataset
ehr_data<- read.csv("health_data.csv")
```

```
# Convert dates to Date type
ehr_data<- ehr_data %>%mutate(Date_of_Admission = as.Date
(Date.of.Admission, format = "%Y-%m-%d"),Discharge_Date = as.
Date(Discharge.Date, format = "%Y-%m-%d"))

# Create a variable for New_Cases by aggregating data
per day
daily_cases<- ehr_data %>%group_by(Date_of_Admission)
%>%summarise(New_Cases = n())

# Merge New_Cases back into the original dataset
ehr_data<- ehr_data %>% left_join(daily_cases, by = c("Date_
of_Admission" = "Date_of_Admission"))

# Encode categorical variables
ehr_data$Gender <- as.factor(ehr_data$Gender)
ehr_data$Blood_Type<- as.factor(ehr_data$Blood.Type)
ehr_data$Medical_Condition<- as.factor(ehr_data$Medical.
Condition)
ehr_data$Admission_Type<- as.factor(ehr_data$Admission.
Type)

# Check for any missing values and handle them if necessary
ehr_data<- na.omit(ehr_data)

 # Fit a Poisson regression model
model <- glm(New_Cases ~ Age + Gender + Blood.Type + Medical
.Condition + Admission.Type, data = ehr_data, family =
"poisson")
# Summary of the model
summary(model)
```

The strategy entails choosing a suitable regression model, such as Poisson regression, for count data, such as newly diagnosed cases of a disease. While accounting for the effects of other variables, the model calculates the impact of each predictor variable. Based on their statistical significance (usually determined using p-values), significant factors are determined to have a high correlation with the dependent variable. In order to properly focus interventions, allocate resources, and devise strategies to prevent the spread of diseases, healthcare providers need to have access to this information. The results obtained after regression analysis are displayed in Figure 7.8.

Research Question 4: Medical problems affect the way that sickness spreads

Visualize the propagation of disease based on various medical problems. Figure 7.9 explains how sickness spreads according to a patient's health.

```
    Min       1Q   Median       3Q      Max
-4.0070  -0.8231  -0.0857   0.7566   3.0430

Coefficients:
                                Estimate Std. Error z value Pr(>|z|)
(Intercept)                    3.450e+00  3.610e-03 955.699   <2e-16
Age                            2.936e-05  3.853e-05   0.762    0.446
GenderMale                    -5.508e-04  1.511e-03  -0.365    0.715
Blood.TypeA+                   3.942e-04  3.017e-03   0.131    0.896
Blood.TypeAB-                  3.363e-03  3.016e-03   1.115    0.265
Blood.TypeAB+                  3.386e-03  3.016e-03   1.123    0.262
Blood.TypeB-                  -1.546e-03  3.020e-03  -0.512    0.609
Blood.TypeB+                   1.060e-03  3.018e-03   0.351    0.726
Blood.TypeO-                   2.243e-03  3.025e-03   0.742    0.458
Blood.TypeO+                  -1.629e-04  3.022e-03  -0.054    0.957
Medical.ConditionAsthma        1.806e-05  2.617e-03   0.007    0.994
Medical.ConditionCancer       -2.333e-03  2.615e-03  -0.892    0.372
Medical.ConditionDiabetes     -1.606e-04  2.609e-03  -0.062    0.951
Medical.ConditionHypertension  2.196e-03  2.611e-03   0.841    0.400
Medical.ConditionObesity      -1.107e-03  2.614e-03  -0.423    0.672
Admission.TypeEmergency        1.381e-03  1.852e-03   0.745    0.456
Admission.TypeUrgent           2.954e-04  1.845e-03   0.160    0.873
---
Signif. codes:  0 '***' 0.001 '**' 0.01 '*' 0.05 '.' 0.1 ' ' 1

(Dispersion parameter for poisson family taken to be 1)

    Null deviance: 64262  on 55499  degrees of freedom
Residual deviance: 64253  on 55483  degrees of freedom
AIC: 357168

Number of Fisher Scoring iterations: 4
```

FIGURE 7.8
Regression analysis on health data.

```
condition_trends<- ehr_data %>% mutate(Month = floor_
date(Date_of_Admission, "month")) %>% group_by(Month, Medical.
Condition) %>% summarise(New_Cases = n()) %>% ungroup()

# Plot new cases by medical condition
ggplot(condition_trends, aes(x = Month, y = New_Cases, color
=Medical.Condition)) +geom_line() + labs(title = "Spread of
Illness by Medical Condition",x = "Month",y = "Number of New
Cases",color = "Medical Condition") +theme_minimal()
```

Research Question 5: Influence of features such as age and gender on disease transmission

```
# Count new cases by age group and gender
age_gender_trends<- ehr_data %>% mutate(Age_Group = cut
(Age, breaks = c(0, 18, 35, 50, 65, 100), labels = c("0-18",
"19-35", "36-50", "51-65", "66+")), Month = floor_date
(Date_of_Admission, "month")) %>% group_by(Month, Age_Group,
Gender) %>% summarise(New_Cases = n()) %>% ungroup()
```

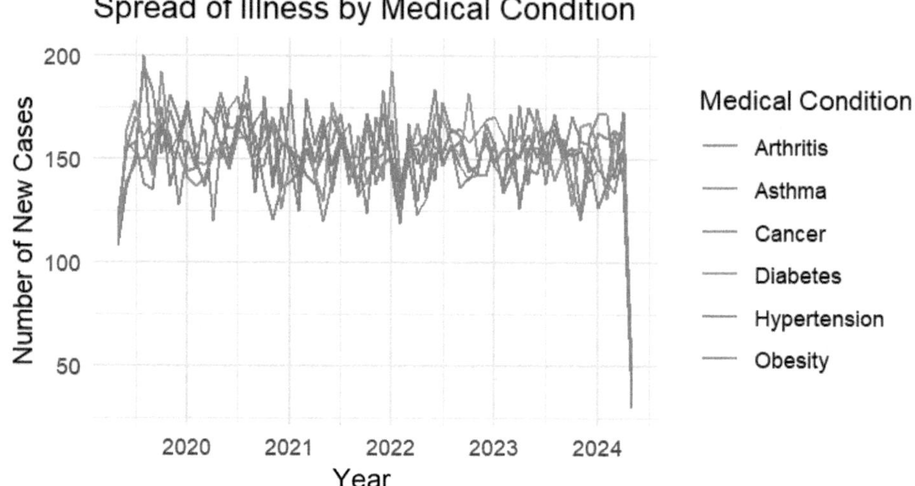

FIGURE 7.9
Disease spread based on a patient's medical history.

```
# Plot new cases by age group and gender
ggplot(age_gender_trends, aes(x = Month, y = New_Cases,
color = Gender)) + geom_line() + facet_wrap(~ Age_Group) +
labs(title = "Impact of Age and Gender on the Spread of
Illness", x = "Month", y = "Number of New Cases", color =
"Gender") + theme_minimal()
```

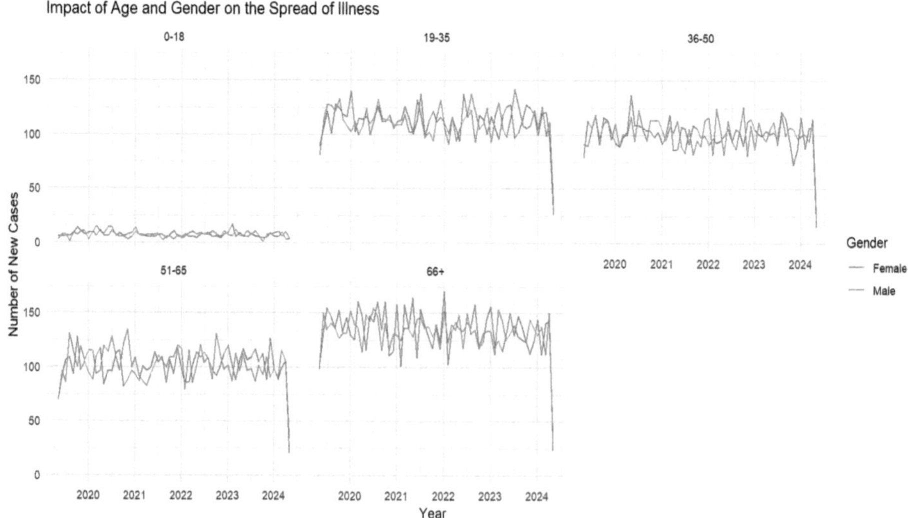

FIGURE 7.10
Spread of disease based on demographic features (age, gender).

Regression analysis, time series models, visualizations, and EHR data are used in this investigation to forecast the spread of sickness and address important research concerns, as visualized in Figure 7.10. The results of this analysis can aid healthcare professionals in better resource planning and allocation, future demand forecasting, and demographically based intervention customization.

7.3 R in Scientific Research

R is a strong and adaptable programming language that is frequently used in statistical modelling, data analysis, and visualization in scientific research (Goodman & Royall, 1988). It is especially well suited for managing complicated data sets and carrying out sophisticated statistical analysis because of its vast ecosystem of packages and tools. R offers reliable techniques for exploratory data analysis, data cleansing, and manipulation that enable scientists to extract patterns and insights from their data (Pearson, 2018).

Additionally, a multitude of statistical methods, such as time series analysis, machine learning, and linear and nonlinear modelling, are supported by R, allowing researchers to construct prediction models and test hypotheses. With the help of programs like ggplot2, R's visualization skills may be used to create high-quality, publication-ready graphs that clearly convey research findings. Furthermore, analyses are transparent and repeatable thanks to R's integration with reproducible research techniques like R Markdown and dynamic report creation, which promotes cooperation and raises the standard of scientific research. All things considered, R is a vital tool for the scientific community, stimulating creativity and advancement in a wide range of disciplines. Because of R's strong analytical and visualization skills, it is widely employed in many scientific research domains. The following are five essential uses:

- Bioinformatics and Genomics
- Environmental Science
- Social Science
- Clinical Trial and Epidemiology
- Behavioural Data Analysis

Large-scale genomic data analysis and interpretation are done with R. Tools for genome annotation, differential expression analysis, and sequence analysis are provided by packages such as Bioconductor, which help researchers comprehend genetic variants and their effects on health and disease. R is a language that ecologists use to examine ecological networks, model species distributions, and evaluate the effects on the ecosystem. R's spatial analysis tools, such as raster and sp, allow scientists to simulate and visualize ecological processes at various spatial scales.

R is used in social science research for text mining, social network analysis, and statistical analysis of survey data. Programs such as tidyverse make data analysis and manipulation easier, and igraph makes it easier to investigate social networks and their connections. R is indispensable for planning and evaluating clinical trials, determining the effectiveness of treatments, and simulating the development of disease. Packages like epitools for epidemiological computations and survival for survival analysis help researchers make well-informed judgements about public health.

Psychometrics researchers create and evaluate psychological exams and evaluations using R. factor analysis, structural equation modelling, and other statistical methods essential for comprehending psychological concepts and behaviour are provided by packages such as psych and lavaan.

Beneath are detailed case studies showcasing R's uses in social and environmental science. We examined earthquake data in the first case study and analysed possible locations for the earthquake forecast zone. In the second case study, we looked at election-related occurrences and conducted sentiment analysis on election data.

7.3.1 Seismic Data Analysis (Case Study)

Seismic data analysis is used to investigate patterns, causes, and effects of earthquakes. This procedure is essential for forecasting earthquakes, evaluating risks, and creating mitigation plans. Statistical modelling, data cleansing, exploratory data analysis (EDA), and visualization are frequently included in the analysis. The earthquake data set includes a number of attributes that offer comprehensive details on every seismic event that has been observed. Table 7.2 explains each attribute.

Aiming to glean significant insights from the data, the analysis of earthquake data entails multiple critical processes. This is a comprehensive instruction explaining how to analyse earthquake data using the included R code.

Step I: Loading of Required Packages and Data

```
# Load necessary libraries
library(tidyverse)
library(lubridate)
library(ggplot2)
library(sf)
library(leaflet)
# Load the dataset
earthquake_data<- read.csv("earthquake_data.csv")
# Preview the dataset
head(earthquake_data)
```

TABLE 7.2

Earthquake Data Description

Attribute_Name	Description
title	The name or title of the seismic event
magnitude	The earthquake's magnitude, which measures the amount of energy released. The Richter scale is frequently used to measure this.
date_time	The time and date of the earthquake, usually expressed in a conventional date and time format.
cdi	Value recorded by individuals reflecting the perceived shaking and damage on the Community Internet Intensity Map.
mmi	Modified Mercalli intensity (MMI), which is frequently based on human observations, is a qualitative indicator of how an earthquake has affected the environment and structures.
alert	A seismic authority-issued alarm level denotes the magnitude and immediacy of the earthquake.
tsunami	A binary (0 or 1) indicator shows whether a tsunami was caused by the earthquake.
sig	A pertinent value serves as a metric employed to evaluate the overall importance of an earthquake by combining several variables such as magnitude, proximity to populated areas, and impact.
net	The seismic network that captures the incident of the earthquake.
nst	The quantity of seismic monitoring stations that record the tremor.
dmin	The minimal distance that separates the epicentre of the earthquake from the closest seismic station at the time of recording.
gap	The highest azimuthal gap (measured in degrees) between the epicentre of the earthquake and the spread of seismic stations.
magtype	The kind of magnitude measurement that was applied.
depth	The focus (or hypocentre) of an earthquake's depth below the surface of the Earth, expressed in kilometres.
latitude	The earthquake's epicentre's latitude coordinate.
longitude	The earthquake's epicentre's longitude coordinate.
location	A recorded description of the earthquake's location that frequently includes the closest city or area.
continent	The continent where the seismic event was recorded.
country	The nation where the seismic event took place.

Step II: Data Cleaning and Summary Statistics

```
# Convert date_time to Date type
earthquake_data<- earthquake_data %>% mutate(date_time
=as_datetime(date_time))

# Check for missing values
colSums(is.na(earthquake_data))
# Basic summary statistics
summary(earthquake_data)
```

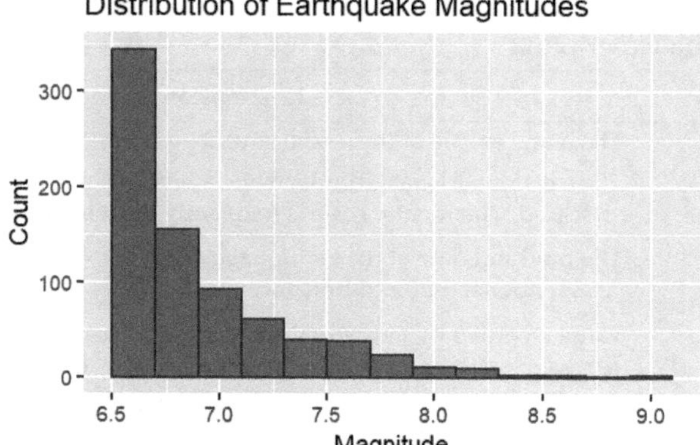

FIGURE 7.11
Distribution analysis of earthquake magnitude.

Inaccurate analysis cannot be ensured if duplicates, missing numbers, or inconsistencies in the raw data are not corrected. The lubridate package's as_datetime() function is used to transform the date_time column into a date-time format. Implement a summary of the statistics to determine the magnitude, depth, and frequency of occurrences for properties by computing basic statistics such as mean, median, standard deviation, and range.

Step III: Exploratory Data Analysis

Analyse the distribution of earthquake depths, magnitudes, and other pertinent characteristics in order to spot trends and anomalies. The distribution of earthquake magnitudes is displayed in Figure 7.11 graphically using a histogram.

```
# Distribution of earthquake magnitudes

ggplot(earthquake_data, aes(x = magnitude)) + geom_
histogram(binwidth = 0.2, fill = "red", color = "blue") +
labs(title = "Distribution of Earthquake Magnitudes",
x = "Magnitude", y = "Count")
```

The link between depth and magnitude is visualized in Figure 7.12 using a scatter plot.

```
# Scatter plot of depth vs magnitude
ggplot(earthquake_data, aes(x = depth, y = magnitude)) +
geom_point(alpha = 0.6) + labs(title = "Depth vs Magnitude",
x = "Depth (km)", y = "Magnitude")
```

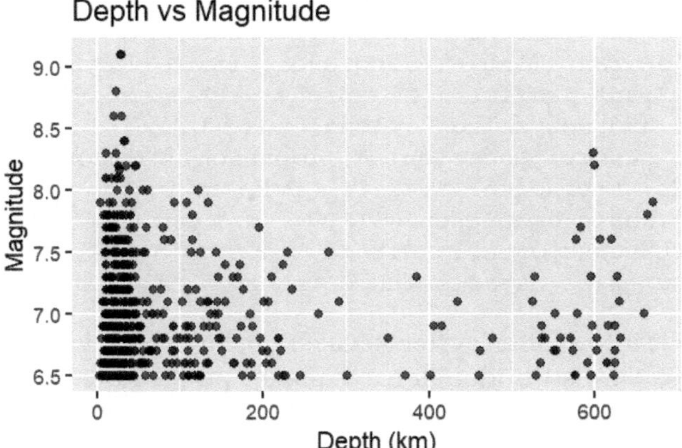

FIGURE 7.12
Association between depth and magnitude.

Step IV: GeoSpatial Analysis

Examine the temporal distribution of earthquakes to find periodicities, trends, or seasonal patterns in the activity. For GeoSpatial Analysis, utilize latitude and longitude values to see geographical distribution of earthquakes. It is useful in locating fault lines and seismic hot spots. Find out seismic activity clusters in order to comprehend regional seismic behaviour and possible danger areas. Generate interactive maps that enable in-depth investigation of seismic data using R packages such as leaflet. The st_as_sf() function from the sf package is used to transform the data into a spatial format for map visualization (Pebesma & Bivand, 2023). Ensure that the columns for latitude and longitude are interpreted correctly.

Mapping and spatial analysis require transforming the data set into a spatial data frame for geographical data, such as earthquake occurrences. This entails creating a data structure that can be plotted on a map using spatial packages. The full code to create seismic zone hot spots is written below:

```
# Map visualization of earthquake locations
# Convert to spatial data frame

earthquake_sf<- st_as_sf(earthquake_data, coords =
c("longitude", "latitude"), crs = 4326, remove = FALSE)
# Check the structure of the spatial data frame
str(earthquake_sf)
```

```
# Create the map visualization
leaflet(data = earthquake_sf) %>% addTiles() %>%
addCircleMarkers(lng = ~longitude, lat = ~latitude,
color = "red", radius = 2, stroke = FALSE, fillOpacity =
0.6) %>% addPopups(lng = ~longitude, lat = ~latitude,
popup = ~paste0("Location: ", location, "<br>Magnitude: ",
magnitude, "<br>Date: ", date_time),options =
popupOptions(closeButton = TRUE)) %>% setView(lng =
mean(earthquake_data$longitude, na.rm = TRUE), lat =
mean(earthquake_data$latitude, na.rm = TRUE), zoom = 2)
```

Geographic information systems (GISs) can be used to visualize and analyse the spatial distribution of earthquakes pursuant to the conversion of spatial data. To better understand earthquake locations and information, add pop-ups and markers. Centre the map at the mean longitude and latitude of the data set, with a zoom level of 2, providing a global view of the earthquake locations. It enables more in-depth and regionally particular information, as seen in Figure 7.13. The global spatial distribution of earthquakes is represented by the map. Geographic coordinates are used to calculate the position of each red circle, which indicates an earthquake. A pop-up window containing comprehensive details on the earthquake, such as its location, magnitude, and date, appears when you click on any marker. Key data points can be quickly accessed with this function. Users can locate places with regular seismic activity or tectonic plate borders by

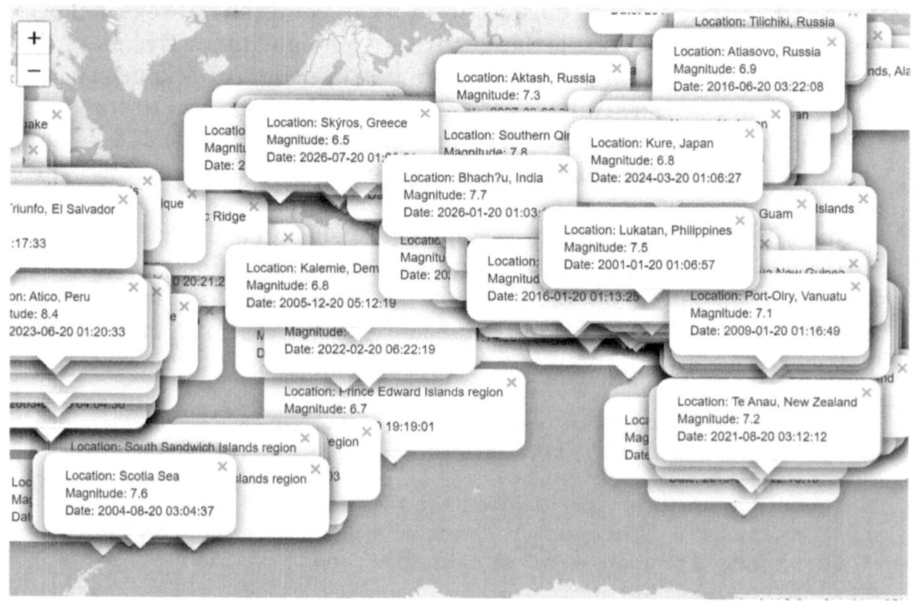

FIGURE 7.13
Spatial distribution of earthquakes.

looking for clusters of earthquake activity on the map. The popups reveal details on the date and magnitude of each earthquake, adding to the map's primary focus on spatial distribution. This information helps to explain the patterns of seismic occurrences' temporal and severity.

Step V: Insights

Determine and describe statistical measures like the average magnitude and the number of earthquakes per continent. Provide an in-depth understanding of the seismic data by summarizing the results. A summary of the total number of earthquakes on each continent is provided, along with the mean magnitude of earthquakes.

```
# Insights and statistics
# - Mean magnitude of earthquakes

mean_magnitude<- mean(earthquake_data$magnitude, na.rm =
TRUE)
print(paste("Mean Magnitude:", mean_magnitude))
```

```
[1] "Mean Magnitude: 6.94112531969309"
```

```
# - Number of earthquakes by continent
earthquakes_by_continent<- earthquake_data %>%group_
by(continent) %>%summarise(count = n())
print(earthquakes_by_continent)
```

```
# A tibble: 7 × 2
  continent        count
  <chr>            <int>
1""                  576
2"Africa"              3
3"Asia"              100
4"Europe"             10
5"North America"      34
6"Oceania"             4
7"South America"      55
```

With adherence to these theoretical procedures, scientists can methodically examine seismic data in order to identify trends, comprehend the consequences of seismic occurrences, and aid in initiatives related to earthquake readiness and mitigation.

7.3.2 Sentiment Analysis of Incidents Pertaining to Elections (Case Study)

Analysing public opinion and occurrences recorded during elections with sentiment analysis can yield insightful information. How to do sentiment analysis on election data is explained below, along with examples of R code

TABLE 7.3
Election Data Description

Attribute_Name	Description
Date	The triggering event's date
State	Specifies the location of the incident
Loksabha seat name	The parliamentary constituency's name
Incident	A summary of the reported incident
Action taken	Actions implemented in reaction to the occurrence
Incident location	Geographical details regarding the incident
Coordinates	Location of the incident's geographic coordinates (latitude and longitude)
Source news article	URL or citation for the news story about the incident

and instructions for setting up the data and formulating research questions (Ozili, 2023). Table 7.3 summarizes a data set for election sentiment analysis. The objective is to use text data from incident descriptions to study the sentiment of election-related occurrences over time, across states, and across Lokshabha seats.

Research Questions

1. What is the general opinion about occurrences connected to the election in various states?
2. What fluctuations in emotions do people have during an election season?
3. Which Lok Sabha constituencies have the highest number of unfavourable reports?
4. What are the most typical responses to unfavourable events?
5. Does the sentiment of occurrences related to elections follow a particular pattern in geography?

The fundamental objective of the election data sentiment analysis is to extract meaning from the data through a number of crucial procedures. In order to obtain insights from data, all research questions should be appropriately answered using the steps mentioned below.

Step I: Load Libraries and Data

```
# Load necessary libraries
library(tidyverse)
library(tidytext)
library(textdata)
library(leaflet)
library(sf)
```

```
# Load the dataset
election_data<- read.csv("election_data.csv")
# Inspect the data
str(election_data)
```

Step II: Preprocessing of Data

```
# Remove rows with missing values
election_data<- election_data %>%filter(complete.cases(.))

# Ensure Date column is of Date type
election_data$Date<- as.Date(election_data$Date)
```

Step III: Perform Sentiment Analysis

```
# Load NRC sentiment lexicon
nrc<- get_sentiments("nrc")

# Tokenize and perform sentiment analysis
election_sentiment<- election_data %>% unnest_tokens(word,
Incident.Description) %>% inner_join(nrc, by = "word",
relationship = "many-to-many") %>% group_by(Date, State,
Lokshabha.Seat.Name, sentiment) %>% summarize(sentiment_
count = n(),.groups = 'drop') %>% spread(sentiment,
sentiment_count, fill = 0) %>% mutate(net_sentiment =
positive - negative) %>% ungroup()

# Merge net_sentiment back into election_data
election_data<- election_data %>% left_join(election_
sentiment %>% select(Date, State, Lokshabha.Seat.Name,
net_sentiment), by = c("Date", "State", "Lokshabha.Seat.
Name"))
```

Step IV: Exploring Research Problems

Research Question 1: What is the general opinion about occurrences connected to the election in various states?

The motive of this question is to identify which states have more positive or negative incidences by adding up sentiment scores at the state level.

```
state_sentiment<- election_sentiment %>% group_by(State) %>%
summarize(overall_sentiment = mean(net_sentiment, na.rm =
TRUE),.groups = 'drop')
print(state_sentiment)
```

```
# A tibble: 8 × 2
  State           overall_sentiment
  <chr>                       <dbl>
1 Gujrat                         -1
2 Karnataka                      -1
3 Madhya Pradesh                 -1
4 Maharasthra                     2
5 Rajasthan                       0
6 Telangana                      -1
7 Uttar Pradesh                  -1
8 West Bengal                     0
```

The result displays the sentiment analysis results, which provide light on the general tone of election-related incidents in various states. The average net sentiment of incidents recorded within a state is reflected in the 'overall_sentiment' score assigned to that state.

States with an overall sentiment score of –1 include Uttar Pradesh, Gujarat, Karnataka, Madhya Pradesh, and Telangana. This suggests that there were more negative terms than positive ones in the descriptions of the incidents, indicating that, on average, incidents recorded in these states had a net negative emotion. Such a pattern can point to unfavourable incidents involving the election process in these areas or popular discontent.

West Bengal and Rajasthan both have a sentiment score of zero. This shows that opinions were split evenly between the positive and negative, suggesting a neutral position on the incidences that were recorded. A neutral attitude could point to a largely consistent or inconsistent public perception of the events surrounding the elections in these states. Maharashtra receives a sentiment score of two. This suggests that there was generally a more positive mood surrounding incidences in this state. A favourable feeling could be the result of more people having positive opinions about the election process in Maharashtra or fewer unfavourable incidences.

Research Question 2: What fluctuations in emotions do people have during an election season?

This research task looks at sentiment patterns over time to find any notable variations in public opinion throughout the course of the election.

```
date_sentiment<- election_sentiment %>% group_by(Date) %>%
summarize(daily_sentiment = mean(net_sentiment, na.rm =
TRUE),.groups = 'drop')
print(date_sentiment)
```

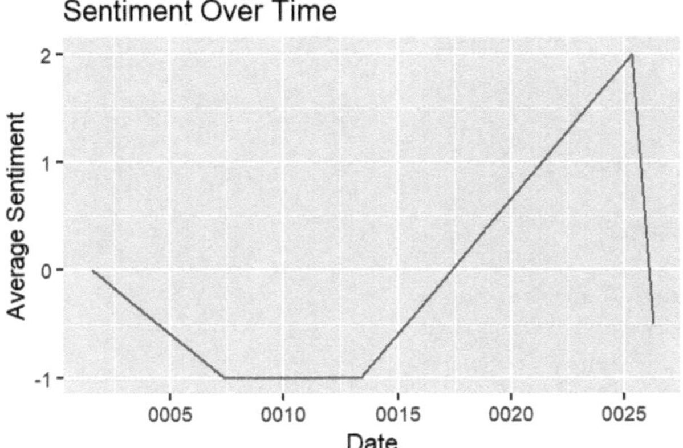

FIGURE 7.14
Sentiment patterns over time.

```
# A tibble: 5 × 2
  Date          daily_sentiment
  <date>        <dbl>
1 0001-06-20     0
2 0007-05-20    -1
3 0013-05-20    -1
4 0025-05-20     2
5 0026-04-20    -0.5
```

```
# Plot sentiment over time
ggplot(date_sentiment, aes(x = Date, y = daily_sentiment)) +
geom_line(color = "red") + labs(title = "Sentiment Over
Time", x = "Date", y = "Average Sentiment")
```

Dates associated with notable negative feeling (e.g., May 20, 0007, and May 20, 0013) might be associated with pivotal moments concerning election-related matters. Stakeholders can better understand when and why there was negative public sentiment by identifying these periods. A sneak peek at variations in sentiment towards election-related incidents throughout time can be discovered in Figure 7.14.

Research Question 3: Which Lok Sabha constituencies have the highest number of unfavourable reports?

Identifying particular Loksabha seats that have the strongest unfavourable sentiment aid in identifying local areas where election-related problems are most prevalent.

```
seat_sentiment<- election_sentiment %>% group_by(Lokshabha.
Seat.Name) %>% summarize(mean_sentiment = mean(net_sentiment,
na.rm = TRUE),.groups = 'drop') %>% arrange(mean_sentiment)
print(seat_sentiment)
```

```
# A tibble: 12 × 2
   Lokshabha.Seat.Name  mean_sentiment
   <chr>                <dbl>
 1 Aonla                            -1
 2 Bhopal                           -1
 3 Chamarajanagar                   -1
 4 Dahod                            -1
 5 Hyderabaad                       -1
 6 Jadavpur                         -1
 7 Jaynagar                         -1
 8 Moradabad                        -1
 9 Ajmer                             0
10 Barasat                           1
11 Mathurapur                        1
12 Beed                              2
```

Research Question 4: What are the most typical responses to unfavourable events?

The efficiency of responses can be determined by examining the actions made in response to situations that elicit unfavourable attitude.

```
negative_incidents<- election_data %>% inner_join(election_
sentiment %>% filter(net_sentiment< 0) %>% select(Date,
State, Lokshabha.Seat.Name), by = c("Date", "State",
"Lokshabha.Seat.Name")) %>% count(Action.taken, sort =
TRUE)
print(negative_incidents)
```

```
Action.taken   n
1              4
2    Repolling 2
3     Arrested 1
4   Legal Case 1
```

The types and rates of responses to incidents related to elections are indicated by the results. This information sheds light on the frequency of various behaviours and how authorities handle situations. Observing the frequency of various activities taken assists in comprehending the standard handling of incidents associated with elections. It offers an insight into the customs and procedures that the authorities adhere to throughout the election season.

Research Question 5: Does the sentiment of occurrences related to elections follow a particular pattern in geography?

This task investigates spatial patterns to figure out if certain regions are more impacted by negative sentiment than others.

```
# Merge net_sentiment back into election_data
election_data<- election_data %>% left_join(election_
sentiment %>% select(Date, State, Lokshabha.Seat.Name,
net_sentiment), by = c("Date", "State", "Lokshabha.Seat.
Name"))

# Convert Coordinates to numeric values
election_data<- election_data %>%mutate(longitude =
as.numeric(sub(".*,", "", Coordinates)),latitude =
as.numeric(sub(",.*", "",  Coordinates)))

# Remove rows with missing coordinates
election_data<- election_data %>%filter(!is.na(longitude) &
!is.na(latitude))
# Convert to spatial data
election_sf<- st_as_sf(election_data, coords = c("longitude",
"latitude"), crs = 4326, remove = FALSE)

# Create a map visualization
leaflet(data = election_sf) %>%addTiles()
%>%addCircleMarkers(~longitude, ~latitude, color =
~ifelse(net_sentiment< 0, "red", "green"), radius = 3,
stroke = FALSE, fillOpacity = 0.7) %>%addPopups(~longitude,
~latitude, ~paste0("Location: ", Location.of.Incident,
"<br>Sentiment: ", net_sentiment), options =
popupOptions(closeButton = TRUE)) %>%setView(lng =
mean(election_data$longitude, na.rm = TRUE), lat
=mean(election_data$latitude, na.rm = TRUE), zoom = 5)
```

Sentiment values are represented by the colour coding, which highlights areas that are primarily positive or negative. Pop-ups that are interactive increase user engagement by giving customized information about each place. Effective sentiment strength and distribution visualization is made possible by the use of circle markers in different colours and opacities. Figure 7.15 provides a clear representation of the geographic distribution and sentiment of the election data, with the map centred on the average coordinates of the data points.

Sentiment analysis of incidents associated with elections can offer insightful information about the sociopolitical environment. Through the identification of sentiment trends and patterns across temporal and spatial dimensions, stakeholders can make well-informed decisions and implement suitable measures. Understanding public opinion, identifying problem regions, and enhancing response tactics during elections can all be aided by this data.

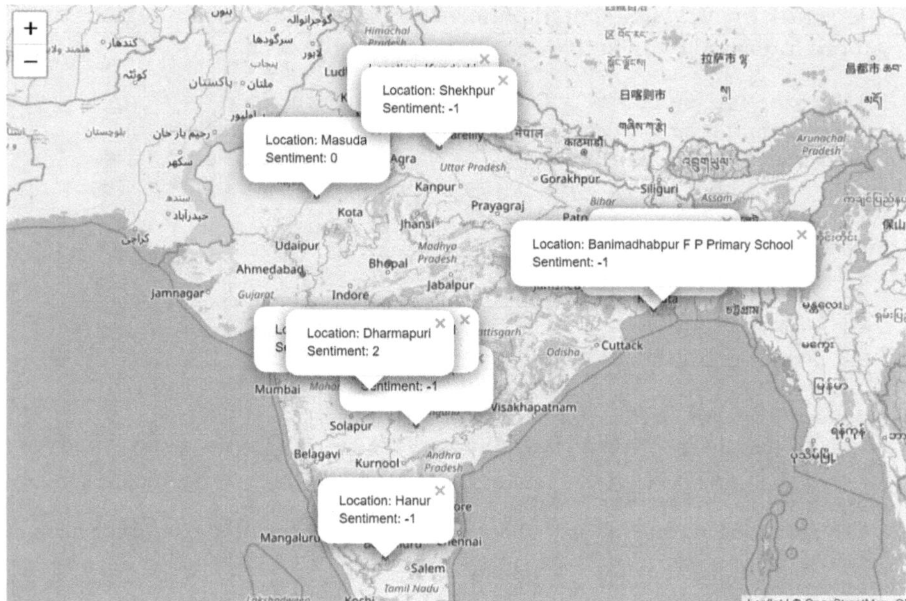

FIGURE 7.15
Spatial sentiment patterns.

7.4 Collaborative Projects

In collaborative projects, a diversity of expertise can be put together to address challenging data analysis problems. Teams can effectively manage big data sets, carry out complex analyses, and produce informative visualizations by utilizing R's robust statistical and graphical features. Version control systems, such as Git, are frequently used in collaboration to manage and synchronize code, guaranteeing that contributions from many users are smoothly merged. Through peer review and the sharing of expertise, this collaborative approach not only shortens project schedules but also improves the accuracy and depth of the analysis. Additionally, R's ability to integrate with R Markdown and Shiny allows for interactive data exploration and clear documentation, which helps teams share findings and improve their methods. In the end, group projects in R create a vibrant and effective atmosphere that stimulates creativity and yields more thorough outcomes. Collaborative projects in R harness collective skills and resources, leading to more comprehensive and reliable data analyses. Here are some advantages of working on a collaborative project:

- Improved Productivity: By substantially allocating tasks to team members according to their areas of expertise, analysis and development can be completed more quickly.

- Enhanced Quality: Robust and precise outcomes are achieved through peer review and the sharing of experience.
- Knowledge Sharing: By exchanging strategies and insights, collaborators promote learning and the growth of skills.
- Reproducibility: The work is made more credible by allowing for the reproduction and verification of analyses by others. This is achieved through the use of common R scripts and version control.

Here are several ways collaborative projects can be applied (Theußl & Zeileis, 2009):

- Projects Integrating Data Analysis: Using R packages such as dplyr, ggplot2, and tidyverse, teams examine intricate data sets including survey or environmental data.
- Research Studies: Using R scripts and markdown documents, researchers work together to examine clinical or experimental data, guaranteeing repeatability and transparency.
- Visualization of Information: Using tools like Shiny or Plotly, collaborative efforts create interactive representations that let many stakeholders interact with data in real time.

7.4.1 Team-Based Analysis

In collaborative initiatives, team-based analysis entails several people cooperating to handle challenging analytical tasks. This method makes use of a variety of viewpoints and skill sets, enabling more thorough data investigation and interpretation. Teams can work more effectively on increasingly complex problems by allocating tasks based on members' areas of competence, such as data cleansing, statistical modelling, and visualization. Continuous feedback and methodological improvement are encouraged by collaborative analysis, which produces more dependable and solid results. Because diverse perspectives foster a deeper grasp of the data and its consequences, the team's synergy improves the analysis's overall quality.

7.4.2 Version Control

A technique called version control keeps track of modifications made to documents and code, enabling several people to work on a project at once without running into problems. Version control systems such as Git offer an organized approach to coordinate and manage contributions from various team members when it comes to collaborative data analysis (Tsitoara, 2024). This system keeps track of all the modifications made, allowing teams to review and understand the project's growth over time and roll back to earlier iterations as needed. By combining updates from different contributors, settling disputes, and guaranteeing that the most recent version of the project is always available, it also makes collaboration easier. Version control improves reproducibility and transparency, which is essential for preserving the analysis's integrity and authority. Here's an illustration of how to use Git version control with RStudio:

Step I: Set Up a Git Repository

Launch your project in RStudio. Click on the 'Tools' menu, then choose 'Version Control' → 'Project Setup.' Select 'Git' and then press 'Create Project.' Your project's Git repository is initialized as a result.

Step II: Build and Alter Files

Create R scripts in RStudio (e.g., VersionExample.R) and other documents.

Step III: Transitions Between Stages

Navigate to the 'Git' tab located in the top-right pane of RStudio. A list of the files that have been added or changed will appear. For each file you wish to stage, check the boxes next to them.

Step IV: Commit Modifications

On the 'Git' tab, click the 'Commit' button. Commit a message outlining the modifications. To save your modifications, click 'Commit.'

Step V: Establish a Branch

Click the 'Branch' drop-down menu in the 'Git' tab, and then choose 'New Branch.' After giving your new branch a name (e.g., VersionExample-branch), click 'Create.' You are now working on this new branch.

Step VI: Make and Adopt Changes

Make any necessary edits to your files. For example, add additional code to analysis.R. Verify the changed files in the 'Git' tab, type a commit message (e.g., 'Added new data visualization'), and then select 'Commit.'

Step VII: Transfer to Remote Repository

First, establish a connection to a remote repository (e.g., GitHub). Click 'More' → 'Shell' on the 'Git' tab, and then use Git commands to add a remote repository:

```
git remote add origin <repository-URL>
git push -u origin main
```

In RStudio, post your commits to the remote repository by using the 'Push' button found in the 'Git' page. Git integration within RStudio simplifies version control, facilitating change management, teamwork, and project history tracking.

7.5 R Shiny

To create interactive web apps straight from R, RStudio created the open-source framework R Shiny (Sievert, 2020). Creating dynamic web apps that can analyse data, produce visualizations, and process user input in real time is made possible. R Shiny leverages the capabilities of R for data analysis and visualization to enable users to create complex and interactive online apps with little to no experience in web programming. Here's a breakdown of the key concepts and components:

1. Reactive Programming: Reactive programming, in which modifications to outputs are automatically triggered by changes in input values, is the foundation of Shiny. Reactive expressions, which are blocks of R code that automatically re-evaluate as their dependents change, are used.
2. User Interface: The application's design and layout are specified by the user interface. It consists of components such as output screens, input areas, buttons, and sliders. Shiny offers tools like mainPanel, sidebarLayout, and fluidPage to build interactive and responsive user interfaces.
3. Server Function: The logic and calculations for the application are contained in the server function. It computes, generates outputs, and handles user inputs. Reactive expressions and observers are used by the server function to dynamically adjust the user interface in response to user interactions.
4. Reactive Expressions: R code blocks known as reactive expressions rerun automatically when their input dependencies change. They are used to produce outputs dependent on user inputs or other reactive values, and they are defined using the reactive function.
5. Input: Inputs, which include text fields, drop-down menus, and sliders, are User Interface (UI) components that let users enter data or choose options. Reactive expressions and outputs are updated by the server function through access to these inputs.
6. Output: The graphic elements that show the outcomes of calculations or data visualizations are called outputs. RenderTable, RenderText,

and RenderPlot are examples of common output functions. Utilizing the reactive expressions as a basis, these routines are utilized to change the user interface.

7. Styling and Layout: Flexible layouts and styling are supported by Shiny using HTML and CSS. By using CSS styles and directly integrating HTML in the user interface, users can alter the app's look.
8. Interactivity: Interactive features like real-time updates, dynamic filtering, and user-driven inputs are made possible with Shiny. This interactive feature makes the experience more user-friendly and increases user engagement.
9. Deployment: Shiny apps can be installed on a number of platforms, such as RStudio Connect, ShinyApps.io, and private servers. Users can utilize web browsers to share their applications with others after deployment.

The following is a basic illustration of an R Shiny application that uses interactive components to view election data. Users of the program will be able to view sentiment scores and filter election results by region on a map (Moraga, 2019).

```
# Load necessary libraries
library(shiny)
library(leaflet)
library(dplyr)

# Sample election data
election_data<- data.frame(
  location = c("Location A", "Location B", "Location C"),
  latitude = c(37.7749, 34.0522, 40.7128),
  longitude = c(-122.4194, -118.2437, -74.0060),
net_sentiment = c(-0.3, 0.5, -0.1),
  region = c("North", "South", "East")
)

# Define UI for application
ui<- fluidPage(
titlePanel("Election Results Visualization"),

sidebarLayout(
sidebarPanel(
selectInput("region", "Select Region:",
                  choices = unique(election_data$region),
                  selected = unique(election_data$region)[1])
    ),
```

```
mainPanel(
leafletOutput("map")
    )
  )
)

# Define server logic
server <- function(input, output) {
filtered_data<- reactive({
election_data %>%
      filter(region == input$region)
  })

output$map<- renderLeaflet({
    data <- filtered_data()

    leaflet(data) %>%
addTiles() %>%
addCircleMarkers(~longitude, ~latitude, color = ~ifelse(net_
sentiment< 0, "red", "green"),
              radius = 5, stroke = FALSE, fillOpacity = 0.7) %>%
addPopups(~longitude, ~latitude, ~paste0("Location: ",
location, "<br>Sentiment: ", net_sentiment),
              options = popupOptions(closeButton = TRUE)) %>%
setView(lng = mean(data$longitude), lat = mean(data$latitude),
zoom = 5)
  })
}

# Run the application
shinyApp(ui = ui, server = server)
```

The Shiny app's user interface is configured by the ui function. It has a sidebar Layout for the sidebar and main panel, a title Panel for the app title, and a leafletOutput for the map. Users are able to filter the data by region using the selectInput widget in the sidebar. The application's server-side logic is managed by the server function. It has a reactive expression called filtered_data that is updated each time the chosen region shifts. The map is then rendered using this reactive data. Using the leaflet package, the renderLeaflet function generates an interactive map. Base map tiles are added using addTiles; sentiment-indicating markers are placed on the map with addCircleMarkers; and when a marker is clicked, addPopups displays comprehensive information. Using the mean coordinates of the filtered data, setView centres the map. Dynamic interaction is possible with the Shiny app (Li & Bilal, 2021). Figure 7.16 illustrates how changing the selected region in the drop-down changes the map in real time. The user's ability to examine election results based on various locations is improved by this interaction.

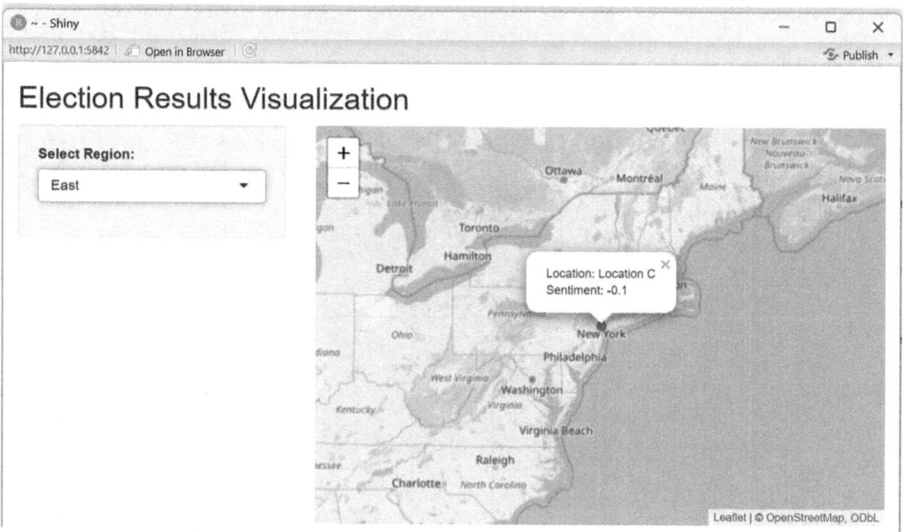

FIGURE 7.16
Interactive web application using Shiny.

7.6 Summary

- R facilitates the integration of real-time data streams, enabling groups to use packages like shiny and plotly to analyse and visualize real-time data.
- R scripts and notebooks can be shared via GitHub or RStudio Cloud, allowing simultaneous access and editing, facilitating team collaboration on data analysis.
- Version management in R projects is made easier by Git and GitHub, which enable teams to monitor changes, control code versions, and settle disputes while maintaining a consistent and repeatable workflow.
- Tasks like data cleansing, analysis, and visualization are divided up among team members according to their areas of expertise in team-based projects, which improves productivity and specialization.
- The shiny package in R facilitates the creation and implementation of interactive dashboards for real-time data visualization and reporting, hence enhancing cooperative decision-making within teams.
- Collaboration and version control solutions facilitate data sharing and guarantee reproducible analysis by keeping track of modifications and updates.

- Version control systems are used in collaborative projects to enable code reviews, during which team members check the quality of the code and offer input before merging changes.
- By utilizing version control, a team may efficiently manage contributions and merge updates while working on separate project branches concurrently.
- Transparency between stakeholders and team members is ensured in collaborative projects by using R Markdown or other documentation tools to document processes and results.

EXERCISES

I. Multiple-Choice Questions

1. How does one construct a reactive expression in Shiny that updates on its own whenever its inputs change?
   a. observe()
   b. output()
   c. reactive()
   d. output$()
2. Which Git command is used in a repository to start a new branch?
   a. git checkout -m<branch-name>
   b. git branch -m <branch-name>
   c. git branch -b<branch-name>
   d. git checkout -b <branch-name>
3. Why does Shiny include the renderPlot() function?
   a. For creating and displaying a plot with reactive data
   b. To change the server function's reactive values
   c. Both a and b
   d. For the UI to display HTML content
4. What does the .gitignore file serve as in the context of Git version control?
   a. To enumerate directories and files that Git ought to ignore and not monitor
   b. To indicate which files Git should monitor
   c. To provide commit messages for various modifications
   d. None of the above

5. For a project, what kind of collaborative technique guarantees that different team members can add to it without erasing the work of one another?
   a. Having a single shared file for all of the team's work
   b. Separating changes using branches and merging them when finished
   c. Avoiding using any version control software
   d. Adding fresh contributions to the project on a regular basis
6. Which package is required in order to publish Shiny apps online?
   a. shinyApp
   b. R shiny
   c. shiny
   d. shinytest
7. What is the main advantage of version control systems for team projects?
   a. It imparts team members a platform for instantaneous communication.
   b. It automatically synchronizes data among the team members.
   c. It enables conflict-free simultaneous work on the same file by team members.
   d. It facilitates efficient version management, change monitoring, and collaboration.
8. The impact of air quality on respiratory health is being examined by a public health department. They possess a data set comprising daily measurements of air quality and reported occurrences of respiratory problems in various areas. Which R function can be applied to a time series study to determine how respiratory problems have changed over time in connection to air quality measurements?
   a. lm()
   b. arima()
   c. anova()
   d. cor.test()
9. A Shiny application is being developed by a research lab to visualize the outcomes of a survey regarding participants' health behaviours. They want the application to dynamically display corresponding visualizations (such as bar charts and histograms) and allow users to choose various health metrics (including physical activity, diet, and sleep) from a drop-down menu. Which Shiny component has to be utilized to build a drop-down menu within the application that allows users to choose health metrics?

a. Input()
b. textInput()
c. InputSlider()
d. selectInput()

10. A group of scientists is looking into how environmental conditions affect the development of plants. Over several months, they have gathered information on plant height, soil moisture content, and temperature. The group's goal is to ascertain whether plant height and temperature are significantly correlated. Which R function is suitable for carrying out a correlation study?
    a. t.test()
    b. aov()
    c. cor()
    d. ks.test()

II. Brief Questions

1. What are some advised practices for starting and overseeing analysis projects involving R teams? How does R make it easier for team members to share and discuss analysis results?
2. Why is documentation crucial for initiatives involving team-based data analysis?
3. How can R work with other apps and platforms to improve data analysis? Enumerate the package names needed to connect to external platforms/apps.
4. How is version control used and what part does it play in managing R projects? How can modifications in data analysis projects be managed using R and version control systems such as Git?
5. What are the advantages of conducting collaborative analysis with R Markdown? In what ways does R help with analysis repeatability in group projects?
6. How can R be used with databases to improve workflows for data analysis?
7. How can R Shiny application management and collaboration be facilitated by version control systems like Git, and what are the advantages of incorporating these technologies during the development process?

8. What are the most significant benefits of employing Git for version control in software development projects, and how does it help developers work together and manage code?
9. What best practices should be followed to efficiently organize and record the analysis process in a scientific research project that requires analysing large-scale genomic data? How may R be used to assure data integrity and reproducibility of the study?
10. What are some efficient ways to handle and analyse this kind of data when using R packages in a research project that requires analysing longitudinal data to evaluate changes in social behaviours over time? Which software tools are essential for organizing, displaying, and analysing longitudinal data, and how can they help guarantee the precision and breadth of the analysis?

III. Coding Exercise

1. Your job is to analyse worldwide electric vehicle (EV) data in order to identify patterns and trends across various geographical areas. Aspects including region, category, parameter, mode, power train, year, unit, and value are included in your data set. The intention is to offer information that businesses and politicians can use to make well-informed decisions about the adoption of EVs and the expansion of infrastructure. Provide a response to the following research questions.
   a. What patterns have you noticed in the last ten years in the sales of EVs in various regions?
   b. What is the regional difference in the market share of BEVs (Battery Electric Vehicle) versus PHEVs (Plug-in Hybrid Electric Vehicle)?
   c. Which areas are adopting commercial EVs at the fastest rate?
   d. How have EV sales changed over time in the passenger car segment?

2. Version control in research and development projects is handled via Git. You have the following tasks to complete. From the main branch, create a new branch called feature-visualization. Write an R script in this new branch that uses ggplot2 to plot the 'mtcars' data set. Add a message stating that the plot script has been added and commit your modifications. Merge the feature-visualization branch back into the main branch by switching back to it. Transfer the modified master branch to the distant repository. For every stage, include the R code and commands.

3. You own a data set that includes details about different parts of the world. The following attributes are included in the data set: Location, Latitude, Longitude, Category, and Value. Develop an interactive Shiny application that lets users see this geographic data represented on a map. The app should allow users to see a scatter plot with the Value for each location and filter the data by Category.
4. You are given a data set with information about the spread of a disease across different geographical locations. The data set includes the following attributes: Location, Latitude, Longitude, Cases, and Date. The main task is to develop an R Shiny application that will allow you to track and examine the disease's spread. The application ought to perform the following tasks:
   a. Display a map where the number of instances at each place is indicated by markers.
   b. Provide an option to choose a time range and see the number of cases changing over that range.
   c. Provide possibilities to filter data by date or location to carry out interactive filtering.

References

Abdelaziz, A. I., Hanson, K. A., Gaber, C. E., & Lee, T. A. (2024). Optimizing large real-world data analysis with parquet files in R: A step-by-step tutorial. Pharmacoepidemiology and drug safety, 33(3), e5728.

Alexander, R. (2023). Telling stories with data: With applications in R. Chapman and Hall/CRC.

Angelone, A. M., & Vittorini, P. (2020). The automated grading of R code snippets: preliminary results in a course of health informatics. In Methodologies and Intelligent Systems for Technology Enhanced Learning, 9th International Conference (pp. 19–27). Springer International Publishing.

Combi, C., Facelli, J. C., Haddawy, P., Holmes, J. H., Koch, S., Liu, H., ... & Yang, C. C. (2023). The IHI Rochester report 2022 on healthcare informatics research: Resuming after the Covid-19. Journal of healthcare informatics research, 7(2), 169–202.

DeVore, C., & Kaukis, N. (2003). Linear regression analysis. STAT, 3013(11/16), 15.

Goodman, S. N., & Royall, R. (1988). Evidence and scientific research. American journal of public health, 78(12), 1568–1574.

Kanagarathinam, K., Ponkumar, G., & Kumar, S. S. (2024). Performance analysis of autoregressive integrated moving average (ARIMA) and 'earlyR' statistical models for predicting epidemic outbreaks: A case study on COVID-19 data in India. Trends in sciences, 21(1), 7246–7246.

Kramar, V., & Alchakov, V. (2023). Time-series forecasting of seasonal data using machine learning methods. Algorithms, 16(5), 248.

Li, R., & Bilal, U. (2021). Interactive web-based data visualization with R, plotly, and shiny (Carson Sievert). Chapman and Hall/CRC.

Moraga, P. (2019). Geospatial health data: Modeling and visualization with R-INLA and shiny. Chapman and Hall/CRC.

Ozili, P. K. (2023). The acceptable R-square in empirical modelling for social science research. In Social Research Methodology and Publishing Results: A Guide to Non-Native English Speakers (pp. 134–143). IGI Global.

Pearson, R. K. (2018). Exploratory data analysis using R. Chapman and Hall/CRC.

Pebesma, E., & Bivand, R. (2023). Spatial data science: With applications in R. Chapman and Hall/CRC.

Prasad, L., Mishra, P., Gupta, S. K., & Aslan, Ş. (2023, March). Dynamics of business and opportunities for employability using predictive modeling tools of Machine Learning and R programming. In 2023 International Conference on Device Intelligence, Computing and Communication Technologies (DICCT) (pp. 223–228). IEEE.

Ratnayake, I., Pepper, S., Anderson, A., Alsup, A., & Mudaranthakam, D. P. (2024). An R Shiny Application (SDOH) for predictive modeling using regional social determinants of health survey responses. International journal of social determinants of health and health services, 54(1), 21–27.

Sievert, C. (2020). Interactive web-based data visualization with R, plotly, and shiny. Chapman and Hall/CRC.

Sillero, N., Campos, J. C., Arenas-Castro, S., & Barbosa, A. M. (2023). A curated list of R packages for ecological niche modelling. Ecological modelling, 476, 110242.

Tang, A. S., Woldemariam, S. R., Miramontes, S., Norgeot, B., Oskotsky, T. T., & Sirota, M. (2024). Harnessing EHR data for health research. Nature medicine, 30(7), 1847–1855.

Theußl, S., & Zeileis, A. (2009). Collaborative software development using R-Forge. Special invited paper on "The Future of R". The R journal, 1(1), 9–14.

Tsitoara, M. (2024). Beginning Git and GitHub: Version control, project management and teamwork for the new developer. Springer Nature.

Yu, Y., Ouyang, Y., & Yao, W. (2018). shinyCircos: An R/Shiny application for interactive creation of Circos plot. Bioinformatics, 34(7), 1229–1231.

Answer Keys

Chapter 1

I. Multiple-Choice Answer Keys:

1. (a) 2. (c) 3. (d) 4. (d) 5. (b) 6. (b) 7. (a) 8. (c) 9. (c) 10. (b)

II. Practice Coding Exercise

1.

```
> A<-c(20,45,50)
> A
[1] 20 45 50
> B<-c(10,20,30)
> B
[1] 10 20 30
> C<A*B
```

2. Array of Odd numbers:

```
> x<-(1:50)*2-1
> x
 [1]  1  3  5  7  9 11 13 15 17 19 21 23 25 27 29 31
[17] 33 35 37 39 41 43 45 47 49 51 53 55 57 59 61 63
[33] 65 67 69 71 73 75 77 79 81 83 85 87 89 91 93 95
[49] 97 99
> A<-array(x,dim=c(5,5))
> A
     [,1] [,2] [,3] [,4] [,5]
[1,]    1   11   21   31   41
[2,]    3   13   23   33   43
[3,]    5   15   25   35   45
[4,]    7   17   27   37   47
[5,]    9   19   29   39   49
```

Array of Even numbers:

```
> x<-(1:50)*2
> x
 [1]   2   4   6   8  10  12  14  16  18  20  22  24
[13]  26  28  30  32  34  36  38  40  42  44  46  48
[25]  50  52  54  56  58  60  62  64  66  68  70  72
[37]  74  76  78  80  82  84  86  88  90  92  94  96
[49]  98 100
```

```
> A<-array(x,dim=c(5,5))
> A
     [,1] [,2] [,3] [,4] [,5]
[1,]    2   12   22   32   42
[2,]    4   14   24   34   44
[3,]    6   16   26   36   46
[4,]    8   18   28   38   48
[5,]   10   20   30   40   50
```

3. Matrix of size 4×3

```
> M<-matrix(c(1:12),nrow=4,ncol=3,byrow=FALSE)
M
[,1] [,2] [,3]
[1,]    1    5    9
[2,]    2    6   10
[3,]    3    7   11
[4,]    4    8   12
```

4.

```
> V<-c(10,23,34,55,78,6,88,90,9,5)
> V
[1] 10 23 34 55 78  6 88 90  9  5
> which(V%%2==0)
[1] 1 3 5 6 7 8
```

5.

```
> X<-c(10,20,30)
> Y<-c(1,2,3,4,5,6)
> X+Y
[1] 11 22 33 14 25 36
```

6. Data Frame “Customer”

```
> Cid<-c(101,102,103,104,105)
> Cname<-c("Atharv","Sneha","Misha","Nisha","Aryan")
> Age<-c(20,30,45,56,77)
> Address<-c("Noida","Delhi","Hyderabad","Bangalore",
"Mumbai")
> Income<-c(100000,200000,345677,788888,888888)
> D<-data.frame(Cid,Cname,Age,Address,Income)
> D
  Cid  Cname Age   Address Income
1 101 Atharv  20     Noida 100000
2 102  Sneha  30     Delhi 200000
3 103  Misha  45 Hyderabad 345677
4 104  Nisha  56 Bangalore 788888
5 105  Aryan  77    Mumbai 888888
```

7. Execute the given command: *matrix(10:10,nrow=2,byrow=TRUE*

```
> M<-matrix(10:10,nrow=2,byrow=TRUE)
> M
     [,1]
[1,]   10
[2,]   10

a. Data Elements count=2
b. Column count=1
c. No
```

matrix(25:15,nrow=2,byrow=TRUE)

```
> M<-matrix(25:15,nrow=2,byrow=TRUE)
Warning message:
In matrix(25:15, nrow = 2, byrow = TRUE) :
data length [11] is not a sub-multiple or multiple of
the number of rows [2]
To remove warning change the data elements as shown in
the executable command:
> M<-matrix(25:26,nrow=2,byrow=TRUE)
> M
     [,1]
[1,]   25
[2,]   26
```

8.

```
A<-c("aa","bb","cc","dd","ee")
 > A[2:4]
            [1] "bb" "cc" "dd"
```

Chapter 2

I. Multiple-Choice Answer Keys:

1. (b) 2. (d) 3. (c) 4. (b) 5. (c) 6. (a) 7. (c) 8. (c) 9. (d)
10. (c) 11. (a) 12. (b) 13. (c) 14. (d) 15. (b)

II. Coding Exercise:

1.

```
> seq(1,30,by=2)
[1]  1  3  5  7  9 11 13 15 17 19 21 23 25 27 29
> seq(2,30,by=2)
[1]  2  4  6  8 10 12 14 16 18 20 22 24 26 28 30
```

2.

```
> x<-c(3,67,89,90,23,10,56,4,7,100,9)
  > min(x)
  [1] 3
  > max(x)
  [1] 100
  > which.min(x)
  [1] 1
  > which.max(x)
  [1] 10
```

3.

```
> #Bubble sort Program in R
    > bbsort<-function(n)
    + {
    + x<-length(n)
    + for(i in 1:(x-1)){
    + for(j in (i+1):x){
    + if(n[i]>n[j]){
    + t=n[i]
    + n[i]=n[j]
    + n[j]=t
    + }
    + }
    + }
    + return(n)
    + }
    #Function Calling
    > n<-c(56,67,3,4,7,89,10,23)
    > bbsort(n)
    [1]  3  4  7 10 23 56 67 89
```

4.

```
> weekday<-function(d,mon,year)
     + {
     + mon<-mon-2
     + if(mon<=0){
     + mon<-mon+12
     + year<-year-1
     + }
     + cc<-year%%100
     + year<-year%%100
     +
     tmp<-floor(2.6*mon-0.2)+d+year+year%%4+cc%%4-2*cc
     + c("sunday","Monday","Tuesday","Wednesday","thursday",
     "friday","Saturday")[1+tmp%%7]
     + }
     > weekday(12,1,2010)
     [1] "friday"
```

5.

```
> qsort<-function(n)
 + {
 + if(length(n)<=1)return(n)
 + pivot<-n[1]
 + rem<-n[-1]
 + rem_1<-rem[rem<pivot]
 + rem_2<-rem[rem>=pivot]
 + return(c(rem_1,pivot,rem_2))
 + }
 > n<-c(1,6,8,9,10,2,3,-9,56)
 > qsort(n)
 [1] -9  1  6  8  9  10  2  3 56
```

6.

```
> customer<-function(CData){
+ print("Customer 1 Details")
+ print(CData[1,])
+ print("Customer 2 Details")
+ print(CData[2,])
+ print("Customer Details")
+ print(CData[c(1,4)])
+ }
   C<-data.frame(CName=c("Maria","John","Nicky","Serah",
   "Ryan"),Gender=c("F","M","F","F","M"),jobprofile=c
   ("Manager","IT Engineer","Doctor","Teacher","Resea
   rch Assistant"),cid=c(101,102,103,104,105))
> customer(C)
[1] "Customer 1 Details"
  CName Gender jobprofile cid
1 Maria      F    Manager 101
[1] "Customer 2 Details"
  CName Gender  jobprofile cid
2  John      M IT Engineer 102
[1] "Customer Details"
  CName cid
1 Maria 101
2  John 102
3 Nicky 103
4 Serah 104
5  Ryan 105
```

7.

```
> raise_f<-function(n,x=3)
+ {
+ i=1
+ r=1
+ while(i<=x){
+ r=r*n
```

```
+ i=i+1
+ }
+ print(r)
+ }
> raise_f(4)
[1] 64
> raise_f(4,5)
[1] 1024
> c_find<-function(a)
+ {
+ print(class(a))
+ }
> data1<-c(1,2,3,4,5)
> c_find(data1)
[1] "numeric"
> data2<-c("Hi","hello","bye")
> c_find(data2)
[1] "character"
```

8.

```
> v<-paste("D",1:5,sep="")
> v
[1] "D1" "D2" "D3" "D4" "D5"
> d<-v[c(1,2:4)]
> d
[1] "D1" "D2" "D3" "D4"
> d<-v[c(-3,-4)]
> d
[1] "D1" "D2" "D5"
```

9.

```
> f<-function(x){
+ n<-length(x)
+ (x[-c(n-1,n)]+x[-c(1,n)]+x[-c(1,2)])/3
+ }
> f(c(1:3,4:1))
[1] 2.000000 3.000000 3.333333 3.000000 2.000000
```

10.

```
> quadeq<-function(a,b,c){
+ if(!identical(a,0)){
+ x=b^2-4*a*c
+ if(identical(x,0)){return(-b/2*a)}
+ if(x<0){return("No real roots")}
+ if(x>0){return(c((-b+c(-sqrt(x),sqrt(x)))/(2*a)))}
else{
+ stop("AS a is zero, not quardatic")
+ }
+ }
+ }
```

```
> quadeq(4,-1,10)
[1] "No real roots"
> quadeq(5,-1,-8)
[1] -1.168858  1.368858
```

Chapter 3

I. Multiple-Choice Answer Keys:

1. (c) 2. (d) 3. (a) 4. (c) 5. (c) 6. (b) 7. (d) 8. (d) 9. (b) 10. (b) 11. (c) 12. (a) 13. (b) 14. (b) 15. (d)

II. Coding Exercise

1.

a.

```
> a<-c(10:30)
> a
[1] 10 11 12 13 14 15 16 17 18 19 20 21 22 23 24 25
[17] 26 27 28 29 30
```

b.

```
> b<-c(30:10)
> b
[1] 30 29 28 27 26 25 24 23 22 21 20 19 18 17 16 15
[17] 14 13 12 11 10
```

c.

```
> r<-c(10:30,29:1)
> r
[1] 10 11 12 13 14 15 16 17 18 19 20 21 22 23 24 25
[17] 26 27 28 29 30 29 28 27 26 25 24 23 22 21 20 19
[33] 18 17 16 15 14 13 12 11 10  9  8  7  6  5  4  3
[49]  2  1
```

d.

```
> temp<-c(14,16,10)
> temp
[1] 14 16 10
```

e.

```
> e<-rep(temp,10)
> e
[1] 14 16 10 14 16 10 14 16 10 14 16 10 14 16 10 14
[17] 16 10 14 16 10 14 16 10 14 16 10 14 16 10
```

f.

```
temp<-c(14,16,10)
rep(temp,times=c(11,10,11))
[1] 14 14 14 14 14 14 14 14 14 14 14 16 16 16 16 16
[17] 16 16 16 16 16 10 10 10 10 10 10 10 10 10 10 10
```

2.

```
> s<-seq(2,4,by=0.2)
> exp(s)*cos(s)
[1]  -3.074932  -5.311231  -8.128421 -11.536926
[5] -15.494514 -19.884531 -24.490697 -28.969238
[9] -32.819775 -35.357194 -35.687732
```

3.

a.

```
> temp<-20:90
> sum(temp^4+3*temp^2)
[1] 1214199116
```

b.

```
> temp<-10:20
> sum((3^temp)/temp+4^temp/(temp^2))
[1] 4074042809
```

4.

```
> set.seed(50)
> x<-sample(0:500,100,replace = T)
> y<-sample(0:500,100,replace = T)
```

5.

a.

```
x[x>400]
y[y>400]
```

b.

```
which(x>400)
which(y>400)
```

c. > sum(x>min(x)-300)
d. > sum(y%%4==0)
e. > y[order(x)]

6.

```
> M<-matrix(c(1,10,-5,10,-5,1,10,10,1),nr=3)
```

a. > M[,1]<-M[,1]+M[,3]
b. >t(M)%*%M OR crossprod(M)
c. > M[abs(col(M)-row(M))==1]<-2

7.

a. outer(1:5,1:5,"-")
b. outer(1:5,1:5,"+")
c. outer(1:5,1:5,"+")%%2

8.

a.

```
> set.seed(100)
> M<-matrix(sample(90:100,size=36,replace = T),nr=6)
> M
     [,1] [,2] [,3] [,4] [,5] [,6]
[1,]   93   98   93   93   94   95
[2,]   92   94   94   97   91  100
[3,]   96   96   98   95   98   93
[4,]   90   91   97   97   99  100
[5,]   95   96   92   95   96   97
[6,]   95   99   93   98   93   99
```

b.

```
> apply(M, 1, function(f){sum(f>99)})
[1] 0 1 0 1 0 0
```

c.

```
> which( apply(M,1,function(x){sum(x==95)==2}) )
[1] 5
```

d.

```
> c<-colSums(M)
> l<-outer(c,c,"+")>500
> l[lower.tri(l,diag = T)]<-F
> which(l,arr.ind = T)
```

9.

```
> Dayweek<-function(day,mon,year)
+ {
+ mon<-mon-2
+ if(mon<=0){
+ mon<-mon+12
+ year<-year-1
+ }
+ y<-year%%100
+ year<-year%%100
+ t<-floor(2.6*mon-0.2)+day+year+year%%4+y%%4-2*y
+ c("Sun","Mon","Tue","Wed","Thu","Fri","Sat")[1+t%%7]
+ }
> Dayweek(12,02,2010)
[1] "Mon"
> c(Dayweek(23,09,1998),Dayweek(13,05,2010),
Dayweek(25,06,1983),Dayweek(14,01,2015))
[1] "Wed" "Sun" "Sun" "Thu"
```

10.

```
> upper<-sample(LETTERS,10)
> upper
[1] "A" "C" "F" "Q" "Z" "L" "N" "X" "G" "M"
> lower<-sample(letters,10)
> lower
[1] "z" "t" "n" "k" "h" "i" "e" "c" "v" "y"
> v<-c(upper,lower)
> sort(v)
[1] "A" "c" "C" "e" "F" "G" "h" "i" "k" "L" "M" "n" "N"
"Q" "t" "v"
[17] "X" "y" "z" "Z"
```

11.

a.

```
> v<-list(a=10:15,b=20:26,c=30:37)
> sapply(v, length)
a b c
6 7 8
```

b.

```
> sapply(v, sum)
a   b   c
75 161 268
```

c.

```
> v$newcol<-c(90:100)
> v
$a
[1] 10 11 12 13 14 15

$b
[1] 20 21 22 23 24 25 26

$c
[1] 30 31 32 33 34 35 36 37

$newcol
[1]  90  91  92  93  94  95  96  97  98  99 100
```

12.

```
> gender=factor(c("M","M","F","F","F","M","M","F","F"))
> table(gender)
gender
F M
5 4
> cumsum(table(gender))
F M
5 9
```

13.

a.

```
> Cid=c('C01','C02','C03','C04')
> Cname=c("Amit","Neha","Aman","Sneha")
> Income=c(56000,80000,45000,56000)
> City=c('Delhi','Noida','Noida','Hyderabad')
> df<-data.frame(Cid,Cname,Income,City)
> df
      Cid Cname Income      City
    1 C01  Amit  56000     Delhi
    2 C02  Neha  80000     Noida
    3 C03  Aman  45000     Noida
    4 C04 Sneha  56000 Hyderabad

> df[2:3]
      Cname Income
    1  Amit  56000
    2  Neha  80000
    3  Aman  45000
    4 Sneha  56000
```

b.

```
> df[c(1,2),c(3,4)]
      Income  City
    1  56000 Delhi
    2  80000 Noida
```

c.

```
> df$Age<-c(34,56,45,30)
> df
      Cid Cname Income      City Age
    1 C01  Amit  56000     Delhi  34
    2 C02  Neha  80000     Noida  56
    3 C03  Aman  45000     Noida  45
    4 C04 Sneha  56000 Hyderabad  30
> s<-subset(df,df$Age==max(Age))
> s
      Cid  Cname Income  City Age
    2 C02   Neha  80000 Noida  56
```

d.

```
> subset(df,df$Income>50000 & df$Age>30)
      Cid Cname Income  City Age
    1 C01  Amit  56000 Delhi  34
    2 C02  Neha  80000 Noida  56
```

14.

a.

```
> l<- list(X=c(10:20),Y=LETTERS[10:15],Z=matrix(1:9,nr
ow=3,byrow = TRUE))
> l
```

```
$X
[1] 10 11 12 13 14 15 16 17 18 19 20

$Y
[1] "J" "K" "L" "M" "N" "O"

$Z
      [,1] [,2] [,3]
[1,]    1    2    3
[2,]    4    5    6
[3,]    7    8    9
> l[1]
$X
[1] 10 11 12 13 14 15 16 17 18 19 20
```

b.

```
> l$Z[2,2]
[1] 5
> l$Z[,2]
[1] 2 5 8
```

c.

```
> sapply(l, length)
X  Y  Z
11  6  9
> all(sapply(l, length)==9)
[1] FALSE
```

d.

```
> l$rnum<-rnorm(5)
> l
$X
[1] 10 11 12 13 14 15 16 17 18 19 20

$Y
[1] "J" "K" "L" "M" "N" "O"

$Z
      [,1] [,2] [,3]
[1,]    1    2    3
[2,]    4    5    6
[3,]    7    8    9

$rnum
[1] -0.1611622  1.6935238 -0.8113666
[4]  0.1787969 -0.5466957
```

Chapter 4

I. Multiple-Choice Answer Keys:

1. (a) 2. (b) 3. (c) 4. (b) 5. (d) 6. (b) 7. (a) 8. (d) 9. (a) 10. (c) 11. (b) 12. (c)

II. Coding Exercise

1.

```
> s1 <- "R Programming for beginners RString@example.com"
> email_pattern <- "[a-zA-Z0-9._%+-]+@[a-zA-Z0-9.-]+\\.
[a-zA-Z]{2,}"
> emails_base <- regmatches(s1, gregexpr(email_pattern,
s1))[[1]]
> emails_base
[1] "RString@example.com"
> library(stringr)
> emails_stringr <- str_extract_all(s1, email_pattern)
[[1]]
> emails_stringr
[1] RString@example.com
```

2.

```
> phone_numbers <- c("(123) 456-7890", "123.456.7890",
"123-456-7890")
> clean_numbers <- str_replace_all(phone_numbers,
"[^0-9]", "")
> formatted_numbers <- str_replace(clean_numbers, "
(\\d{3}) (\\d{3})  (\\d{4})",  "\\1-\\2-\\3")
> clean_numbers
[1] "1234567890" "1234567890" "1234567890"
> formatted_numbers
[1] "123-456-7890" "123-456-7890" "123-456-7890"
```

3.

a.

```
> library(stringi)
> # Extract substring
> str <- "Statistical Programming Language"
> substring <- stri_sub(str, 5, 11)
> substring
          [1] "istical"
```

b.

```
> # Replace all occurrences of a pattern
> new_str <- stri_replace_all_fixed("I like R and
Python Programming", "Python", "R")
> new_str
[1] "I like R and R Programming"
```

c.

```
> # Split string with regex
> split_str <- stri_split_regex("Data, analytics;
science", "[,; ]+")[[1]]
> split_str
        [1] "Data" "analytics" "science"
```

4.

```
> # Load the stringr packagelibrary(stringr)
> library(stringr)
> # Create a sample dataset of product reviews
> reviews <- c("I recently bought PROD123456 and it
exceeded my expectations.","The product PROD654321 was
not up to the mark.","I recommend checking out
PROD000111 for its amazing features.","Beware of
duplicates like PROD222333 and PROD444555 in the
market.")
> # Define the pattern for product IDs
> pattern <- "PROD\\d{6}"
> # Extract all product IDs from the reviews
> extracted_ids <- str_extract_all(reviews, pattern)
> extracted_ids
> # Replace all product IDs with "[PRODUCT ID]"
> anonymized_reviews <- str_replace_all(reviews,
pattern, "[PRODUCT ID]")
> anonymized_reviews
```

5.

```
> #Create Sample Data
> comments <- c("The service was excellent! Contact us
at service@example.com for more info.","Visit our
website http://example.com for details on prices and
quality.","Email support@example.org for any queries
regarding product quality and price.","The product's
price is justified given its exceptional quality and
service.")
> # Extract Email Addresses
> # Load the stringr package
> library(stringr)
> # Define the pattern for email addresses
> email_pattern <- "[a-zA-Z0-9._%+-]+@[a-zA-Z0-9.-]+\\.
[a-zA-Z]{2,}"
> # Extract email addresses from the comments
> email_addresses <- str_extract_all(comments,
email_pattern)
> # Print the extracted email addresses
> print(email_addresses)
> # Count Keyword Occurrences
> keywords <- c("service", "price", "quality")
> # Function to count occurrences of each keyword in a
string
```

```
> count_keywords <- function(text, keywords)
{sapply(keywords, function(keyword) str_count(text,
keyword))}
> keyword_counts <- t(sapply(comments, count_keywords,
keywords))
> colnames(keyword_counts_df) <- keywords
> print(keyword_counts_df)
> # Remove URL's
> url_pattern <- "http[s]?://[a-zA-Z0-9./?=_-]+"
> cleaned_comments <- str_replace_all(comments,
url_pattern, "")
> print(cleaned_comments)
> # Normalize the Comments
> normalized_comments
<- str_trim(tolower(cleaned_comments))
> print(normalized_comments)
```

6.

```
> library(stringr)
> vec <- c("R Programming", "Java",
"Python Prigramming", "Data Science","HTML")
> # str_detect
> result <- str_detect(vec, "a")
> vec <- c("R Programming", "Java", "Python Programming",
"Data Science","HTML")
> # str_detect
> result <- str_detect(vec, "a")
> result
[1]  TRUE  TRUE  TRUE  TRUE FALSE
> indices <- str_which(vec, "a")
> indices
[1] 1 2 3 4
```

7.

```
> reverse_words <- function(str) {
+     sapply(strsplit(str, " "), function(word)
sapply(word, function(w) paste(rev(strsplit(w, "")
[[1]]), collapse = ""))) |>
+         paste(collapse = " ")
+ }
> print(reverse_words("Data Science"))
[1] "ataD ecneicS"
```

8.

```
> extract_numbers <- function(str) {
+     as.numeric(unlist(regmatches(str, gregexpr("\\b\\
      d+\\b", str))))
+ }
> print(extract_numbers("There are 12 apples for $15"))
[1] 12 15
```

9.

```
> is_palindrome <- function(str) {
+     cleaned_str <- tolower(gsub("[^a-z]", "", str))
+     cleaned_str == rev(cleaned_str) |>
+         paste(collapse = "") |>
+         identical(cleaned_str)
+ }
> print(is_palindrome("A man, a plan, a canal,
Panama"))
[1] FALSE
```

10.

```
> library(tm)
> library(wordcloud)
> library(RColorBrewer)
> reviews <- readLines(file.choose())
> corpus <- Corpus(VectorSource(reviews))
> corpus <- tm_map(corpus, content_
transformer(tolower)) # Convert to lowercase
> corpus <- tm_map(corpus, removePunctuation) # Remove
punctuation
> corpus <- tm_map(corpus, removeNumbers) # Remove
numbers
# Remove common stopwords
> corpus <- tm_map(corpus, removeWords,
stopwords("english"))
> dtm <- TermDocumentMatrix(corpus)
> matrix <- as.matrix(dtm)
# Calculate word frequencies
> word_freqs <- sort(rowSums(matrix), decreasing=TRUE)
> data <- data.frame(word=names(word_freqs), freq=word_
freqs) # Convert to data frame
> set.seed(1234) # For reproducibility
> wordcloud(words = data$word, freq = data$freq, min.
freq = 5, max.words=200,random.order=FALSE, rot.
per=0.35, colors=brewer.pal(8, "Dark2"))
```

Chapter 5

I. Multiple-Choice Answer Keys:

1. (a) 2. (d) 3. (c) 4. (d) 5. (a) 6. (b) 7. (d) 8. (d) 9. (a) 10. (c)
11. (b) 12. (b) 13. (d) 14. (c)

II. Coding Exercise

1.

```
> #User defined function to clear range
> library(openxlsx)
> clearRange <- function(wb, sheet, startRow, endRow,
startCol, endCol) {
+     for (row in startRow:endRow) {
+         for (col in startCol:endCol) {
+             writeData(wb, sheet = sheet, x = "",
startCol = col, startRow = row, colNames = FALSE)
+         }
+     }
+ }
> wb <- loadWorkbook("emp.xlsx")
> d1 <- read.xlsx(wb, sheet = "Sheet1")
> cl<-clearRange(wb,sheet="Sheet1",3,5,6,7)
> d1
```

2.

```
> library(xml2)
> df <- data.frame(name = c("Item 1", "Item 2"),price =
c(10.00, 20.00),stringsAsFactors = FALSE)
>
> df
    name price
1 Item 1    10
2 Item 2    20
> # Create the root node
> root <- xml_new_root("root")
> # Add items to the root
> for (i in 1:nrow(df)) {
+     item <- xml_add_child(root, "item")
+     xml_add_child(item, "name", df$name[i])
+     xml_add_child(item, "price", as.character
      (df$price[i]))
+ }
>
> # Convert the XML document to a string
> xml_string <- as.character(root)
> print(xml_string)
[1] "<?xml version=\"1.0\" encoding=\"UTF-8\"?>\
n<root>\n  <item>\n    <name>Item 1</name>\n
<price>10</price>\n  </item>\n  <item>\n    <name>Item 2
</name>\n    <price>20</price>\n  </item>\n</root>\n"
```

3.

```
> library(xml2)
> #List Creation
> my_list <- list(item1 = list(name = "Item 1", price =
10.00),item2 = list(name = "Item 2", price = 20.00))
```

```
>
> #Root node creation
> root <- xml_new_root("root")
> #Add elements to root node
> for (item in names(my_list)) {
+     node <- xml_add_child(root, "item")
+     xml_add_child(node, "name", my_list[[item]]$name)
+     xml_add_child(node, "price", as.character
      (my_list[[item]]$price))
+ }
> # Convert XML to String
> xml_string <- as.character(root)
> print(xml_string)
[1] "<?xml version=\"1.0\" encoding=\"UTF-8\"?>\
n<root>\n  <item>\n    <name>Item 1</name>\n
<price>10</price>\n  </item>\n  <item>\n    <name>Item 2
</name>\n    <price>20</price>\n  </item>\n</root>\n"
```

4.

```
>library (dplyr)
> #D #Compare chest pain type (cp) with cholesterol
level (chol)
< cp_vs_chol <- hdata %>%group_by(cp) %>%
summarise(avg_chol = mean(chol))
>
> cp_vs_chol
# A tibble: 4 × 2
     cp avg_chol
  <int>    <dbl>
1     0     250.
2     1     245.
3     2     243.
4     3     237.

> #A)
> hdata <- read.csv("heart.csv", header = TRUE,
stringsAsFactors = FALSE)
> head(hdata)
  age sex cp trtbps chol fbs restecg thalachh exng oldpeak slp caa thall output
1  63   1  3    145  233   1       0      150    0     2.3   0   0     1      1
2  37   1  2    130  250   0       1      187    0     3.5   0   0     2      1
3  41   0  1    130  204   0       0      172    0     1.4   2   0     2      1
4  56   1  1    120  236   0       1      178    0     0.8   2   0     2      1
5  57   0  0    120  354   0       1      163    1     0.6   2   0     2      1
6  57   1  0    140  192   0       1      148    0     0.4   1   0     1      1
> #B) # Calculate statistics
> num_records <- nrow(hdata)
> num_records
[1] 303
```

```
> summary_stats <- summary(hdata[c("age", "chol",
"thalachh")])
> summary_stats
      age             chol            thalachh
Min.   :29.00   Min.   :126.0   Min.   : 71.0
1st Qu.:47.50   1st Qu.:211.0   1st Qu.:133.5
Median :55.00   Median :240.0   Median :153.0
Mean   :54.37   Mean   :246.3   Mean   :149.6
3rd Qu.:61.00   3rd Qu.:274.5   3rd Qu.:166.0
Max.   :77.00   Max.   :564.0   Max.   :202.0
> #C) #Calculate proportion of patients with heart
disease
> proportion_heart_disease <- mean(hdata$output == 1) *
100
> cat("\nStep 3: Proportion of Patients with Heart
Disease:\n")

Step 3: Proportion of Patients with Heart Disease:
> cat("Percentage of patients with heart disease:",
round(proportion_heart_disease, 2), "%\n")
Percentage of patients with heart disease: 54.46 %
> library(dplyr)
> #D #Compare chest pain type (cp) with cholesterol
level (chol)
> cp_vs_chol <- hdata %>%group_by(cp) %>%
summarise(avg_chol = mean(chol))
>
> cp_vs_chol
# A tibble: 4 × 2
     cp  avg_chol
  <int>     <dbl>
1     0      250.
2     1      245.
3     2      243.
4     3      237.
```

5.

a.

```
> library(xml2)
> library(dplyr)
>
> # Load XML file
> xml_file <- "patient_data.xml"
> doc <- read_xml(xml_file)
> doc
> #Extract patient data
> patients <- xml_find_all(doc, "patient_info")
```

b.

```
> # B # Determine and print the total number of
 patients in the dataset
> num_patients <- nrow(patient_data)
> num_patients
[1] 5
```

c.

```
> # C# Identify most common medical condition
 diagnosed
> common_diagnosis <- patient_data %>%count(diagnosis,
 sort = TRUE)
> most_common_condition <- head(common_
 diagnosis$diagnosis, 1)
> num_patients_diagnosed <- head(common_diagnosis$n, 1)
> cat("Step 3: Most Common Medical Condition
 Diagnosed:\n")
Step 3: Most Common Medical Condition Diagnosed:
> cat("Condition:", most_common_condition, "\n")
Condition: Cancer
> cat("Number of Patients Diagnosed:", num_patients_
 diagnosed, "\n\n")
Number of Patients Diagnosed: 2
```

d.

```
> # D # Calculate average age by gender
> average_age_gender <- patient_data %>% group_
 by(gender) %>% summarise(avg_age = mean(age), num_
 patients = n())
>
> cat("\nStep 5: Bonus - Additional Analysis\n")

Step 5: Bonus - Additional Analysis
> cat("Average Age by Gender:\n")
Average Age by Gender:
> print(average_age_gender)
# A tibble: 2 × 3
  gender avg_age num_patients
  <chr>    <dbl>        <int>
1 Female    42.5            2
2 Male      59.3            3
```

6.

a.

```
> # 1 # Load Data
> # Install and load the jsonlite package
> library(jsonlite)
>
```

```
> # Read products data
> jdata <- fromJSON("products.json")
> print(jdata)
```

b.

```
> #2 # Convert JSON to Data Frame
> products_df <- as.data.frame(jdata)
> products_df
```

c.

```
> #3 # Calculate Total Stock Value for Each Categor
> # Calculate stock value for each product
> products_df$stock_value <- products_df$price *
products_df$stock
> # Summarize total stock value by category
> stock_value_by_category <- aggregate(stock_value ~
category, data = products_df, sum)
> print("Total Stock Value by Category:")
[1] "Total Stock Value by Category:"
> print(stock_value_by_category)
     category stock_value
1 Electronics    19425000
2   Furniture      100000
```

d.

```
> # 4 # Identify the product with the highest stock
value
> # Find the product with the highest stock value
> highest_stock_value_product <- products_df[which.
max(products_df$stock_value), ]
> print("Product with Highest Stock Value:")
[1] "Product with Highest Stock Value:"
> print(highest_stock_value_product)
  id name    category  price stock stock_value
4  4   TV Electronics 380000    35    13300000
```

e.

```
> # 5 # Create Summary Table
> summary_table <- aggregate(. ~ category, data =
products_df[, c("category", "price")], FUN =
function(x) c(total = length(x), average = mean(x)))
> ## Format the summary table
> summary_table <- do.call(data.frame, summary_table)
> colnames(summary_table) <- c("Category", "Total_
Products", "Average_Price")
> print("Summary Table:")
[1] "Summary Table:"
> print(summary_table)
     Category Total_Products Average_Price
1 Electronics              3        165000
2   Furniture              1          4000
```

7.

a. read.csv
b. mean
c. data.frame, average_new.csv
d. students$Grade
e. aggregate, mean

Chapter 6

I. Multiple-Choice Answer Keys:

1. (d) 2. (c) 3. (a) 4. (a) 5. (b) 6. (d) 7. (c) 8. (b) 9. (c) 10. (a) 11. (c)

II. Coding Exercise

1.

```
>library(ggplot2)
> library(dplyr)
> library(tidyr)
> # Example dataset creation
> data <- data.frame(
+     date = rep(seq.Date(as.Date("2020-01-01"),
as.Date("2020-12-31"), by="days"), 3),
+     region = rep(c("Region A", "Region B", "Region
C"), each = 366),
+     cases = sample(0:100, 1098, replace = TRUE),
+     recoveries = sample(0:100, 1098, replace = TRUE),
+     deaths = sample(0:10, 1098, replace = TRUE)
+ )
>
> # Transform data into long format for ggplot2 using
pivot_longer
> data_long <- data %>%
+     pivot_longer(cols = c(cases, recoveries, deaths),
names_to = "type", values_to = "count")
>
> # Plot
> ggplot(data_long, aes(x = date, y = count, color =
type)) +
+     geom_line() +
+     facet_wrap(~region, scales = "free_y") +
+     labs(title = "COVID-19 Trends Over Time by Region",
+          x = "Date",y = "Count",color = "Type")
+theme_minimal()
```

2.

```
> # Load necessary library
> library(lattice)
>
> # Example dataset creation
> set.seed(123)  # For reproducibility
> dates <- seq.Date(as.Date("2023-01-01"),
as.Date("2023-12-31"), by="week")
> categories <- c("Category A", "Category B",
"Category C")
>
> data <- data.frame(
+     date = rep(dates, each = length(categories)),
+     category = rep(categories, times =
length(dates)),
+     sales = sample(50:200, length(dates) *
length(categories), replace = TRUE)
+ )
>
> # Create the line plot using lattice
> xyplot(sales ~ date | category, data = data, type = "l",
+        layout = c(1, 3), col = "blue",xlab = "Date",
ylab = "Sales",
+        main = "Weekly Sales Trends by Product
Category")
```

3.

```
> # Load necessary libraries
> library(ggplot2)
> library(dplyr)
>
> # Example dataset creation
> set.seed(123)  # For reproducibility
> teams <- c("Team A", "Team B", "Team C")
> modules <- c("Module 1", "Module 2", "Module 3")
> data <- data.frame(
+     team = sample(teams, 300, replace = TRUE),
+     module = sample(modules, 300, replace = TRUE),
+     score = rnorm(300, mean = 75, sd = 10)
+ )
>
> # Plot
> ggplot(data, aes(x = module, y = score, fill = team))
+
+     geom_violin(trim = FALSE) +
+     geom_jitter(shape = 16, position = position_
jitter(0.2), alpha = 0.4) +
```

```
+     labs(title = "Distribution of Performance Scores
by Module and Team",
+          x = "Module",y = "Score",fill = "Team")
+theme_minimal()
```

4.

```
> # Load necessary library
> library(leaflet)
>
> # Example dataset creation
> data <- data.frame(
+     city = c("New York", "Los Angeles", "Chicago",
"Houston", "Phoenix"),
+     attraction = c("Statue of Liberty", "Hollywood
Sign", "Willis Tower", "Space Center", "Desert
Botanical Garden"),
+     latitude = c(40.6892, 34.1341, 41.8789, 29.5505,
33.4606),
+     longitude = c(-74.0445, -118.3215, -87.6359,
-95.0973, -111.9440),
+     visits = c(4000000, 1500000, 1200000, 900000,
700000)
+ )
>
> # Create the leaflet map
> leaflet(data) %>%
+     addTiles() %>%
+     addCircleMarkers(
+         ~longitude, ~latitude,
+         radius = ~sqrt(visits) / 1000,# Adjust the
size relative to the visits
+         color = ~ifelse(visits > 2000000, "red",
"blue"),
+         stroke = FALSE, fillOpacity = 0.7,
+         label = ~paste0(attraction, ": ", visits, "
visits")
+     ) %>%
+     addLegend("bottomright", colors = c("blue",
"red"), labels = c("<= 2M visits", "> 2M visits"),
title = "Visit Counts")
```

5.

```
> # Load necessary library
> library(leaflet)
>
> # Example dataset creation with schools in a city in
India
> data <- data.frame(
+     school = c("School A", "School B", "School C",
"School D", "School E"),
```

```
+     latitude = c(28.6139, 28.7041, 28.5355, 28.4595,
28.4082), # Example coordinates for New Delhi area
+     longitude = c(77.2090, 77.1025, 77.3910, 77.0266,
77.0636),
+     students = c(500, 800, 1200, 600, 400)
+ )
>
> # Create the leaflet map
> leaflet(data) %>%
+     addTiles() %>%  # Add default OpenStreetMap tiles
+     addCircleMarkers(
+         ~longitude, ~latitude,
+         radius = ~sqrt(students) / 10,  # Adjust the
size relative to the number of students
+         color = "red",
+         stroke = FALSE, fillOpacity = 0.7,
+         label = ~paste0(school, ": ", students, "
students")
+     ) %>%
+     addLegend("bottomright", colors = "blue", labels
= "Number of Students", title = "School Enrollments")
```

6.

```
> # Load necessary libraries
> library(treemap)
> library(dplyr)
>
> # Example dataset creation
> data <- data.frame(
+     category = c("Electronics", "Electronics",
"Furniture", "Furniture", "Clothing", "Clothing"),
+     subcategory = c("Smartphones", "Laptops",
"Sofas", "Tables", "Shirts", "Jeans"),
+     region = c("North", "South", "North", "South",
"North", "South"),
+     sales = c(120000, 150000, 80000, 70000, 60000,
40000)
+ )
>
> # Create a treemap
> treemap(
+     data,
+     index = c("category", "subcategory"),
+     vSize = "sales",
+     vColor = "sales",
+     draw = TRUE,
+     palette = "Blues",
+     title = "Sales Distribution by Product Category
and Sub-Category"
+ )
```

Chapter 7

I. Multiple-Choice Answer Keys:

1. (c) 2. (d) 3. (a) 4. (a) 5. (b) 6. (c) 7. (d) 8. (b) 9. (d) 10. (c)

II. Coding Exercise

1.

```
library(dplyr)
library(ggplot2)
library(tidyr)
# Assuming the data is in a CSV file named 'ev_data.
csv'
ev_data <- read.csv("ev_data.csv")
# Convert necessary columns to appropriate data types
if required
ev_data$Year <- as.integer(ev_data$Year)

# a) Analyze Trends in EV Sales Across Different
Regions
> # Filter data for passenger vehicles and sales
> passenger_sales <- ev_data %>% filter(category ==
"Historical" & parameter == "Sales")
> # Summarize sales by region and year
> sales_trends <- passenger_sales %>% group_by(region,
year) %>% summarize(Total_Sales = sum(value))
ggplot(sales_trends, aes(x = year, y = Total_Sales,
color = region)) + geom_line() + labs(title = "EV Sales
Trends by Region", x = "Year", y = "Total Sales")

# b) Compare Market Share of BEVs and PHEVs in
Different Regions
# Filter data for market share
market_share <- ev_data %>% filter(parameter == "EV Stock
Share")

# Summarize market share by region and powertrain
market_share_summary <- market_share %>% group_
by(region, powertrain) %>% summarize(Average_Market_
Share = mean(value))
# Plot the market share comparison
> ggplot(market_share_summary, aes(x = region, y =
Average_Market_Share, fill = powertrain)) + geom_
bar(stat = "identity", position = "dodge") + labs
(title = "Market Share Comparison of BEVs and PHEVs
by Region", x = "Region", y = "Average Market Share")
```

```
# c) Identify Regions Leading in Commercial EV Adoption
# Filter data for commercial vehicles
commercial_ev <- ev_data %>% filter(category ==
"Historical" & parameter == "EV Stock Share")
# Summarize market share by region
commercial_summary <- commercial_ev %>% group_by(region)
%>% summarize(Average_Market_Share = mean(value))
# Plot the regions leading in commercial EV adoption
ggplot(commercial_summary, aes(x = reorder(region,
-Average_Market_Share), y = Average_Market_Share)) +
geom_bar(stat = "identity", fill = "red") + labs(title
= "Regions Leading in Commercial EV Adoption",
x = "Region", y = "Average Market Share")

# d) Evolution of EV Sales in Passenger Vehicle
Category
# Filter data for passenger vehicle sales
passenger_sales_trend <- ev_data %>% filter
(category == "Passenger" & parameter == "Sales")
# Summarize sales by year
annual_sales <- passenger_sales_trend %>% group_
by(year) %>% summarize(Total_Sales = sum(value))
# Plot the evolution of sales
ggplot(annual_sales, aes(x = year, y = Total_Sales)) +
geom_line(color = "red") + labs(title = "Evolution of
EV Sales in Passenger Vehicle Category", x = "Year",
y = "Total Sales")
```

2.

```
Create a new branch named feature-visualization from
the main branch:
git checkout main
git pull origin main
git checkout -b feature-visualization
#R script that generates a plot of the mtcars dataset
using ggplot2
# plot_mtcars.R
library(ggplot2)
# Load mtcars dataset
 data(mtcars)
# Create a scatter plot of mpg vs. wt
ggplot(mtcars, aes(x = wt, y = mpg)) + geom_point()
+labs(title = "Scatter Plot of MPG vs. Weight",
x = "Weight",y = "Miles per Gallon") +theme_minimal()
#Commit Changes
git add plot_mtcars.R git commit -m "Added plot script
for mtcars dataset using ggplot2"
Switch back to the main branch and merge the feature-
visualization branch:
git checkout main
```

```
git pull origin main
git merge feature-visualization
git push origin main
```

3.

```
library(shiny)
library(leaflet)
library(ggplot2)
# Sample Data Creation
set.seed(123)
data <- data.frame(Location = paste("Location",
1:20),Latitude = runif(20, min = 34, max = 42),
Longitude = runif(20, min = -118, max = -74),
Category = sample(c("Park", "Museum", "Restaurant"), 20,
replace = TRUE),Value = runif(20, min = 10, max = 100))
# Define UI
ui <- fluidPage(
  titlePanel("Geographical Data Visualization"),

  sidebarLayout(
    sidebarPanel(
      selectInput("category", "Select Category:",
                  choices = unique(data$Category),
                  selected = unique(data$Category)[1])
    ),

    mainPanel(
      leafletOutput("map"),
      plotOutput("scatterPlot")
    )
  )
)

# Define server logic
server <- function(input, output) {

  # Reactive expression to filter data based on
selected category
  filteredData <- reactive({
    subset(data, Category == input$category)
  })

  # Render Leaflet map
  output$map <- renderLeaflet({
    leaflet(filteredData()) %>%
      addTiles() %>%
      addMarkers(~Longitude, ~Latitude,
                 popup = ~paste("<b>Location:</b>",
Location, "<br>",
                                "<b>Value:</b>", Value))
  })
```

```
  # Render Scatter Plot
  output$scatterPlot <- renderPlot({
    ggplot(filteredData(), aes(x = Longitude,
y = Latitude, size = Value)) +
     geom_point() +
     labs(title = paste("Scatter Plot of", input$category,
"Locations"),
          x = "Longitude",
          y = "Latitude",
          size = "Value") +
      theme_minimal()
  })
}
# Run the application
shinyApp(ui = ui, server = server)
```

4.

```
library(shiny)
library(leaflet)
library(ggplot2)
library(dplyr)
# Sample Data Creation
set.seed(42)
data <- data.frame(Location = paste("Location",
1:30),Latitude = runif(30, min = 34, max =
42),Longitude = runif(30, min = -118, max = -74),
Cases = sample(1:100, 30, replace = TRUE),Date = seq.
Date(from = as.Date("2024-01-01"), by = "week",
length.out = 30))
# Define UI
ui <- fluidPage(
  titlePanel("Disease Spread Analysis"),

  sidebarLayout(
    sidebarPanel(
      dateRangeInput("dateRange", "Select Date Range:",
                     start = min(data$Date),
                     end = max(data$Date)),
      selectInput("location", "Select Location:",
                  choices = c("All", unique(data$Location)),
                  selected = "All")
    ),

    mainPanel(
      leafletOutput("map"),
      plotOutput("timeSeriesPlot")
    )
  )
)
```

```
# Define server logic
server <- function(input, output) {

  # Reactive expression to filter data based on
selected date range and location
  filteredData <- reactive({
    data %>%
      filter(Date >= input$dateRange[1] & Date <=
input$dateRange[2]) %>%
      filter(if (input$location == "All") TRUE else
Location == input$location)
  })

  # Render Leaflet map
  output$map <- renderLeaflet({
   leaflet(filteredData()) %>%
    addTiles() %>%
    addCircleMarkers(~Longitude, ~Latitude,
            radius = ~sqrt(Cases) * 0.5,  # Adjust radius
            for better visibility
            popup = ~paste("<b>Location:</b>", Location,
            "<br>",
                    "<b>Date:</b>", Date, "<br>",
                    "<b>Cases:</b>", Cases),
            color = "red", fillOpacity = 0.7)
  })

  # Render Time Series Plot
  output$timeSeriesPlot <- renderPlot({
    filteredData() %>%
      group_by(Date) %>%
      summarize(TotalCases = sum(Cases)) %>%
      ggplot(aes(x = Date, y = TotalCases)) +
      geom_line() +
      geom_point() +
      labs(title = "Disease Cases Over Time",
           x = "Date",
           y = "Total Cases") +
      theme_minimal()
  })
}

# Run the application
shinyApp(ui = ui, server = server)
```

Index

Note: **Bold** page numbers refer to **tables** and *Italic* page numbers refer to *figures.*

D

For Product Safety Concerns and Information please contact our EU representative GPSR@taylorandfrancis.com Taylor & Francis Verlag GmbH, Kaufingerstraße 24, 80331 München, Germany

Batch number: 10397790

Printed by Printforce, the Netherlands